fundamentals of
Modern Drafting

SECOND EDITION

PAUL ROSS WALLACH

CENGAGE
Learning®

Australia • Brazil • Japan • Korea • Mexico • Singapore • Spain • United Kingdom • United States

CENGAGE Learning

Fundamentals of Modern Drafting, Second edition
Paul Ross Wallach

VP, General Manager, Skills and Planning: Dawn Gerrain

Product Manager: Daniel Johnson

Sr. Director of Development: Marah Bellegarde

Managing Editor: Larry Main

Content Developer: Richard Hall

Product Assistant: Kaitlin Schlicht

Vice President, Marketing: Jennifer Baker

Marketing Director: Deborah Yarnell

Senior Market Development Manager: Erin Brennan

Brand Manager: Kay Stefanski

Senior Production Director: Wendy Troeger

Production Manager: Mark Bernard

Senior Content Project Manager: William Tubbert

Senior Art Director: Bethany Casey

Technology Project Manager: Joe Pliss

Media Editor: Debbie Bordeaux

Cover & Chapter Opener image(s): © Kuzma/Veer.

Library of Congress Control Number: 2013935859

ISBN-13: 978-1-133-60362-7

ISBN-10: 1-133-60362-9

Cengage Learning
200 First Stamford Place, 4th Floor
Stamford, CT 06902
USA

Cengage Learning is a leading provider of customized learning solutions with office locations around the globe, including Singapore, the United Kingdom, Australia, Mexico, Brazil, and Japan. Locate your local office at: **international.cengage.com/region**

Cengage Learning products are represented in Canada by Nelson Education, Ltd.

To learn more about Cengage Learning, visit **www.cengage.com**

Purchase any of our products at your local college store or at our preferred online store **www.cengagebrain.com**

Notice to the Reader

Printed in the United States of America
1 2 3 4 5 6 7 17 16 15 14 13

CONTENTS

A Word from the Author

With the increasing development of industry and new breakthroughs in technology, there will always be a large demand for qualified engineers and drafters. With this demand comes the need for qualified schools and instructional materials to provide students with the skills they need to compete in today's workplace.

I would like to welcome you to this student-oriented engineering drawing textbook. *Fundamentals of Modern Drafting* provides the basic information and skill-building procedures of modern design and drafting techniques. The many drafting and design exercises in this text may be performed with either manual drafting, CAD, or freehand sketching techniques. Each chapter is designed to teach basic drafting concepts and skills in a logical order using the latest ASME conventions. The concepts and drawing exercises for each chapter progress from simple to complex. This will ensure beginning students a degree of success and offer sufficient materials for more advanced students. Many of the drafting concepts are presented with visual step-by-step illustrations.

The last chapter contains concise but complete instructional materials and exercises to design and draw residential working drawings.

Note to Instructors

It is critical that the students understand the design and drawing concepts of engineering and architectural drawing *before* they start creating drawings with a CAD system. Designing and drawing with a CAD system without this basic knowledge will not create any good-quality or useful working drawings.

Supplements

The Instructor's Companion Web Site to Accompany Fundamentals of Modern Drafting offers free resources for instructors to enhance the educational experience. The Web site contains the following features:

• Slides created in PowerPoint, which outline key concepts from each chapter

• Test bank to evaluate student learning

Features of this Text

Fundamentals of Modern Drafting fulfills the need for an instructional drafting text that will teach the fundamentals of engineering drawing through sketching, instrument drafting, and introductory CAD skills. Some of the special features include the following:

- Each chapter opens with objectives and sets the stage for clear and concise learning.
- Over 1,300 illustrations and photographs help clarify the content and aid students in reading two-dimensional and three-dimensional working drawings.
- Step-by-step procedural illustrations take the students through the concepts of drafting, design, and layouts.
- Key terms are highlighted within the text and listed at the end of each chapter to reinforce important concepts and terminology.
- Each chapter will develop and strengthen specific technical concepts, allowing the student to develop proficiency in solving drafting problems.
- All chapters are organized in a logical sequence; however, each chapter may be used as a stand-alone unit of instruction.
- No previous drafting knowledge is required to use this textbook.
- Exercises at the end of each chapter start with simple concepts and become progressively more complex.
- Students use the following methods to solve end-of-chapter exercises:
  - Freehand sketching
  - Instrument drafting
  - CAD system drawings
  - Special design exercises
  - Engineering change orders (ECOs)
  - Inch-decimal, inch-fraction, and metric units of measure

Chapter Overview

Chapter 1 presents an introduction to modern industry. Specific careers related to drafting also give the students an insight into occupational options.

Chapters 2 through 7 give the student the background needed to learn and draw the basic drafting concepts with: sketching, instruments, drafting supplies, lettering, formats, conventions, and an overview of CAD. The latest ASME standards are used throughout the text. Chapter 5 offers instruction on how students can use their creativity and drafting skills in a design team.

Chapters 8 through 15 teach the students the concepts of mechanical drafting required to design and draw finished multiview drawings, dimensioning, tolerancing, sectional drawings, auxiliary drawings, revolutions, descriptive geometry, development drawings, and pictorial drawings.

Chapters 16 through 24 teach students how to prepare finished working drawings that are required for production: fasteners, drafting systems, working drawings, welding drawings, gear drawings, cam drawings, piping drawings, electronics drawings, jig and fixture drawings, green planning in industry, and architectural drawings.

Standards

The language of drafting is a uniform and standardized system that is used throughout the world. The standards for the U.S. Customary system are developed by the American Society of Mechanical Engineers (ASME). The standards for the metric system are developed by the International Organization for Standardization (ISO). Careful attention was given to the dimensioning and tolerancing chapters (ASME Y14.5).

ACKNOWLEDGMENTS

I would like to express my appreciation and gratitude to the following individuals who took the time to offer their expertise and wisdom in the development of this textbook.

Chuck Bales
Moraine Valley Community College
Palos Hill, Illinois

John Scheblein
Suffolk Community College
Selden, NY

Keith Bright
Chattanooga Central
Harrison, Tennessee

Ed Wheeler
University of Tennessee at Martin
Martin, Tennessee

"This textbook is dedicated to Mike Robbins"

Introduction to Contemporary Drafting

OBJECTIVES

The student will be able to:

- Relate to the historical development of drafting
- State the importance and need for drafting as a technical communication skill
- State how drafting is used in different industrial fields as the major source of communication
- Identify the roles and responsibilities of various drafting specialists
- Recognize the levels of education, training, and experience required for the various professional and drafting positions

History

Prehistoric people drew crude drawings in the soil and cave walls long before people were able to write (**Figure 1-1**). Drawings have been used throughout history as an art form and a method of communication. As time progressed, drawing instruments and drawing surfaces, such as stone and clay tablets, limestone, wood and **parchment** made from the papyrus plant, were developed and refined. During the years of 3,600 BC. the first Egyptian pyramid was built using detailed construction drawings (**Figure 1-2**). Other historical buildings that used construction drawings for their construction was the **Parthenon** in Greece in 447 BC. (**Figure 1-3**), and the Roman Coliseum in 72 BC. (**Figure 1-4**). These are only a few of the very early structures designed and constructed with detailed drawings on parchment. By the first century AD, Romans were using detailed instrument drawings that were dimensioned for their building projects of roadways, aqueducts, and buildings. It was not until the fifteenth century that two-dimensional working drawings were used to produce products. Note a few of Leonardo da Vinci's (1452–1519) drawings of his inventions from 500 years ago (**Figure 1-5**). Can you tell what their functions are?

Figure 1-1 *Early humans drew crude pictures on cave walls.*

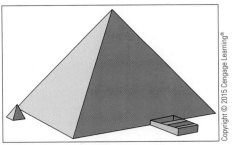

Figure 1-2 *Egyptian Pyramid.*

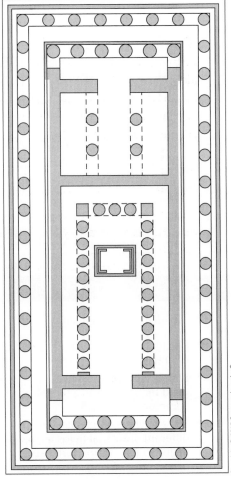

Figure 1-3 *Floor plan of the Greek Parthenon.*

Figure 1-4 *Roman Coliseum.*

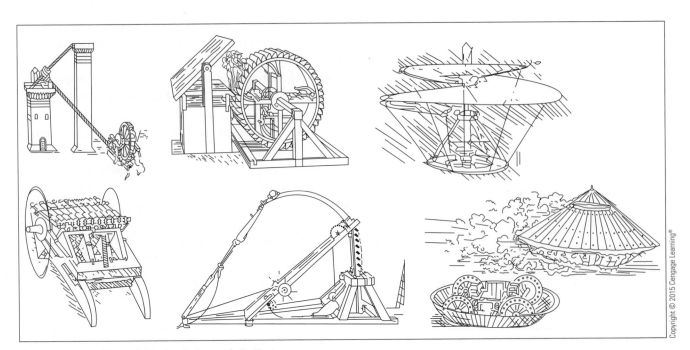

Figure 1-5 *Inventions and drawings by Leonardo Da Vinci.*

The Graphic Language

Human beings routinely communicate with each other through speech, written words, body movements, and an assortment of artistic and technical drawings. Verbal, written, and body language are very effective in communicating personal and social ideas or emotions. Only drawings, not language, are effective in describing the precise shape, size, and form of objects. To illustrate this point, verbally describe the spinning wheel in **Figure 1-6** to a friend. Now show the picture to your friend. How good was your verbal description? This exercise should clearly illustrate the Chinese proverb, "A picture is worth a thousand words."

Drafting is the basic technical form for visual communication. It is the universal language of industry and construction. It will translate the technical ideas, sketches, and the data of engineers and designers into clear, easy-to-read working drawings. Drafting is the basic technical form for visual communication.

Figure 1-6 *The spinning wheel was invented in the 13th Century AD.*

Today's Technical Working Drawings

Most of the aspects of technical graphics (drawings) are common to all the industrial areas of drafting. The major types of drawings are:

- **Multiview drawings** are two-dimensional (2D) views of the object drawn using orthographic projection (See Chapter 8). The complete shape of the item with its dimensions and details is shown with multiview drawings (**Figure 1-7**).

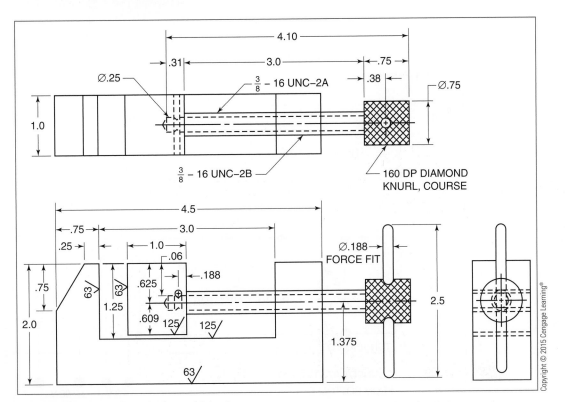

Figure 1-7 *Multiview drawing of a machinist vise.*

- **Pictorial drawings** are three-dimensional (3D) drawings that show an object as in a photograph (**Figure 1-8**). Usually three adjacent surfaces are shown in one drawing (See Chapter 15).

- **Schematic drawings** (See Chapter 22) use symbols and lines to show the flow of energy or fluids (**Figure 1-9**).

- **Block diagrams** are used to show the flow of a working process (**Figure 1-10**).

The use of **American Society of Mechanical Engineers (ASME)** standardized symbols and engineering drawing conventions makes it possible for technical drawings to be interpreted in all countries regardless of the language barriers. All manufactured products and structures, regardless of simplicity (**Figure 1-11**) or complexity (**Figure 1-12**), will still require a technical working drawing. Complex products may require many hundreds of working drawings (**Figure 1-13**).

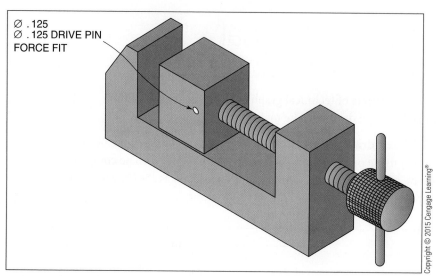

Ø .125
Ø .125 DRIVE PIN
FORCE FIT

Copyright © 2015 Cengage Learning®

Figure 1-8 *Pictorial drawing.*

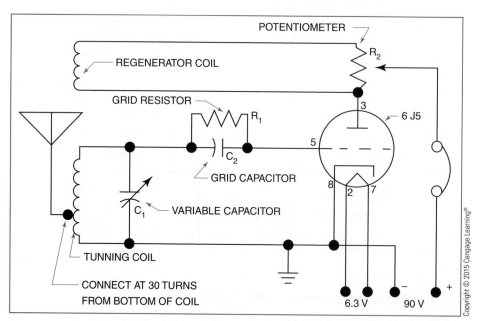

POTENTIOMETER

R₂

REGENERATOR COIL

GRID RESISTOR

R₁

6 J5

3

5

C₂

GRID CAPACITOR

8   2   7

VARIABLE CAPACITOR

C₁

TUNNING COIL

CONNECT AT 30 TURNS
FROM BOTTOM OF COIL

6.3 V   90 V   +   −

Copyright © 2015 Cengage Learning®

Figure 1-9 *An example of an electronic schematic drawing for a small radio.*

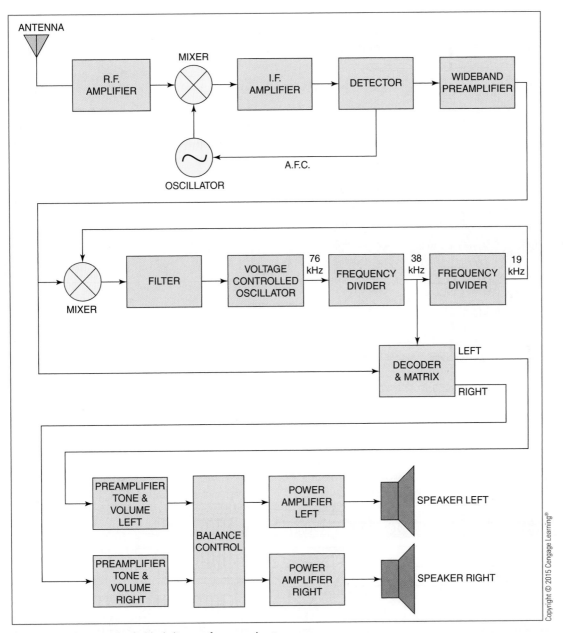

Figure 1-10 *An example of a block diagram for a sound system.*

Today's drafters must possess a broad understanding of the drafting knowledge and skills covered in this text. In addition, drafters must gain the specific knowledge of the manufacturing methods and standards in the specific industry where they work. Regardless of the level of responsibility or the drafting specialization, all drafters must:

- Understand the basics of drafting and design
- Be familiar with the basic types of engineering drawings (**Figure 1-14**).
- Clearly communicate ideas with freehand sketches
- Be proficient with computer-aided drafting and design systems
- Have some skill with manual drafting instruments

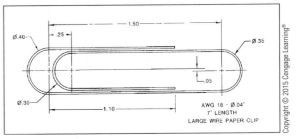

Figure 1-11 *The design and manufacture of a simple paper clip requires technical working drawings.*

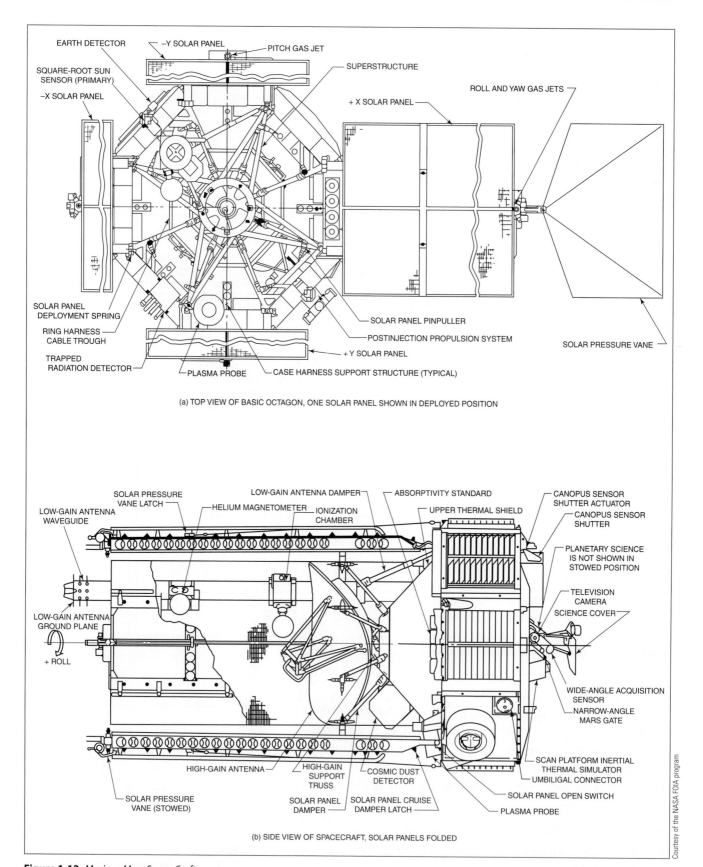

(a) TOP VIEW OF BASIC OCTAGON, ONE SOLAR PANEL SHOWN IN DEPLOYED POSITION

(b) SIDE VIEW OF SPACECRAFT, SOLAR PANELS FOLDED

Figure 1-12 *Mariner-Mars Space Craft.*

The Graphic Creators

Because each industry is highly specialized, the drafters in each industry must have the knowledge to correctly design and draw each part for manufacture. The personnel involved with designing and producing the working drawings may be placed into three general categories according to their formal education, knowledge, creativity, experience, and work ethics. These categories are top-level professionals, midlevel drafters, and intern-level drafters.

Regardless of the level an individual may reach, she or he should concentrate during the high school years on classes in math, physics, science, and drafting. This will provide a good background to start a formal education or an intern-level position in industry.

THE TOP-LEVEL PROFESSIONAL

The formal education for all top-level professionals is a college degree (BA, MA, or PhD). Often an internship and a state license in a specialty is required. These professionals have the responsibility for the successful designing and manufacturing operations. That is why they draw the largest salaries. Following is a brief description of several professional areas:

- The **aeronautical** and **astronautical engineers** perform a variety of work related to the research, planning, designing, manufacturing, and testing of airplanes, satellites, rockets, and spaceships.
- The **architect** does the planning, designing, environmental study, structural engineering, and supervising of the construction for all types of construction.
- The **cartographer** does the planning and drawings for all types of maps.
- The **civil engineer** plans does the planning, designing, and supervision for roads, airports, harbors, dams, tunnels, and most construction systems that are not inhabited buildings.
- The **chemical engineer** plans the research to develop new and improved industrial chemicals for manufacturing processes and production procedures.
- The **developmental engineer** does the data research for the development of new ideas and new products.
- The **drafting/engineering supervisor** coordinates all the workers involved with the production of all the working drawings for specific projects. It is her or his responsibility to get well-designed and error-free working drawings finished on schedule.
- The **electrical/electronics engineer** does the planning, designing, and supervising of the manufacturing of electrical and electronics components such computer systems, and all other types of electrical/electronic systems.
- The **environmental engineer** does the research and study of the materials and the effect a product may have on the environment. Also called green engineering.

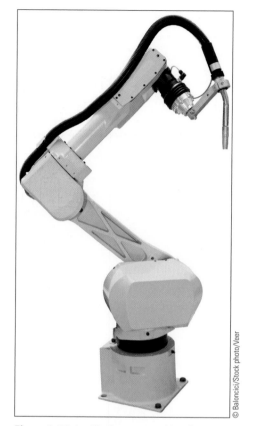

© Baloncici/Stock photo/Veer

Figure 1-13 *It will take many working drawings to accurately manufacture this robot welder.*

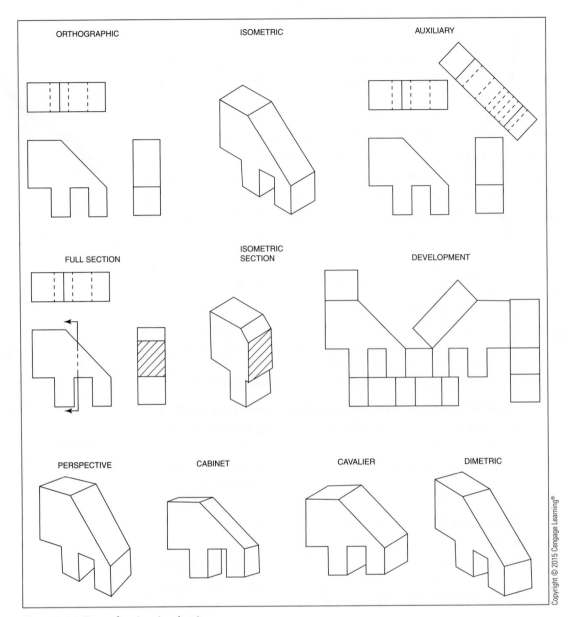

Figure 1-14 *Types of engineering drawings.*

- The **industrial designer** is a creative person who will use new ideas and materials to design a functional and attractive external package for industrial products.

- The **industrial engineer** and **mechanical engineer** works in all areas of industry, applying the math and physics to machines and parts so they will function properly and safely.

- **Instructor of engineering** or **drafting** is a rewarding position for people who enjoy working with helping others. Teaching positions in secondary schools, technical schools, colleges, and universities are open to people with degrees and/or teaching credentials with industrial experience.

- The **molecular** and **nanotechnology engineers** research and develop products on a molecular level that are used in the medical field and often in industrial products.

- The **robotics engineer** specializes in the design of robotic tools for industry and miscellaneous items such as toys.
- The **tool designer** plans and designs the tools used to produce the manufacturing systems, machinery, and tools for all areas of the industrial production.

MIDLEVEL DRAFTERS

Midlevel drafters are the technicians. It is recommended that they obtain an AA degree from a technical school. They should have a broad and diversified job classification, depending on their education, knowledge, creativity, drafting skills, and attitudes. They are classified as semi-professionals. Most will be supervised by the top-level professionals. Some technicians, even though they do not have a college degree, are highly capable and may perform the top-level job activities, but they will not reach the top salary levels.

Following is a brief description of some of the positions for midlevel drafters:

- The **checker** is an experienced drafter whose responsibility is to see that all the drafters' working drawings are properly drawn and are error-free. This is critical, because correcting an error during production is very expensive.
- The **chief /senior drafter** supervises the drafting personnel and sets the parameters for the standards, practices, schedules, and drawing procedures for the project's set of working drawings.
- The **commercial artist** prepares attractive illustrations for magazines, books, posters, and so on, to help promote recognition and sales of the manufactured item.
- The **design technician** combines design skills and drafting ability, usually working from the top-level sketches.
- The **senior detailer** is skilled in understanding the engineer's concepts and can produce complex working drawings needed for manufacture.
- The **technical illustrator** creates three-dimensional drawings from the working drawings. The drawings are used to view a complicated part for a better understanding, or to show an exploded drawing of all the individual parts to simplify its assembly.

INTERN-LEVEL DRAFTERS

Intern-level drafters do not require a formal technical education, but receive on-the-job training. However, it is recommended that they have some training from a two-year or technical college with drafting and CAD training. Many engineers and technicians will have interns working with them.

The following is a brief description of some intern-level drafting positions:

- The **computer-aided drafting operator (CAD)** must have the CAD skills that are important for all designers and drafters. These are an asset in gaining employment and advancement in the workplace.
- The **junior detailer** prepares simple working drawings from the sketches of the senior detailer and corrects drawing errors marked by the checker.

- The **junior drafter** should have good manual drafting and lettering skills and be trainable on a CAD system. He or she starts with simple working drawings and is closely supervised.

- The **printing operator** makes reproductions of working drawings on various types of equipment such as copy machines, cameras, and printers connected to a computer or from a CD.

- The **tracer** is the most basic entry level position. This job requires good drafting and lettering skills because it involves the copying or recopying sketches or quick drawings from the engineers and designers who did not take the time to do a neat drawing.

The Design Process

The design process will involve all levels of personnel involved. Each industry will vary in its design process, but the goal of an efficient product remains the same. A typical design process from beginning to end might be:

1. Recognition of needs
2. Scientific investigation
3. Proposal of concept
4. Layout and development drawings
5. Conceptual design reviews and analysis
6. Component testing of a prototype
7. Final design review
8. Detail design drawings
9. Quality assurance
10. Manufacturing
11. Assembly
12. Final testing
13. Sales
14. Planning for the second-generation upgrade

Nearly every phase of the 14-step design process will require some degree of freehand sketching, manual drafting and CAD for the development of ideas and working drawings.

Conclusion

With today's advancing technologies, we are continually improving existing industries and creating new ones. All industries will require a degree of graphic drawings for manufacturing, assembly, maintenance, and sales.

There are now over one million women and men working in the drafting, design, and related positions. As more products are developed and manufactured, and more buildings are designed and constructed, the need for personnel with high levels of drafting skills and knowledge will continue to increase.

DRAFTING EXERCISES

1. Select one of the professional positions that you find interesting and write a short paper on the education, training, and the type of work it involves.

2. Select one of the midlevel drafting positions that you find interesting and write a short paper on the education, training, and the type of work it involves.

3. If you had to go to work directly from high school, what type of intern position would you prefer? Write a paper on how you would prepare yourself for the internship.

4. Interview a college counselor and list the high school prerequisites needed for acceptance into an college engineering program.

5. Interview a technical or community college counselor and list the recommended high school classes needed for enrollment in a technical program.

6. Talk to high school counselor about applying for a scholarship or educational grant.

7. Practice writing a resume to use to apply for an internship drafting position.

8. Visit an industry using a CAD system. Interview the CAD operator and take notes on her or his background, education, training, and work performed on the CAD system.

9. Observe the Mainer-Mars spacecraft drawing in **Figure 1-12**. List the personnel involved with its design and drawings.

10. Discuss the concepts of the jet's multiview drawings in **Figure 1-15**.

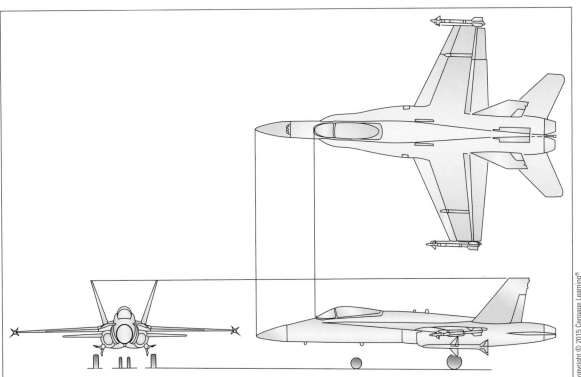

Figure 1-15 *Discuss the concepts of the jet's multiview drawings.*

KEY TERMS

Aeronautical engineer

Architect

Astronautical engineer

American Society of
    Mechanical Engineers
    (ASME)

Block diagram

Cartographer

Checker

Chief/senior drafter

Chemical engineer

Civil engineer

Commercial artist

Computer-aided drafting
    operator

Design technician

Developmental engineer

Drafting/engineering
    supervisor

Electrical/electronics
    engineer

Environmental engineer

Industrial designer

Industrial engineer

Instructor of engineering/
    drafting

Junior detailer

Junior drafter

Mechanical engineer

Molecular/nanotechnology
    engineers

Multiview drawing

Parchment

Parthenon

Pictorial drawing

Printing operator

Robotics engineer

Schematic drawing

Senior detailer

Technical illustrator

Tool designer

Tracer

Drafting Equipment and Supplies

2

The student will be able to:

- Understand the function of manual drafting instruments and supplies
- Complete a pencil drawing using drafting instruments
- Measure with a civil engineer's scale
- Measure with a mechanical engineer's scale
- Measure with a metric scale
- Measure with an architect's scale

Introduction

There are three basic forms of drafting procedures. They are:

1. Freehand sketching (**Figure 2-1**)
2. Manual drafting (**Figure 2-2**)
3. Computer-aided drafting (CAD) (**Figure 2-3**)

The drafter must be familiar with all three methods so he or she can select the best method to fit the needs for the design requirements. Note the variations of drawing times with the three drafting

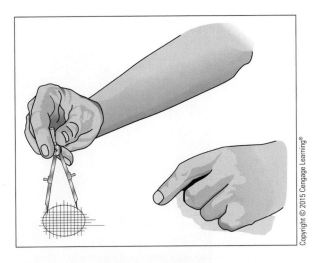

Courtesy of Ann Ross Wallach

Figure 2-1 *Freehand sketching on quarter-inch grid paper.*

© Steven Morris/E+/Getty Images

Figure 2-2 *Manual drafting.*

methods in **Figure 2-4**. The final drawings must be dimensionally correct, accurate, and easily readable using the correct standards. It is, therefore, critical when enrolled in an engineering or architectural drawing and design class to first obtain knowledge of the concepts of the subject, *before,* learning how to operate a CAD system. Without this knowledge, your CAD drawings will be useless in industry.

The successful drafter, designer, engineer, or architect should be skilled with freehand sketching and manual drafting for the following reasons:

- A CAD system may not always be available.
- Not all companies can or will supply CAD systems to every drafter.
- Field drawings and instructional sketches might have to be completed in environments hostile to computers.
- Creative design in many areas is still done in pencil.
- There may be times when the CAD system cannot create a specific type of detail. It would be expedient to make a print of your drawing and complete the detail with manual drafting instruments.

Manual Drafting Supplies

This section described manual drafting supplies (**Figure 2-5**).

Copyright © 2015 Cengage Learning®

Figure 2-3 *This CADD workstation has a computer, keyboard, mouse, and flat-panel monitor.*

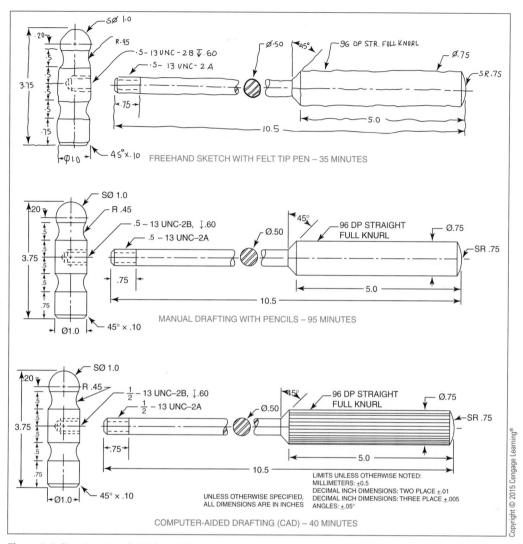

Figure 2-4 *Drawing times for Ball-peen hammer.*

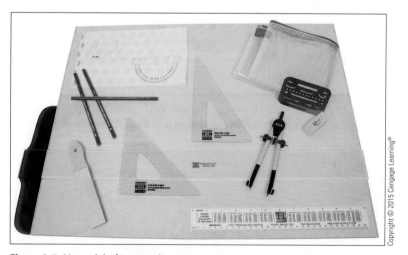

Figure 2-5 *Manual drafting supplies.*

DRAWING BOARDS

Drawing boards come in many different sizes to fit large drawing formats. The typical size for a manual drafting class is 20" × 26" basswood drafting board, as shown in **Figure 2-6**.

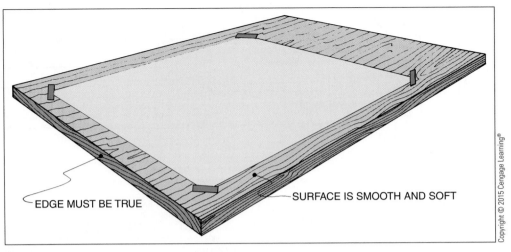

EDGE MUST BE TRUE

SURFACE IS SMOOTH AND SOFT

Copyright © 2015 Cengage Learning®

Figure 2-6 *Drawing board qualities.*

DRAWING PAPERS

Drawing papers include a variety of materials that may be used for drafting:

- **Opaque drawing paper** is a heavy paper that comes in white, buff, and light green.

- **Tracing paper** is a thin, transparent, inexpensive paper that is usually used for sketching ideas. It tears easily when handled or erased. It is not used for permanent type of drawings.

- **Vellum** is a transparent paper that is treated with oils and chemicals to make it heavier and durable.

- Polyester **film** is a transparent and indestructible media. Because of its smooth surface, only a black plastic led pencil or ink can be used to draw on its smooth surface (**Figure 2-7**).

- **Grid paper** is printed with various patterns. The grid paper used for engineering drawing is usually the inch divided into 10 parts or 2 millimeters for metric drawings. Architects use the 1/8" or 1/4" graph paper (**Figure 2-8**).

All drafting papers come in rolls or cut sheets (**Figure 2-9**). Standard cut sheet paper sizes are shown in **Figure 2-10**. Drafting tape is used to hold the paper to the drafting board (**Figure 2-11**).

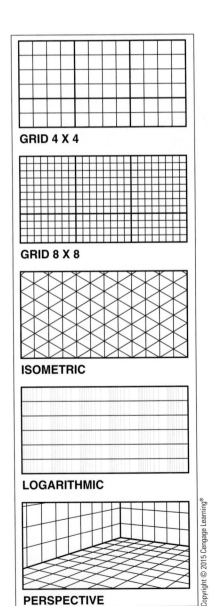

GRID 4 X 4

GRID 8 X 8

ISOMETRIC

LOGARITHMIC

PERSPECTIVE

Copyright © 2015 Cengage Learning®

Figure 2-8 *Examples of grid paper.*

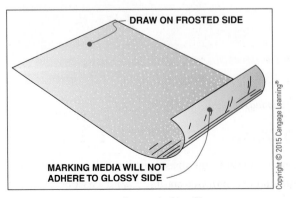

DRAW ON FROSTED SIDE

MARKING MEDIA WILL NOT ADHERE TO GLOSSY SIDE

Copyright © 2015 Cengage Learning®

Figure 2-7 *Drawing on polyester drafting film.*

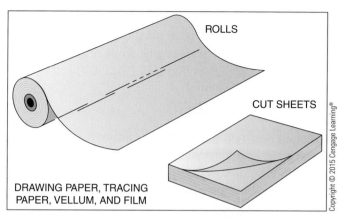

Figure 2-9 *Paper is manufactured in rolls and sheets.*

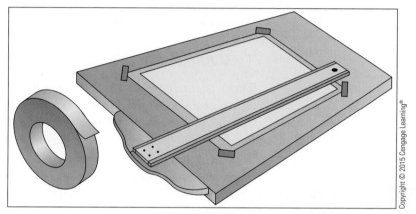

Figure 2-11 *Drafting tape is the most efficient method to attach drawing media to the drawing board's surface.*

LETTER SIZES	STANDARD SIZES	
A	9 × 12	8.5 × 11
B	12 × 18	11 × 17
C	18 × 24	17 × 22
D	24 × 36	22 × 34
E	36 × 48	34 × 44
F	28 × 40	

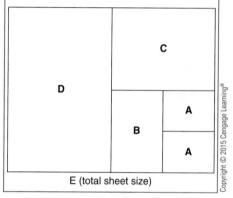

Figure 2-10 *Standard cut paper sizes.*

DRAFTING PENCILS

The two most common **drafting pencils** are wood and thin-lead mechanical pencils. The graphite leads are in various degrees of hardness and softness (**Figure 2-12**). The recommended pencils for drafting are shown in **Figure 2-13**. Sharp points produce fine lines, while round or dull points produce broad lines. Sharpening devices for drafting pencil points include hand or electric rotating sharpeners, cylindrical lead pointers, and sandpaper for hand-pointing leads.

Thin-lead mechanical pencils do not require either sharpening or pointing. Leads for mechanical holders are available in all grades; fine-line leads are available in thicknesses of 0.3, 0.5, 0.7, and 0.9 mm. Drafting leads are made of graphite, not lead. However, some drafting leads are made of plastic specially designed for use on drafting film. After drawing with any graphite lead, remember to brush the surface periodically with a dusting brush to remove the accumulation of foreign matter and eraser leavings.

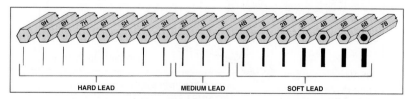

Figure 2-12 *The various degrees of draftiung pencils and their graphite lead widths.*

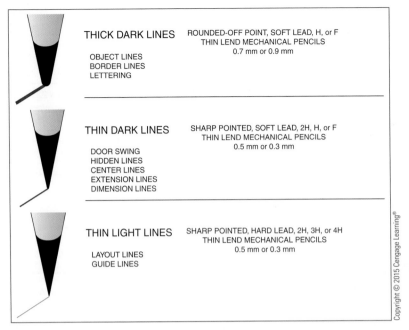

THICK DARK LINES   ROUNDED-OFF POINT, SOFT LEAD, H, or F
THIN LEND MECHANICAL PENCILS
0.7 mm or 0.9 mm

OBJECT LINES
BORDER LINES
LETTERING

THIN DARK LINES   SHARP POINTED, SOFT LEAD, 2H, H, or F
THIN LEND MECHANICAL PENCILS
0.5 mm or 0.3 mm

DOOR SWING
HIDDEN LINES
CENTER LINES
EXTENSION LINES
DIMENSION LINES

THIN LIGHT LINES   SHARP POINTED, HARD LEAD, 2H, 3H, or 4H
THIN LEND MECHANICAL PENCILS
0.5 mm or 0.3 mm

LAYOUT LINES
GUIDE LINES

Figure 2-13 *Recommended pencils for manual drafting.*

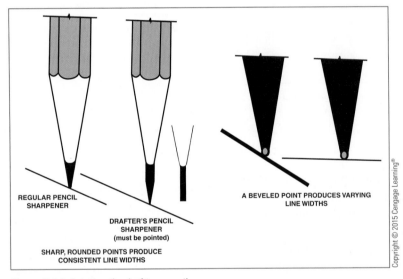

REGULAR PENCIL
SHARPENER

DRAFTER'S PENCIL
SHARPENER
(must be pointed)

SHARP, ROUNDED POINTS PRODUCE
CONSISTENT LINE WIDTHS

A BEVELED POINT PRODUCES VARYING
LINE WIDTHS

Figure 2-14 *Pointing the drafting pencil.*

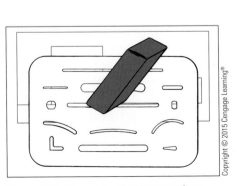

Figure 2-15 *The erasing shield protects lines not to be erased.*

When drawing, it is important to rotate the pencil so the point does not become beveled, because it will produce different line thicknesses (**Figure 2-14**).

ERASERS

An eraser is used to eliminate construction lines, drafting errors, and graphite smudges from the drawing surface. A rubber eraser is best for most line work. A soft vinyl eraser is good for light lines and smudges. When using an eraser it is often helpful to use an erasing shield to ensure that only the desired line area is erased (**Figure 2-15**).

T SQUARES

The **T square** comes in various lengths to fit different sized formats. For the 20" × 26" drawing board, a 24" T square is recommended. Keep the T square firmly against the left side of the drawing board when drawing horizontal lines (**Figure 2-16**).

TRIANGLES

The two drafting **triangles** used with manual drafting are the 30-60-degree and the 45-degree triangles (**Figure 2-17**). They are available in several sizes. The recommended sizes are 10" for the 30-60-degree triangle, and 8" for the 45-degree triangles. Be certain to hold the T square and the triangles firmly with one hand when drawing the lines (**Figure 2-18**).

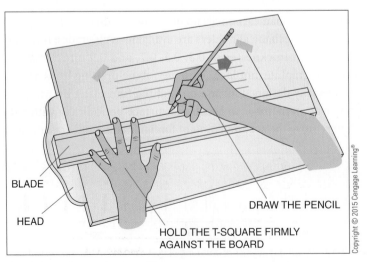

BLADE

HEAD

DRAW THE PENCIL

HOLD THE T-SQUARE FIRMLY
AGAINST THE BOARD

Figure 2-16 *Drawing horizontal lines.*

Incremental angles of 15 degrees may be drawn with the two triangles (**Figure 2-19**). An adjustable triangle (**Figure 2-20**) can be set at any required angle.

COMPASSES

Compasses are used to draw circles and arcs. The **bow compass** is usually used for drafting (**Figure 2-21**). It is important to adjust the shoulder needle and sharpen the graphite lead on a rough surface before using it (**Figure 2-22**).

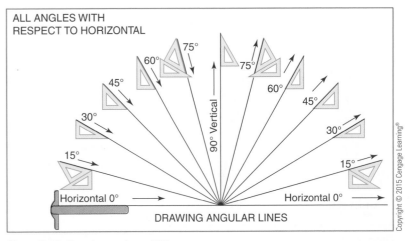

Figure 2-19 *Drawing angles at 15° increments.*

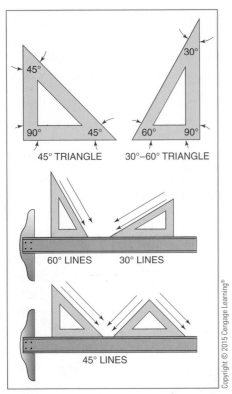

Figure 2-17 *Drawing 45°, 30°, and 60° lines.*

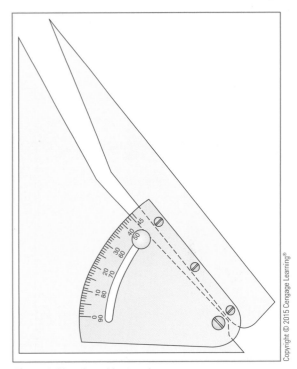

Figure 2-20 *Adjustable triangle.*

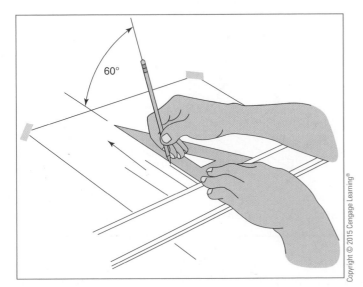

Figure 2-18 *Drawing with a T-square and triangle.*

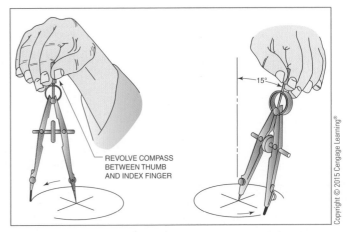

Figure 2-21 *Drawing procedure for handling a bow compass.*

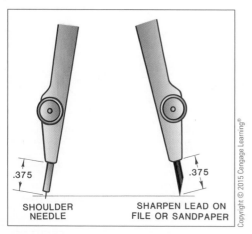

Figure 2-22 *The compass point and the compass lead.*

DIVIDERS

Dividers are similar to compasses except that both tips are pointed. They are very helpful to divide a line or transfer distances from one drawing to another (**Figure 2-23**).

PROTRACTORS

A **protractor** is used to lay out a specific angle on a drawing or measure an existing angle (**Figure 2-24**).

IRREGULAR CURVES

All drawings are composed of straight lines, circles, arcs, and irregular lines. All these lines must be drawn with drafting instruments. An **irregular curve**, also called a French curve, is used to draw the irregular lines (**Figure 2-25**). Only the lettering on a manual drawing is done freehand.

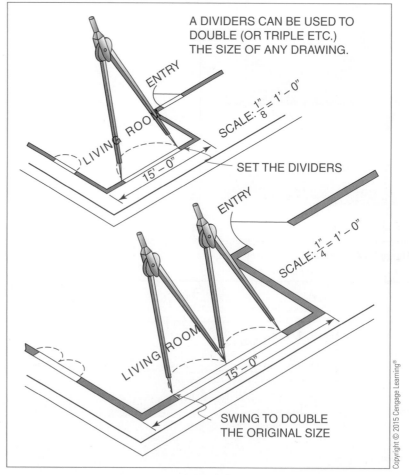

Figure 2-23 *Stepping off dimensions with the dividers.*

TEMPLATES

Templates for drawing symbols and geometric figures are available for all areas of engineering and architectural drawing. An electronics template is shown in **Figure 2-26** and an architectural template in **Figure 2-27**. As small circles are difficult to draw with a compass, a circle template will be required (**Figure 2-28**).

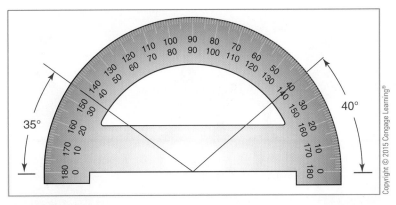

Figure 2-24 *Protractor.*

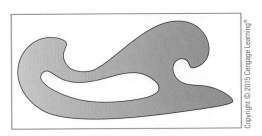

Figure 2-25 *Irregular curve.*

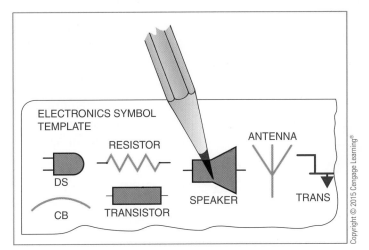

Figure 2-26 *Line up and position the graphic symbol on the schematic drawing and trace around the inside of the cut-out symbol.*

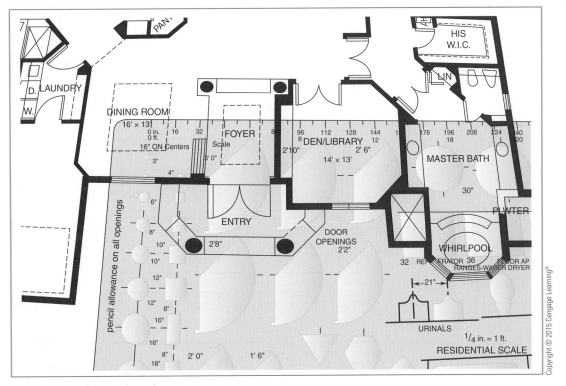

Figure 2-27 *Architectural template.*

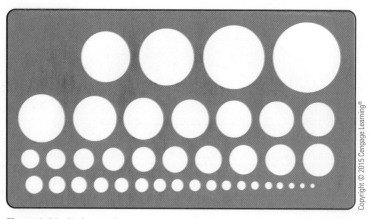

Figure 2-28 *Circle template.*

DRAFTING SCALES

Whether you draw with manual equipment or with a CAD system, an understanding of different scales is essential. Scales are available in several different shapes **(Figure 2-29)**. Scales allow drafters to create accurate drawings that are proportioned to the actual size of the object being drawn. Scales are based on either the **U.S. customary system** or the **metric system**. Scales used in drafting include the mechanical engineer's scale, the civil engineer's scale, the architect's scale, and the metric scale.

The **mechanical engineer's scale** uses a U.S. customary inch-fraction unit of measure **(Figure 2-30)**. Scale subdivisions include ½₂", ⅟₁₆", ⅛", ¼", ⅜", ¾", ½", and 1" units.

The **civil engineer's scale** uses a U.S. customary inch-decimal unit of measure **(Figure 2-31)**. The inch-decimal unit is used by most industries using the U.S. customary system. Civil engineer's scale subdivisions include 10, 20, 30, 40, 50, and 60 parts per inch.

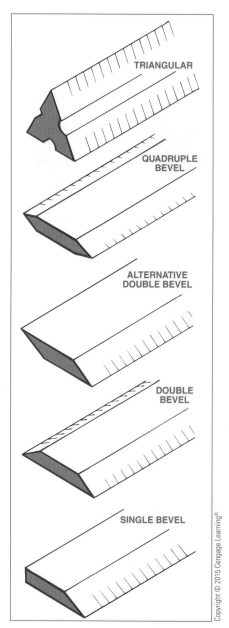

Figure 2-29 *Various scale shapes.*

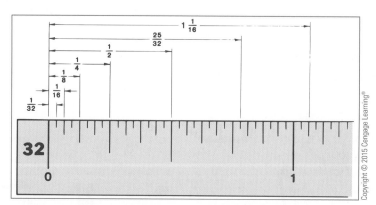

Figure 2-30 *The mechanical engineer's scale uses inch-fraction units of measure.*

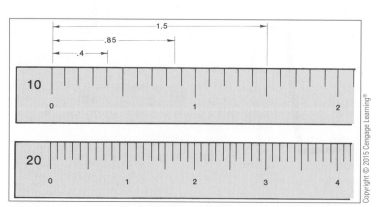

Figure 2-31 *The civil engineer's scale uses inch-decimal units of measure.*

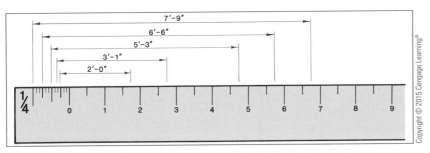

Figure 2-32 *The architect's scale uses foot-inch units of measures. The most often used architect's scale is 1/4" = 1'–0".*

The **architect's scale** is used to prepare plans for structures. The foot is divided into twelve parts, so inches on a drawing can equal the actual inch or feet dimensions of a building **(Figure 2-32)**. For example, most architectural drawings are prepared to a scale ¼" = 1'–0". This means that a ¼" line on a drawing represents one foot on a building. Thus, at ¼" = 1'–0" scale, an 8' wall would appear 2" long on the drawing. The various architect's scales are ³⁄₃₂", ³⁄₁₆", ⅛", ¼", ⅜", ¾", ½", 1", 1½", and 3".

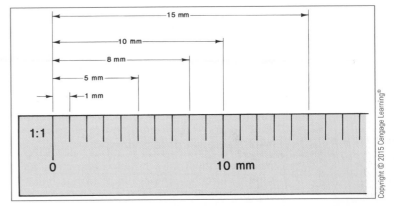

Figure 2-33 *The basic unit of measure for metric scales is the millimeter (mm).*

The basic unit of measure for metric scales is the **millimeter (mm) (Figure 2-33)**. The abbreviation mm is not used in metric drawings since all dimensions are in millimeters. All countries except the United States use the metric system for technical drawing and manufacturing, although we are gradually converting.

Drafting scales are either **open-divided** or **full-divided (Figure 2-34)**. Only one major unit of open-divided scales is graduated with a full-divided unit. It is adjacent to the zero. Full-divided scales contain full subdivision lines throughout the entire length of the scale. In selecting the proper scale for each drawing, the drafter must consider the amount of space available, the readability of the finished drawing, and ease of drawing. **Figure 2-35** provides some basic guidelines for proper scale selection. In selecting a scale, remember that a decimal or fractional part of an inch can be made equal to any unit of measure such as an inch, foot, yard, or mile.

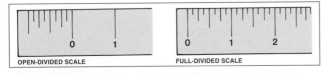

OPEN-DIVIDED SCALE   FULL-DIVIDED SCALE

Figure 2-34 *Drafting scales are either open-divided or full-divided.*

SCALE SELECTIONS FOR ENGINEERING DRAWING			
Types of Drawings	**Mechanical Engineer's Scale**	**Civil Engineer's Scale**	**Metric Scale**
The object to be drawn must be smaller than the drawing format. All the drawings, dimensions, and notations will fit at actual size.	Full Size (1:1)	1:1	1:1
The object to be drawn is larger than the drawing format and must be reduced by half to fit the format.	1/2" = 1"	1:2	1:2
The object to be drawn is much larger than the drawing format and must be reduced eight to ten times in size to fit the drawing format.	1/8" = 1"	1" = 10" 1:10	1:10
The object to be drawn is very large, such as a building, and must be reduced at least fifty times in size to fit the drawing format.	1/4" = 1'–0" (1:48)	1:50	1:50
The object to be drawn is small and cannot easily be drawn full size. Doubling the drawing size makes the drawing easier to draw and interpret.	1" = 1/2" (2:1)	2:1	2:1
The object to be drawn is very small. For ease of drawing and interpreting, the original size must be increased eight or ten times.	1" = 1/8" (8:1)	10:1	10:1
When an industrial product is extremely small, such as circuitry chips, the drawings must be drawn approximately 100 to 500 times larger.	Not used	100:1 500:1	100:1 500:1

Figure 2-35 *Typical scale selection for engineering drawings.*

Manual Drafting versus CAD

As computer-aided drafting becomes more prevalent in schools and industry, a decision must be made as to how many CAD stations and how many manual drafting stations should be installed. The decision will depend on the needs, goals, and budgets of each school, industry, and business.

DRAFTING EXERCISES

Exercises are provided for freehand drawing, manual drafting, and CAD.

1. Divide a "B" (11" × 17" or 12" × 18") drawing format into 11 parts as shown in **Figure 2-36**. Practice sketching the line work with drafting pencils.

2. Divide a "B" drawing format into 11 parts (**Figure 2-36**). Practice drawing with instruments and drafting pencils.

   Measure each line in **Figure 2-37** with these scales:

   full size—inch-decimal

   full size—inch-fraction

   1/2" = 1"

   1" = 10'

   full size—metric (millimeters)

   1/4" = 1'–0"

3. Practice drawing the line work in **Figure 2-38** with drafting instruments, freehand, and with a CAD system.

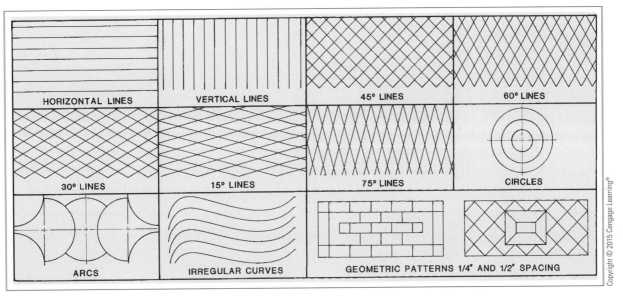

Figure 2-36 *Divide four B-size drawing formats into eleven parts as shown. Practice the line work with pencil sketching, drawing instruments and pencil, drawing instruments and a CAD drawing.*

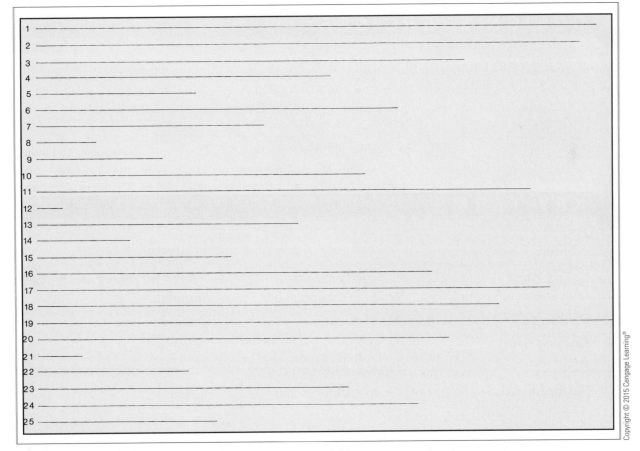

Figure 2-37 *Measure each line with as many different scales as are available.*

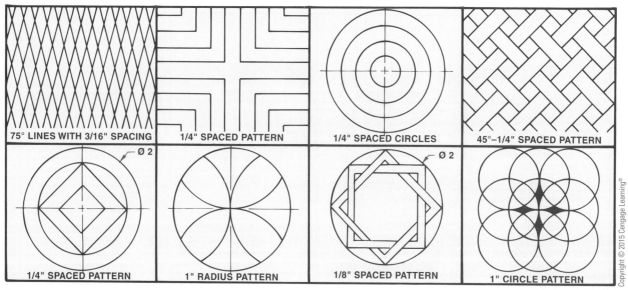

Figure 2-38 *Practice drawing the line work.*

DESIGN EXERCISES

1. With your drafting instruments, make a geometric design of your initials about 3" in height.

2. Design a dart board target with the scoring values.

3. Design a rifle range target.

4. With your drafting instruments, design and cut out a puzzle on stiff paper.

KEY TERMS

Architect's scale

Bow compass

Civil engineer's scale

Dividers

Drafting pencils

Film

Full-divided scale

Grid paper

Irregular curve

Mechanical engineer's scale

Metric system

Millimeter (mm)

Opaque drawing paper

Open-divided scale

Protractor

T square

Template

Tracing paper

Triangle

U.S. customary system

Vellum

Sketching and Lettering

3

OBJECTIVES

The student will be able to:

- Use proper line weights to sketch and letter
- Complete a two-dimensional sketch of simple objects
- Complete a three-dimensional sketch of simple objects
- File a record of sketched ideas
- Select and sharpen sketching and lettering pencils
- Keep sketches in proportion
- Dimension a sketch
- Shade a sketch
- Letter clearly on a sketch

Introduction

Highly developed manual drafting skills are not required for the preparation of CAD drawings (**Figure 3-1**); however, a competent CAD operator must possess a thorough understanding of the principles of drafting and be skilled in technical sketching.

Observe how the sketch (**Figure 3-2**) is used to describe the object shown in **Figure 3-3**. Try describing a pair of pliers to a person who has never seen one. You will now understand the proverb "a picture is worth a thousand words."

Supplies

Sketching supplies consist of a wide variety of pencils, erasers, and drawing media. Soft pencils in the range shown in **Figure 3-4** are used for sketching. Pencil points (**Figure 3-5**) are rounded for sketching compared to the sharp points used for instrument drawing. However, sharp points on soft pencils are used for layout and lettering guide line sketching. Chisel points are used for shading. Sketching pencils are sharpened with a mechanical sharpener, with a knife and file, or with sandpaper to shape the point.

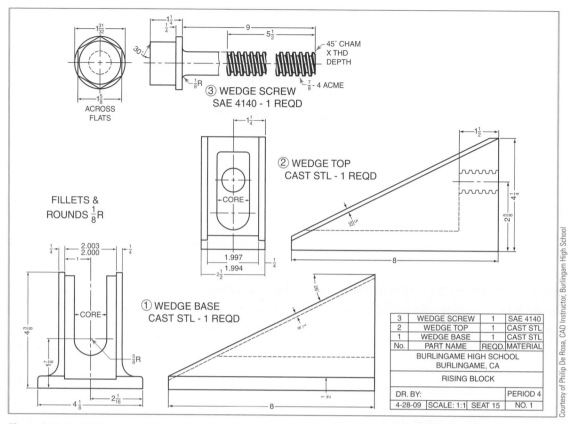

Figure 3-1 *As a CAD system is only another drafting tool, it is important that the CAD operator have a knowledge of engineering drawing.*

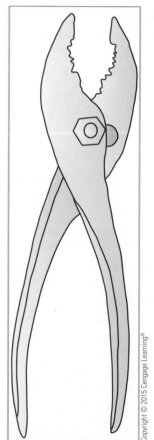

Figure 3-2 *A freehand sketch is worth a thousand words.*

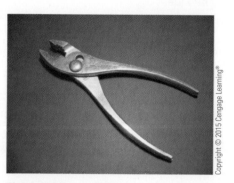

Figure 3-3 *A pair of pliers.*

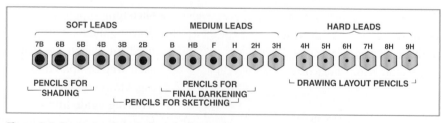

Figure 3-4 *Recommended pencil grades for sketching.*

Although many types of drawing media are used for technical sketching, grid paper is the most practical and easy to use **(Figure 3-6)**. Grids provide **guide lines** for rough scaling and proper alignment of perpendicular and parallel lines. Use light-colored surfaces if sketches are to be photocopied, and use translucent surfaces if diazo prints are to be made. Translucent surfaces are also helpful if progressive design sketches are to be traced.

Working Drawings

Working drawings are sketches used as the major design reference in manufacturing and construction. Sketches are also used as working drawings when time and conditions preclude the preparation of instrument or CAD working drawings. These sketches are usually orthographic multiview drawings, but they are sometimes prepared as pictorial drawings. An **orthographic projection** shows several views of an object on a drawing surface that is perpendicular to both the view and the lines of projection. A **pictorial drawing** shows an object's depth; three sides of an object can be seen in one view.

ORTHOGRAPHIC MULTIVIEW DRAWINGS

Multiview drawings provide the greatest amount of detail for manufacturing and construction. **Figure 3-7** shows the steps recommended to complete a multiview sketch. As with manual drafting (see Chapter 18), blocking in the overall outline before completing the internal details is the key to maintaining correct scales, angles, and proportions.

PICTORIAL DRAWING

Blocking in the basic outline is also the recommended procedure for pictorial drawings. There are three types of pictorial drawings: isometric, oblique, and perspective.

Isometric drawings are prepared by establishing a vertical corner line and projecting receding lines 30° **(Figure 3-8)**. As with multiview drawings, always block in the entire outline of the object before cutting corners or adding surface details such as holes and projections.

Oblique drawings are sketches that recede on only one side of an orthographic view. In preparing an oblique drawing, first draw a front view of the object (**Figure 3-9**), then extend lines from each corner upward at 45° or 30°. Connect the ends of these lines with lines that are parallel to the lines of the front view. Oblique drawings are easy to dimension since the width and length of the front view are drawn to actual scale. Only the depth dimension is foreshortened by the receding lines.

Perspective drawings are sketches that contain receding sides designed to approximate the actual appearance of an object. There are three types of perspectives: one-point, two-point, and three-point.

1. **One-point perspective** sketches are similar to oblique drawings, except the receding lines do not follow a consistent angle. Receding lines are connected to a **vanishing point.** A vanishing point represents the point at which all receding lines appear to come together. It is similar to road or train tracks

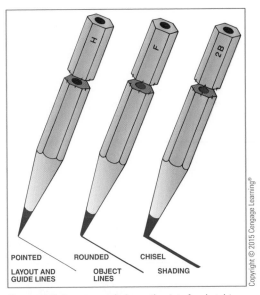

Figure 3-5 *Recommended pencil points for sketching.*

Figure 3-6 *Drawing on grid paper is faster and neater.*

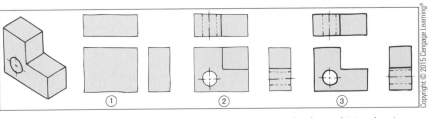

Figure 3-7 *Steps to sketch a multiview drawing.*

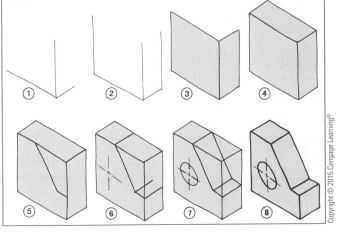

Figure 3-8 *Steps to sketch an isometric drawing.*

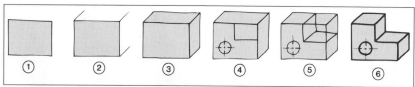

Figure 3-9 *Steps to sketch an oblique drawing.*
Copyright © 2015 Cengage Learning®

disappearing on the horizon. In sketching a one-point perspective, first outline the front view, as in oblique sketching. Establish a vanishing point and sketch lines from each corner to this point **(Figure 3-10)**. Sketch lines representing the back of the object parallel to the lines on the front view. This blocks in the sketch. Now any angular cuts and details can be added.

2. **Two-point perspective** sketches are similar to isometric sketches, except the side lines recede to two vanishing points. First, sketch the front vertical corner **(Figure 3-11)**. Establish two vanishing points above or below the front corner line, and connect the top and bottom of the front corner line to each of these points. Vanishing points on two-point perspective sketches must always be located on a horizontal line representing the horizon. Next, establish the depth of the sides and connect the back corner lines also to the vanishing points. In two-point perspectives, vertical lines are all parallel.

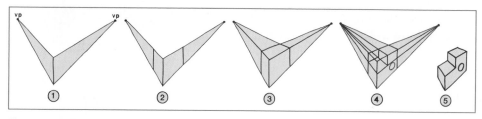

Figure 3-10 *Steps to sketch a one-point perspective drawing.*

3. In **three-point perspectives**, vertical lines are projected to a third vanishing point that is aligned with the vertical front corner line.

Three-point perspectives are usually used for architectural sketches, and rarely for technical drawings.

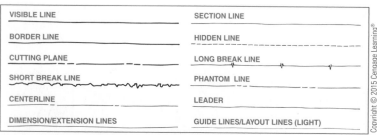

Figure 3-11 *Steps to sketch a two-point perspective drawing.* Copyright © 2015 Cengage Learning®

Sketching Guidelines and Procedures

Line conventions and lettering for technical sketches are similar to those used for instrument drawings **(Figure 3-12)**. Sketching standards differ only in the degree of line raggedness. Lines are dark and wide for object lines, dark and thin for dimension and center lines, and thin and light for layout and guide lines. **Figure 3-13** shows the application of line conventions to a technical sketch.

All sketches are composed of straight lines, circles and arcs, irregular curves, and letters and numerals **(Figure 3-14)**. Sketches cannot be prepared with the precision and accuracy of an instrument or CAD drawing. Care must be taken to ensure that dimensions are relatively

VISIBLE LINE	SECTION LINE
BORDER LINE	HIDDEN LINE
CUTTING PLANE	LONG BREAK LINE
SHORT BREAK LINE	PHANTOM LINE
CENTERLINE	LEADER
DIMENSION/EXTENSION LINES	GUIDE LINES/LAYOUT LINES (LIGHT)

Figure 3-12 *Line conventions for sketching working drawings.*

proportional. If dimensional proportions are grossly inaccurate, the sketch will misrepresent the actual appearance of the object (**Figure 3-15**).

When sketching straight lines, squares, and rectangles, use short strokes. Do not attempt to draw continuous lines. Right-angle lines, unless sketched on grid paper, should be laid out and sketched (**Figure 3-16**).

Circles and arcs can be accurately and symmetrically sketched by following the sequence shown in **Figure 3-17a**. Just as the circle was derived from the square in **Figure 3-17a**, all fillets and rounds should first be blocked in square, as shown in **Figure 3-17b**. By following this procedure, proper proportions and symmetry can be maintained.

The procedure and sequence for sketching ellipses is similar to circle sketching (**Figure 3-18**). Sketching accurate angles, other than right angles (90°), can be difficult without using a triangle or protractor. However, by estimating and dividing a right angle into even angles, you can achieve an acceptable level of accuracy (**Figure 3-19**).

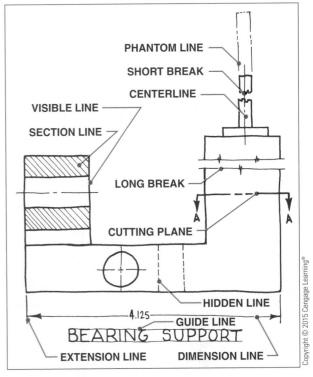

Figure 3-13 *Using line conventions on a working drawing.*

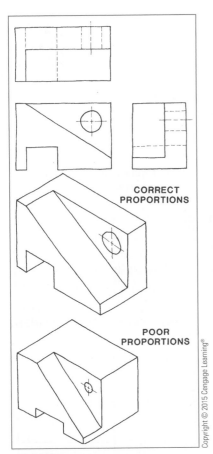

Figure 3-15 *Well-proportioned drawings have superior communications.*

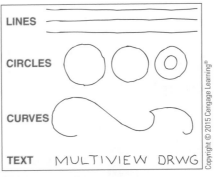

Figure 3-14 *Four basic drawing forms make up all drawings.*

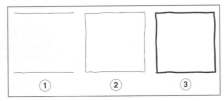

Figure 3-16 *Sketching straight lines, squares, and rectangles.* Copyright © 2015 Cengage Learning®

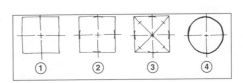

Figure 3-17a *Sketching circles.* Copyright © 2015 Cengage Learning®

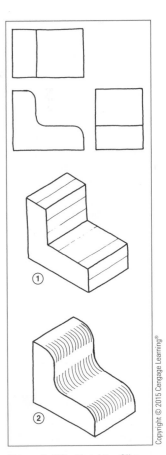

Figure 3-17b *Sketching fillets and rounds.*

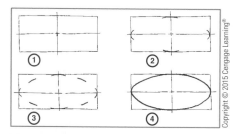

Figure 3-18 *Sketching ellipses.*

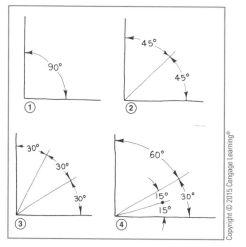

Figure 3-19 *Sketching and estimating angles.*

Figure 3-20 *Use a comfortable pencil grip for sketching.*

Figure 3-23 *Shade (darken) the surfaces opposite the light source.*

When sketching, hold the pencil comfortably (**Figure 3-20**) and pull it (**Figure 3-21**), never push it. To maintain a consistently rounded point and avoid wearing a flat chisel point, rotate the pencil frequently (**Figure 3-22**). When erasing soft pencil sketches, use a good quality medium-soft eraser.

SHADING

Surfaces exposed to a major light source will appear bright. Conversely, surfaces not directly exposed to a light source will be shaded; therefore, adding shading to sketches creates realism (**Figure 3-23**). In **Figure 3-23**, the light source is located above and to the left of the object. The opposite areas are shaded because direct light is blocked from these surfaces. **Figure 3-24** shows techniques for shadowing (shading) these areas.

Surfaces are rarely totally hidden from a light source. Some surfaces are very light, very dark, or appear in a variety of shadow grades depending on the position, intensity, and number of light sources. **Figure 3-25** shows an object with three light levels: light, medium, and dark. In addition, this figure shows a separate shadow cast by the object.

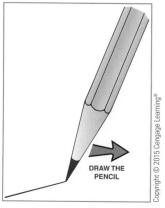

Figure 3-21 *Always pull the pencil—never push it.*

Figure 3-22 *Rotate the pencil frequently for rounded points.*

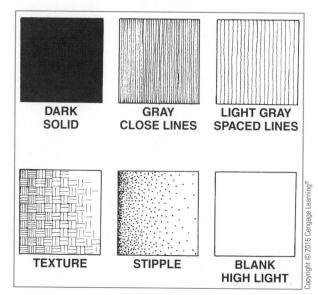

Figure 3-24 *Shading techniques.*

Light travels in a straight line and cannot bend around corners unless reflected; therefore, light intensity on surfaces always changes at the corners of objects. **Figure 3-26** shows several methods of sketching these differences to add realism to a sketch. On objects without corners, such as spheres and cylinders, light and dark areas change gradually. **Figure 3-27** shows several methods of shadowing a cylinder sketch to add realism.

DIMENSIONING

When sketches are used for instruction, manufacturing, or construction, dimensions are usually necessary to adequately describe the object. When sketches are dimensioned, the overall width, depth, and height dimensions are placed on the outside of the location dimensions (**Figure 3-28**). These are known as overall dimensions. Dimension lines are sketched parallel to object lines and connected to the object with extension lines and arrows. Location dimensions show the location of parts of an object. Size dimensions show the size of any hole or projection on the object and are placed between the object and the overall dimension lines. Dimensions are usually shown on multiview sketches. Pictorial sketches (**Figure 3-29**) are normally used only to show the general appearance of products and do not require dimensions.

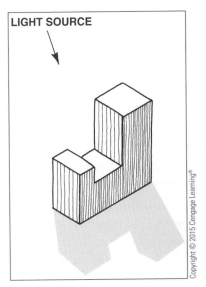

Figure 3-25 *The light source dictates the position of the shadows.*

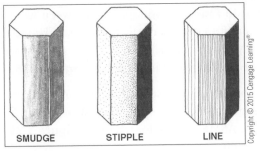

Figure 3-26 *Freehand shading of short corners.*

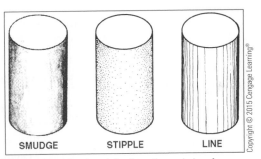

Figure 3-27 *Freehand shading of rounded surfaces.*

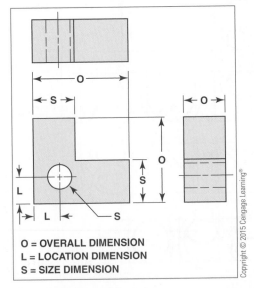

O = OVERALL DIMENSION
L = LOCATION DIMENSION
S = SIZE DIMENSION

Figure 3-28 *Types of dimensions.*

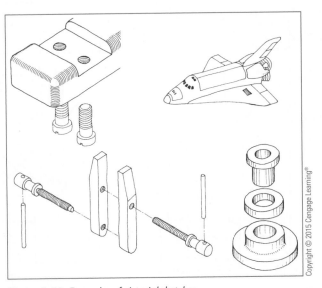

Figure 3-29 *Examples of pictorial sketches.*

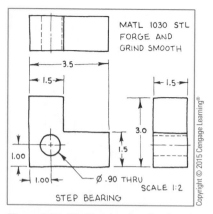

Figure 3-30 *Working drawing sketch with dimensions and notations.*

Figure 3-31 *Single-stroke vertical Gothic lettering (ASME standards).*

LETTERING

Some information is best communicated with words or numerals. Almost every sketch or drawing contains notes, labels, and dimensions that must be legible and consistent. In **Figure 3-30**, the title, material, and hole size are described with a lettered notation. Each numeral used for dimensioning must be distinct from all other numerals. The letter *D* which looks like an *O*, or the numeral *3*, which looks like an *8*, can be easily misread and create costly manufacturing or construction mistakes. For this reason, the **American Society of Mechanical Engineers (ASME)** style of letters and numerals (**Figure 3-31**) is used by most American industries. **Figure 3-32** shows the most efficient and quickest method of making these letters and numerals. A slanted form of this style (**Figure 3-33**) is acceptable but is rarely used or recommended for engineering drawings.

To letter most efficiently and legibly, always use guide lines (**Figure 3-34**). Guide lines keep letters consistent and contained in the area selected. As a general rule, most lettering is ⅛" high with ³⁄₁₆"-high title headings. Stylized lettering may be used for architectural drawings. Several examples are shown in **Figure 3-35**.

Correct spacing between letters is also needed to produce lettering that is most readable. **Figure 3-36** shows lettering that has the area between the letters approximately equal. By contrast, **Figure 3-37** shows wide and inconsistent spacing of letters. Notice how difficult it is to quickly read **Figure 3-37** compared to **Figure 3-36**. Spacing between words and sentences is also important for effective reading. The space between words should be approximately equal to the height of the letters. The space between sentences should be twice the height of the letters (**Figure 3-38**).

Fractions are among the most frequent causes of dimensional errors. If a numerator or denominator is misread as a whole number, the result can be disastrous. For this reason, the ASME spacing for fractions (**Figure 3-39**) should be used to avoid confusion.

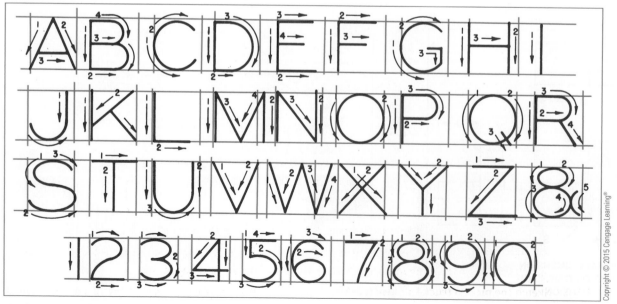

Figure 3-32 *Recommended lettering strokes.*

When all of the procedures and guidelines for sketching are combined, the result can be a readable and functional technical sketch as shown at the left in **Figure 3-40**. If these guidelines are ignored, the confusing series of lines and letters shown at right in **Figure 3-40** can result.

ABCDEFGHIJKLMNOPQRSTUVWXYZ
1234567890

Figure 3-33 *Single-stroke slant (68°) Gothic lettering.* Copyright © 2015 Cengage Learning®

A B C D E F G H I J K L M N O P Q R S T U V W X Y Z

Figure 3-34 *Guide lines will aid freehand lettering.* Copyright © 2015 Cengage Learning®

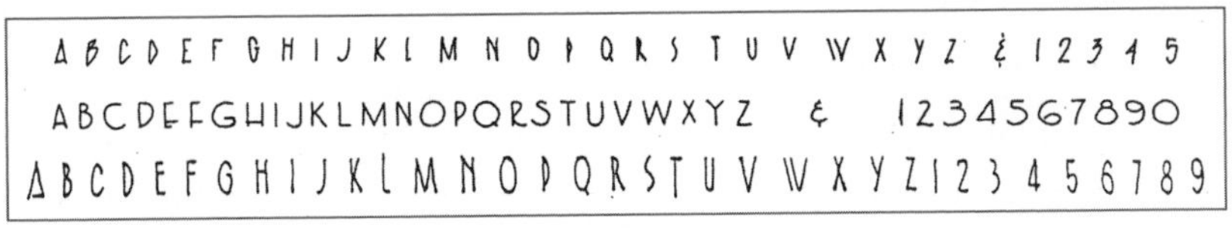

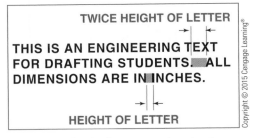

Figure 3-35 *Examples of stylized architectural lettering.* Copyright © 2015 Cengage Learning®

FUNDAMENTALS OF DRAFTING WITH AUTOCAD

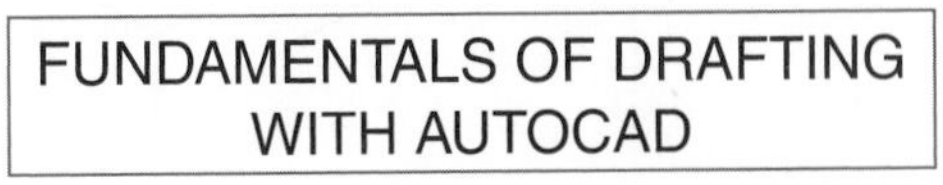

Figure 3-36 *The area between letters should be as equal as possible.* Copyright © 2015 Cengage Learning®

FU NDAM ENTAL S O F DRAF T IN G
WI TH AU TOC A D

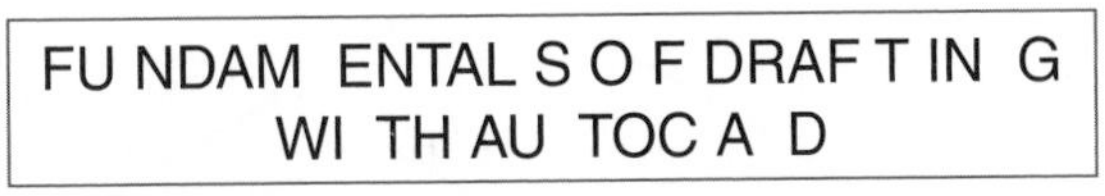

Figure 3-37 *Inconsistent areas between letters create reading difficulties.* Copyright © 2015 Cengage Learning®

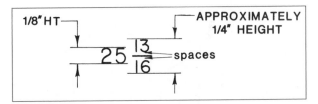

Figure 3-39 *Fractions on the working drawing.* Copyright © 2015 Cengage Learning®

TWICE HEIGHT OF LETTER

THIS IS AN ENGINEERING TEXT FOR DRAFTING STUDENTS. ALL DIMENSIONS ARE IN INCHES.

HEIGHT OF LETTER

Figure 3-38 *Spacing between words and sentences.*

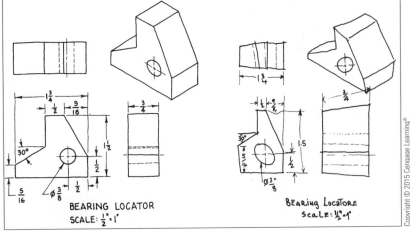

Figure 3-40 *Which sketch gives better communication?*

DRAFTING EXERCISES

1. Sketch the isometric drawings in **Figures 3-41** through **3-47** on isometric grid paper.

2. Sketch the multiview drawings in **Figure 3-48**.

3. Sketch the pickup truck in **Figure 3-49**.

4. Sketch the objects in **Figure 3-50**.

5. Sketch the tool-bit holder in **Figure 3-51**.

6. Practice lettering single-stroke Gothic lettering on $^1/_8$" grid paper.

7. Practice placing various types of fonts in different sizes and angular positions with a CAD system.

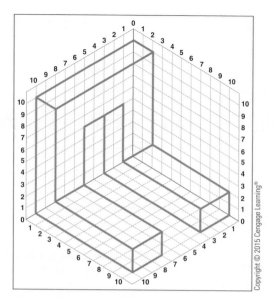

Figure 3-41

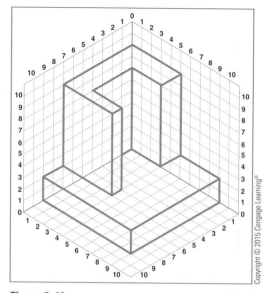

Figure 3-42

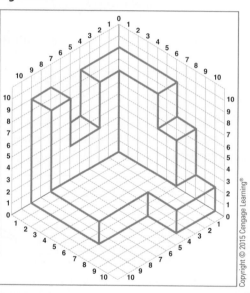

Figure 3-43

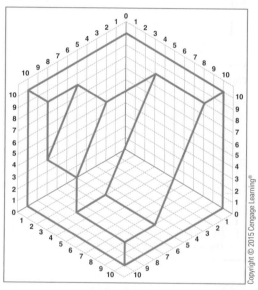

Figure 3-44

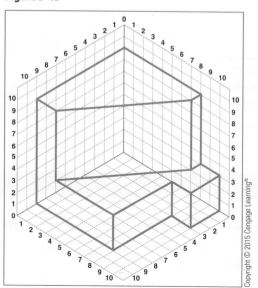

Figure 3-45

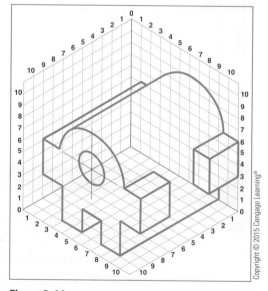

Figure 3-46

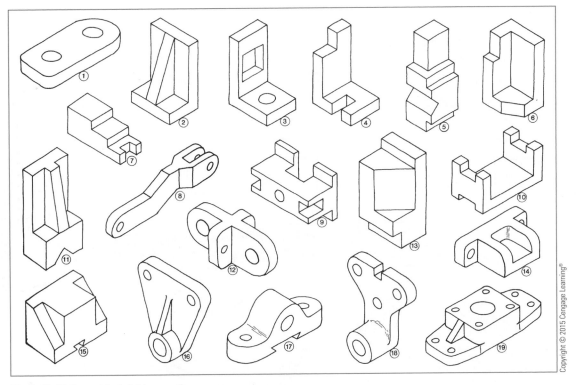

Figure 3-47 *Isometric sketching practice.*

Figure 3-48 *Orthographic multiview sketching practice.*

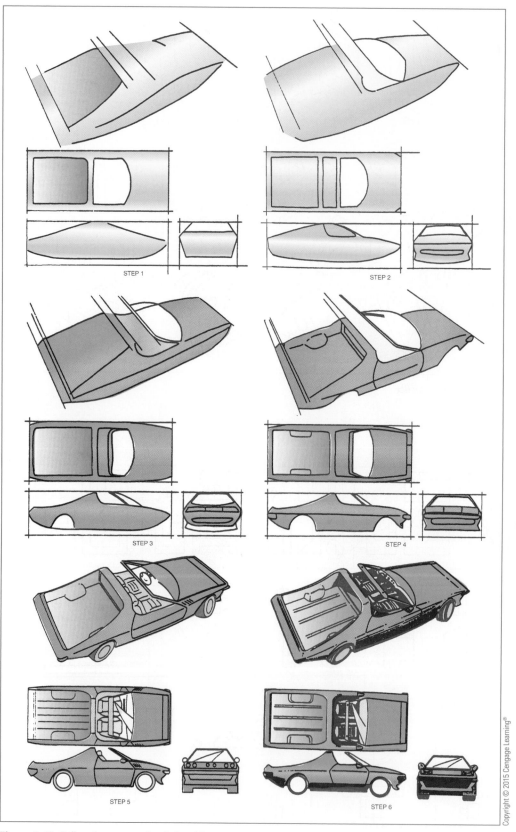

Figure 3-49 *Follow the steps to sketch the pickup.*

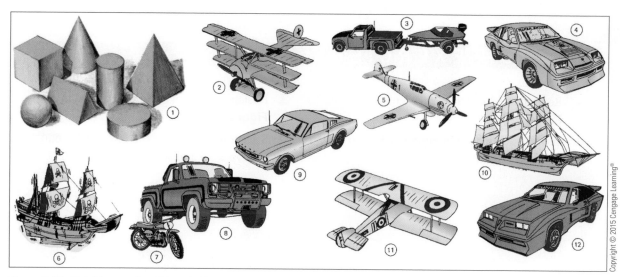

Figure 3-50 *Sketching practice with real objects.*

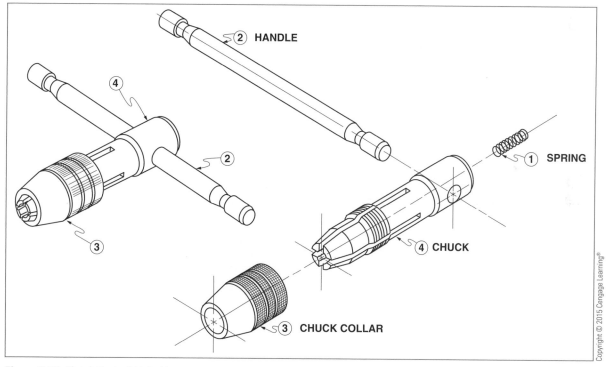

Figure 3-51 *Sketch the tool-bit holder.*

DESIGN EXERCISES

1. Redesign and sketch the hexagon wrench in **Figure 3-52**.

2. Sketch the bookend in **Figure 3-53** and design your stylized initials into the surface.

3. Design a new tent stake as specified in **Figure 3-54**.

4. Select one or more of the industrial products in **Figure 3-55** and sketch a new design.

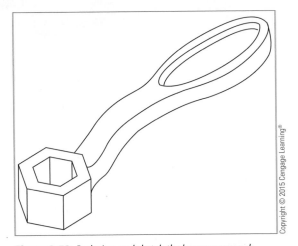

Figure 3-52 *Redesign and sketch the hexagon wrench.*

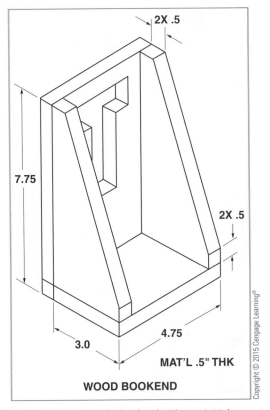

WOOD BOOKEND

Figure 3-53 *Sketch the bookend with your initials.*

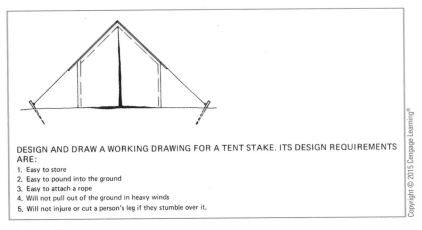

DESIGN AND DRAW A WORKING DRAWING FOR A TENT STAKE. ITS DESIGN REQUIREMENTS ARE:

1. Easy to store
2. Easy to pound into the ground
3. Easy to attach a rope
4. Will not pull out of the ground in heavy winds
5. Will not injure or cut a person's leg if they stumble over it.

Figure 3-54

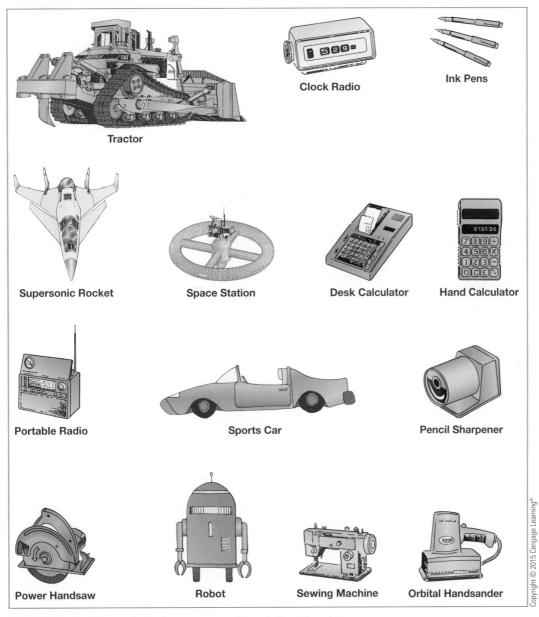

Tractor

Clock Radio

Ink Pens

Supersonic Rocket

Space Station

Desk Calculator

Hand Calculator

Portable Radio

Sports Car

Pencil Sharpener

Power Handsaw

Robot

Sewing Machine

Orbital Handsander

Figure 3-55 *Sketch a new design for one or more of these industrial products.*

 KEY TERMS

American Society of Mechanical Engineers (ASME)

Guide lines

Isometric drawing

Line conventions

Location dimensions

Multiview drawing

Oblique drawing

One-point perspective

Orthographic projection

Overall dimensions

Perspective drawing

Pictorial drawing

Size dimension

Three-point perspective

Two-point perspective

Vanishing point

Working drawing

4

Introduction to Computer-Aided Drafting Systems

The student will be able to:

- Describe the role of CAD in today's industry
- Identify the major hardware components of a CAD system
- Describe the functions of a CAD's system major hardware
- Describe the function and role of engineering software programs
- Describe the advantages of a CAD system over manual drafting

Introduction

In prehistoric times objects were drawn in the soil with a stick. Later, drawings were made on the walls of a cave (**Figure 4-1**). Eventually drawings and manuscripts were cut or scratched into the surfaces of stones. As time passed, drawings were produced on various flexible surfaces with a stylus. Much later in history, paper and drawing instruments were developed as the drawing became more complex, as discussed in Chapter 2. For the past hundred years, drafters have used freehand and manual drafting techniques. The latest development is

This CAD workstation has a computer, keyboard, mouse, and flat-panel monitor.

Copyright © 2015 Cengage Learning®

Figure 4-1 *Early Humans drew on walls.*

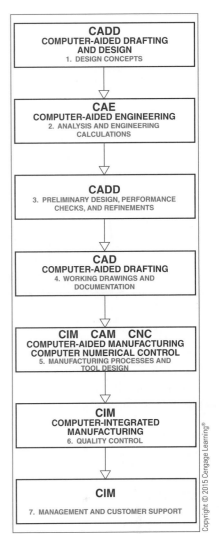

Figure 4-2 *Flow chart for CAD-related processes.*

the **computer-aided drafting** and design systems (CAD/CADD). With these systems, two-dimensional (2D) and three-dimensional (3D) drawings can be generated quickly and accurately. They have also been designed to integrate working drawings into the manufacturing of the drawn item (**Figure 4-2**). Note the relationship of the quality of line work, readability, and the time frame a professional drafter used to produce the drawings in **Figure 4-3a, b, c,** and **d**.

Learning how to use a CAD system and its engineering software program will take some time. Then after becoming proficient with the system, engineering and architectural drawings may be generated at a much greater rate with more accuracy than with manual drafting. The ease and speed with which drawings can be drawn, digitally stored, and retrieved results in fast accurate changes, and revisions is one of CADs greatest features.

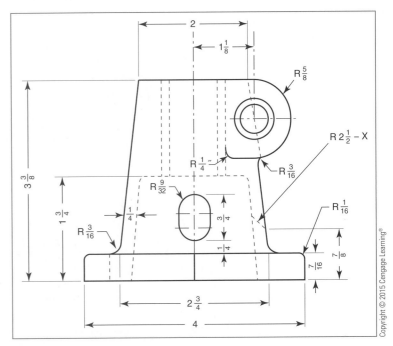

Figure 4-3a *Freehand sketch - 25 minutes.*

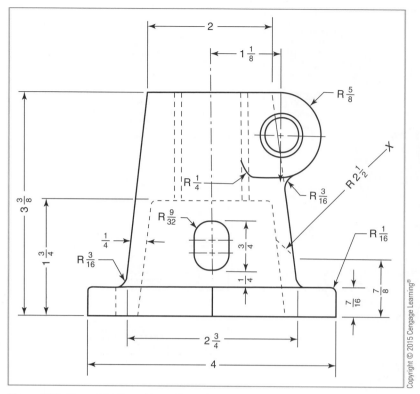

Figure 4-3b *Manual drafting with pencils - 1.5 hours.*

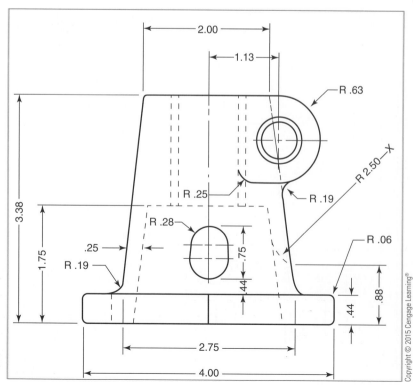

Figure 4-3c *Manual drafting with technical pens - 2.25 hours.*

In addition to generating drawings, a CAD system can produce related documents, mathematical calculations, graphics analyses, and budgets. These documents can be transmitted worldwide over the Internet.

There are two specific types of computer systems: **microcomputer** systems and the mainframe computers. Microcomputers may be **stand-alone** systems (**Figure 4-4**) or **networked** so the operators can share their work and programs with each other. A **mainframe system** (**Figure 4-5**) consists of a large central computer with many workstations that access the mainframe's data.

Before a CAD operators can successfully create working drawings, they must have knowledge of the standards and practices with the subject areas of engineering graphics and/or architectural design. The CAD system is simply an electronic drafting tool. Remember that a CAD system cannot design and generate a working drawing any more than a pencil, pen, or typewriter can write a book.

The Computer System

The **motherboard** is the main component inside the computer case. It is a large rectangular board with the **integrated circuits (ICs)** that connect all the components, plus all the external components with **USB**, **serial**, and/or **parallel ports** (**Figures 4-6** and **4-7**).

The most important part of the computer is the **central processing unit (CPU)**. It is called the brain or engine in a computer system and is about the size of a postage stamp. It controls the speed, power, calculations, and controls all other components.

The **power supply unit (PSU)** converts the alternating current (AC) to low-voltage direct current (DC) and provides the power for the internal components of the computer.

An **internal bus** connects the CPU to various internal components and to expansion cards such as graphics cards and sound cards.

A **modem** is a telecommunications device that allows the sending and receiving over telephone wires, optic cable lines, or a wireless system.

Memory

The storage of memory in a computer comes in two basic categories: long term and short term. The long term memory or **read-only memory (ROM)** is permanent. It cannot be erased or changed. It is the fixed data that the computer uses while it is operating. The short-term memory or **random-access memory (RAM)** operates only when the computer is turned on. The first thing that goes into the RAM is the operating system (**OS**), followed by the software commands. Large software programs will require a large amount of RAM.

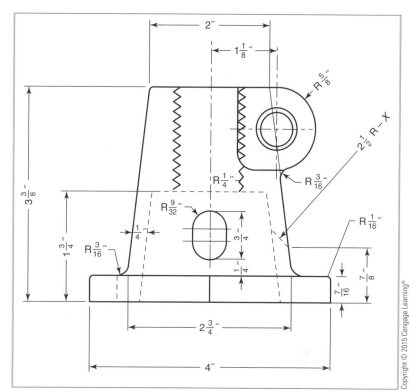

Figure 4-3d *CAD drawing - 35 minutes.*

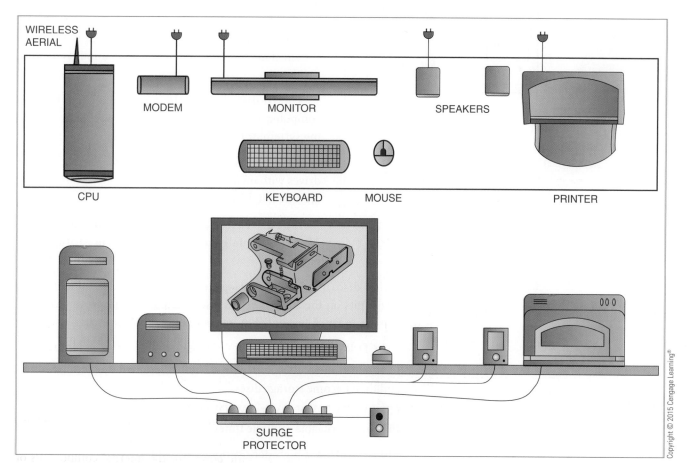

Figure 4-4 *A stand-alone computer workstation.*

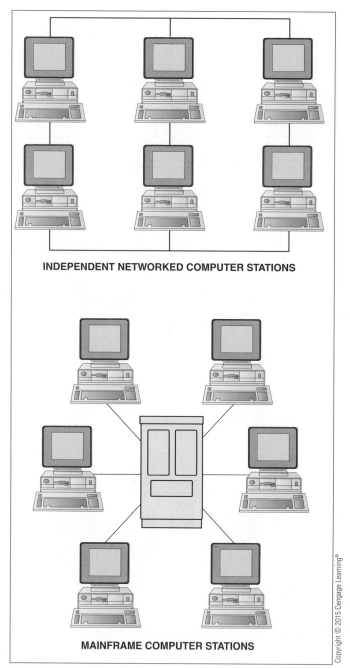

INDEPENDENT NETWORKED COMPUTER STATIONS

MAINFRAME COMPUTER STATIONS

Figure 4-5 *Computer systems.*

The digital computer represents all the data's memory for operating with the **binary system**. The computer's memory is measured in bytes. All forms of information are converted into strings of bits (binary digits with a value of 1 or 0). The most common unit of data is the **byte** that has 8 bits. An example of a binary byte is 00110001. Some common units to measure memory are:

Kilobyte (KB) 1,024 bytes—thousand

Megabyte (MB) 1,048,567 bytes—million

Gigabyte (GB) 1,073,741,824 bytes—billion

Terabyte (TB) 1,099,511,628,000 bytes—trillion

Data Storage Devices

Storage capacity is not the same as memory. Memory enables a computer to function properly. Storage devices are used to store software programs and the work created on the computer in files. There are several categories of digital storage: primary storage, secondary storage, tertiary storage, and offline storage. **Primary storage** is the permanent internal memory, such as ROM, stored in the computer. **Secondary storage** is a mass storage device that is removable. **Tertiary storage** is a robotic system that stores and retrieves large amounts of storage devices.

A **hard disk drive** is the main storage device in a computer. It uses magnetic currents to store the operating system, software, and all the drawing files.

Flash drives (thumb drives) are removable storage drives with memory chips on which data is stored. They are small, lightweight, and rewritable.

Tape drives read and write data on a **magnetic tape**. They are usually used for long-term storage and backups.

Optical discs uses laser technology to write and read data from the long and short pits on the surface of the disc. There are many types of optical discs in use today. A few of the more commonly used discs are:

- CD, CD-ROM, DVD, BD-ROM — These discs are read-only (**Figure 4-8**).
- CD-R, DVD-R, DVD+R, BD-R — These discs are writeable once only and then become storage discs.
- CD-RW, DVD-RW, DVD+RW, DVD-RAM, BD-RE — These are discs that are rewriteable multiple times.

A **Blu-ray disc** is a high-density optical disc for high-definition video. It can store 70 times more data than a **compact disc (CD)**.

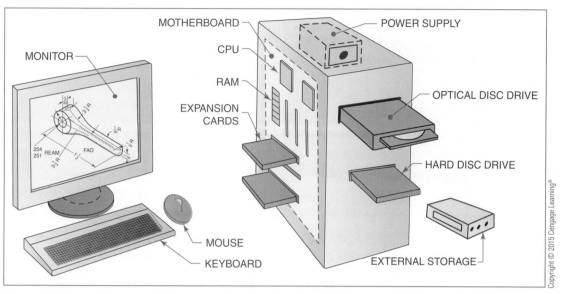

Figure 4-6 *The component parts for a typical computer station.*

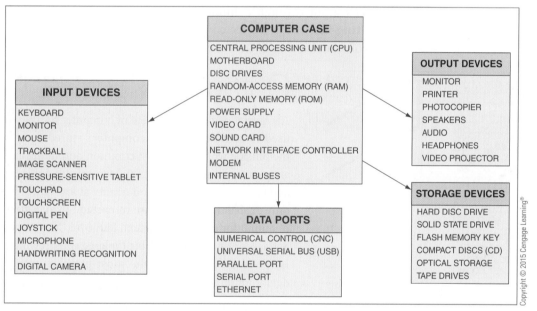

Figure 4-7 *The computer with peripherals.*

Input Devices

Computer **input** is commands and information fed into a computer. Input devices are important because they are the link between the computer and the operator. These devices are housed externally to the computer.

The **keyboard** is similar to a typewriter (**Figure 4-9**). A user inputs data into the computer with a keyboard using specific keys for drawing commands or text. The keyboard may be connected to the computer with a cable or be wireless. The keys are usually spring-based. Smaller keyboards are provided for laptops and PDAs. New versions may use virtual keys, or laser-projected keyboards on a flat surface.

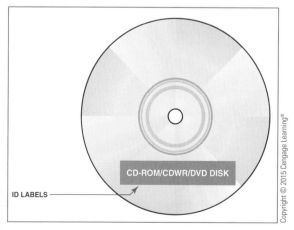

Figure 4-8 *CD-ROM (also CDRW and DVD).*

Figure 4-9 *Typical keyboard.*

The **monitor** resembles a television screen. The cathode ray tube (CRT) monitor has been replaced with the liquid crystal display (LCD) monitor. A newer, but expensive, monitor is the organic **light-emitting diode (LED)** monitor. For CAD work, a large monitor with a high resolution for a sharper line images is recommended. The monitor has the option to be an input and an **output device**. The higher the number of **pixels** per square inch (PSI), The better the line clarity and sharpness of the output on the screen and prints (**Figure 4-10**).

Pointing devices allows the operator to move a **cursor** about the screen and select specific areas. A pointing device may be a **mouse** (**Figure 4-11**), trackball, touch screen, or a stylus pen.

An **optical image scanner** can scan existing drawings and text into the computer's memory. A **voice-activation device** controls the computer's input through commands spoken into a microphone. A **handwriting-activation device** controls the computer's input through written commands on an electronic pad that recognizes handwriting.

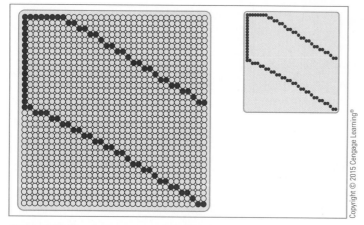

Figure 4-10 *The display on a monitor with a low number of pixels will have low resolution.*

Output Devices

An output device is any piece of computer **hardware** used to communicate the computer's stored data to the outside world. The most common output devices are **printers** and speakers. Desktop **ink-jet printers** and **laser printers** offer good quality at low cost. The only problem is that they are restricted to small sizes of output for drawings and text. Large ink-jet plotters and laser printers are available for large size drawings. Large **photocopiers** are the most popular output device for large drawings.

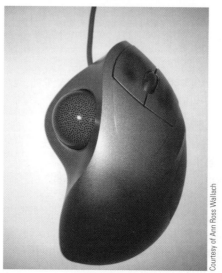

Figure 4-11 *The mouse comes in various shapes.*

CAD Software

CAD software contains programs that instruct the hardware devices to perform specific tasks for the drafter. Regardless of how proficient the drafter is with the software program, without a knowledge of engineering and architectural drawing, the working drawings produced will be worthless. Most CAD programs have many hundreds of drawing commands to create drawings. While there is a long learning curve, once a user is familiar with the drawing task commands, drawings become easier and faster to create, and more accurate.

Some of the many engineering software programs are: AutoCAD, AutoCAD LT, Solid Works, Pro/Engineer, and Inventor. Some architectural programs are: Chief Architect, Envisioneer, Revit, and 3ds MAX.

Most CAD systems will work on one of the major platforms such as Windows, Mac OS X, Linux, or UNIX. These platforms employ a **graphical user interface (GUI)**, which controls the location of a pointer on the monitor.

Computer-aided design is only one of other manufacturing uses by engineers:

- **Computer-aided engineering (CAE)**
- **Computer-aided manufacturing (CAM)**
- **Computer-numerical control (CNC)**
- **Finite-element analysis (FEA)**
- **Product-data management (PDM)**

A new software program called Building Information Model (BIM) has become very popular in industry. BIM permits a seamless integration for the design and manufacturing of a product. Some of its operations:

- Uses parametrics, so when a change is made, every sheet in the set of working drawings is updated with the change
- Allows designing in 3D while the 2D drawings are automatically updated
- Coordinates all the written documentation
- Links together all the detailed data for the drawings and calculations into a database
- Has real-time ordering and pricing for all the materials required for the manufacture of the specific item
- Allows coordination, visualization, analysis, and conception of the final design before manufacturing.

CAD Drawings

CAD systems can produce many types of working drawings in two-dimensions (**2D**) and three-dimensions (**3D**). **Wireframe** drawings are single line see through drawings shown in 3d (**Figure 4-12**). **Surface models** drawings have solid surfaces and show the item in 3D. These drawings may be rotated to show all the surfaces. **Solid models** appear the same as surface models, except that solid models have mass properties allowing the operator to do structural analysis, sectional views, and measure mass, and density. They may also be rendered. **Virtual reality systems** create a computer-generated reality scene in which the user seems to be immersed in the computer images.

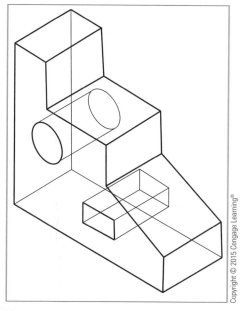

Figure 4-12 *Wire frame drawing.*

The CAD drawing commands shown in **Figures 4-13a, b,** and **c,** are typical for most CAD software programs. With these drawing commands a CAD operator may generate most types of drawings as shown in **Figures 4-14**, **4-15**, **4-16**, and **4-17**.

COMMAND	ICON DIAGRAMS	DESCRIPTION
LINE		DRAWS A LINE FROM ONE SELECTED POINT TO ANOTHER
ARC		DRAWS AN ARC THROUGH THREE SELECTED POINTS
CIRCLE		DRAWS A CIRCLE. THE CENTER IS SELECTED, THEN THE RADIUS. THE RADIUS MAY BE TYPED IN OR SPECIFIED WITH THE PUCK.
ELLIPSE		DRAWS AN ELLIPSE. THE MINOR AND MAJOR DIAMETERS ARE SPECIFIED AND THE ELLIPSE IS DRAWN. IT MAY THEN BE ROTATED AND MOVED.
POLYGON		DRAWS A POLYGON OF A SPECIFIED NUMBER OF SIDES AND SIZE. THE NUMBER OF SIDES IS TYPED IN, AND THE SIZE MAY BE SPECIFIED BY ITS INSIDE OR OUTSIDE DIAMETER.
POINT	SELECT POINT STYLE & SIZE	PLACES A POINT ON THE DRAWING. THE SIZE AND SHAPE OF THE POINT MAY BE SELECTED BY THE OPERATOR. THIS ONE IS A SIMPLE CROSS.
DONUT	SELECT ID & OD	DRAWS 2 CONCENTRIC CIRCLES WITH A SPECIFIED INSIDE AND OUTSIDE DIAMETER.
DIMEN-SION, DIAMETER	⌀0.1642	IF THE DIAMETER DIMENSION IS TOO SMALL TO APPEAR INSIDE THE CIRCLE, IT IS DISPLAYED ON THE OUTSIDE WITH A LEADER.
DIMEN-SION, DIAMETER	⌀0.7232	DIMENSION THE DIAMETER OF A CIRCLE. THE CIRCLE IS SELECTED, THE COMPUTER MEASURES IT AND DISPLAYS THE DIMENSION PRECEDED WITH THE ⌀ SYMBOL FOR DIAMETER.
DIMEN-SION, RADIUS	R 0.2934	DIMENSION AN ARC. THE ARC IS SELECTED, THE COMPUTER MEASURES IT AND DISPLAYS THE DIMENSION PRECEDED BY AN R.
DIMEN-SION, ARC	60°	DIMENSION AN ARC. THE TWO LINES OF THE ARC ARE SELECTED AND THE DIMENSION IS DISPLAYED WITH THE ANGLE SYMBOL.
DIMEN-SION, VERTICAL	0.6106	A VERTICAL DIMENSION IS DISPLAYED BY SELECTING THE BEGINNING OF THE TWO EXTENSION LINES AND THE LOCATION OF THE DIMENSION.

A

Figure 4-13a *Typical CAD software program drawing commands.*

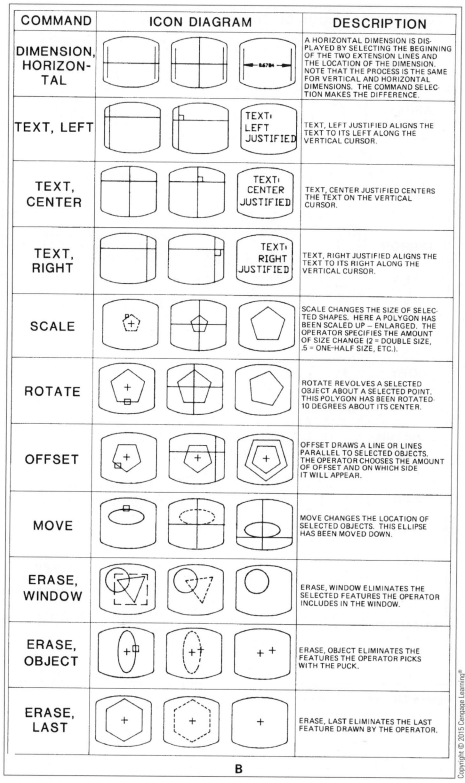

COMMAND	ICON DIAGRAM	DESCRIPTION
DIMENSION, HORIZONTAL		A HORIZONTAL DIMENSION IS DISPLAYED BY SELECTING THE BEGINNING OF THE TWO EXTENSION LINES AND THE LOCATION OF THE DIMENSION. NOTE THAT THE PROCESS IS THE SAME FOR VERTICAL AND HORIZONTAL DIMENSIONS. THE COMMAND SELECTION MAKES THE DIFFERENCE.
TEXT, LEFT		TEXT, LEFT JUSTIFIED ALIGNS THE TEXT TO ITS LEFT ALONG THE VERTICAL CURSOR.
TEXT, CENTER		TEXT, CENTER JUSTIFIED CENTERS THE TEXT ON THE VERTICAL CURSOR.
TEXT, RIGHT		TEXT, RIGHT JUSTIFIED ALIGNS THE TEXT TO ITS RIGHT ALONG THE VERTICAL CURSOR.
SCALE		SCALE CHANGES THE SIZE OF SELECTED SHAPES. HERE A POLYGON HAS BEEN SCALED UP – ENLARGED. THE OPERATOR SPECIFIES THE AMOUNT OF SIZE CHANGE (2 = DOUBLE SIZE, .5 = ONE-HALF SIZE, ETC.).
ROTATE		ROTATE REVOLVES A SELECTED OBJECT ABOUT A SELECTED POINT. THIS POLYGON HAS BEEN ROTATED 10 DEGREES ABOUT ITS CENTER.
OFFSET		OFFSET DRAWS A LINE OR LINES PARALLEL TO SELECTED OBJECTS. THE OPERATOR CHOOSES THE AMOUNT OF OFFSET AND ON WHICH SIDE IT WILL APPEAR.
MOVE		MOVE CHANGES THE LOCATION OF SELECTED OBJECTS. THIS ELLIPSE HAS BEEN MOVED DOWN.
ERASE, WINDOW		ERASE, WINDOW ELIMINATES THE SELECTED FEATURES THE OPERATOR INCLUDES IN THE WINDOW.
ERASE, OBJECT		ERASE, OBJECT ELIMINATES THE FEATURES THE OPERATOR PICKS WITH THE PUCK.
ERASE, LAST		ERASE, LAST ELIMINATES THE LAST FEATURE DRAWN BY THE OPERATOR.

B

Figure 4-13b *Typical CAD software program drawing commands.*

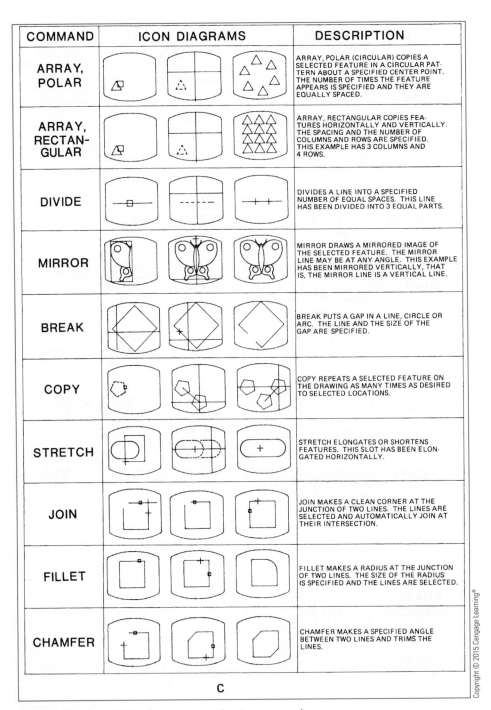

COMMAND	ICON DIAGRAMS	DESCRIPTION
ARRAY, POLAR		ARRAY, POLAR (CIRCULAR) COPIES A SELECTED FEATURE IN A CIRCULAR PATTERN ABOUT A SPECIFIED CENTER POINT. THE NUMBER OF TIMES THE FEATURE APPEARS IS SPECIFIED AND THEY ARE EQUALLY SPACED.
ARRAY, RECTANGULAR		ARRAY, RECTANGULAR COPIES FEATURES HORIZONTALLY AND VERTICALLY. THE SPACING AND THE NUMBER OF COLUMNS AND ROWS ARE SPECIFIED. THIS EXAMPLE HAS 3 COLUMNS AND 4 ROWS.
DIVIDE		DIVIDES A LINE INTO A SPECIFIED NUMBER OF EQUAL SPACES. THIS LINE HAS BEEN DIVIDED INTO 3 EQUAL PARTS.
MIRROR		MIRROR DRAWS A MIRRORED IMAGE OF THE SELECTED FEATURE. THE MIRROR LINE MAY BE AT ANY ANGLE. THIS EXAMPLE HAS BEEN MIRRORED VERTICALLY, THAT IS, THE MIRROR LINE IS A VERTICAL LINE.
BREAK		BREAK PUTS A GAP IN A LINE, CIRCLE OR ARC. THE LINE AND THE SIZE OF THE GAP ARE SPECIFIED.
COPY		COPY REPEATS A SELECTED FEATURE ON THE DRAWING AS MANY TIMES AS DESIRED TO SELECTED LOCATIONS.
STRETCH		STRETCH ELONGATES OR SHORTENS FEATURES. THIS SLOT HAS BEEN ELONGATED HORIZONTALLY.
JOIN		JOIN MAKES A CLEAN CORNER AT THE JUNCTION OF TWO LINES. THE LINES ARE SELECTED AND AUTOMATICALLY JOIN AT THEIR INTERSECTION.
FILLET		FILLET MAKES A RADIUS AT THE JUNCTION OF TWO LINES. THE SIZE OF THE RADIUS IS SPECIFIED AND THE LINES ARE SELECTED.
CHAMFER		CHAMFER MAKES A SPECIFIED ANGLE BETWEEN TWO LINES AND TRIMS THE LINES.

C

Figure 4-13c *Typical CAD software program drawing commands.*

The Cartesian System

The 2D drawings are based on a **Cartesian coordinate system**. All coordinate Cartesian points are based on a X-axis and a Y-axis at right angles (**Figure 4-18**). The intersection point is the origin. The X-axis to the right of the origin is positive. To the left of the origin, the X-axis is negative. The Y-axis going up from the origin is positive, and going down is

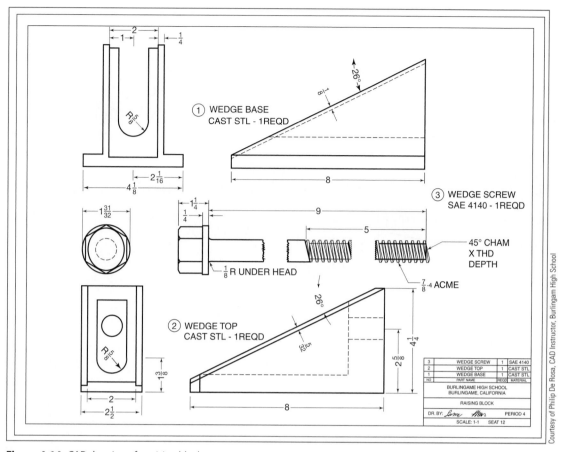

Figure 4-14 *CAD drawing of a raising block.*

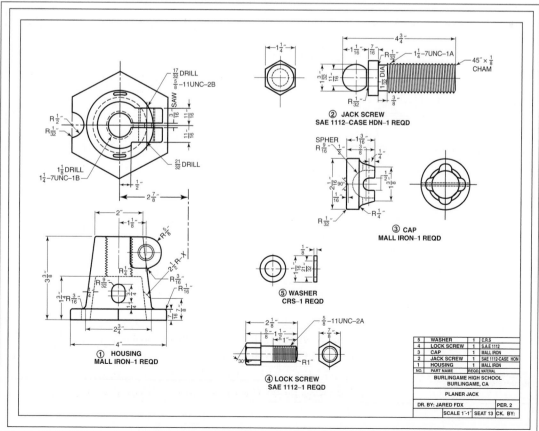

Figure 4-15 *CAD drawing of a planer jack.*

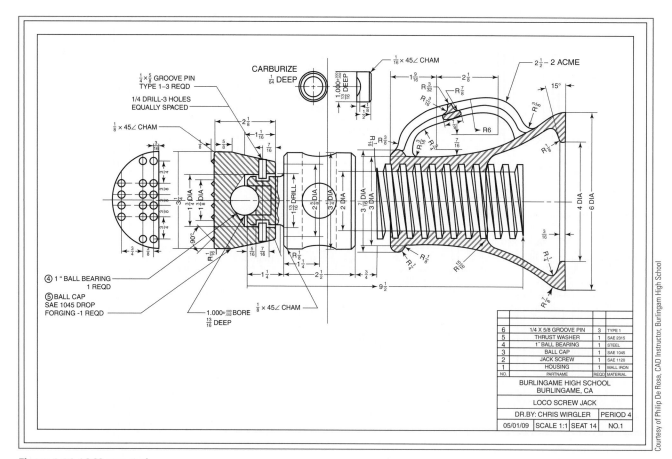

Figure 4-16 *LOCO screw jack.*

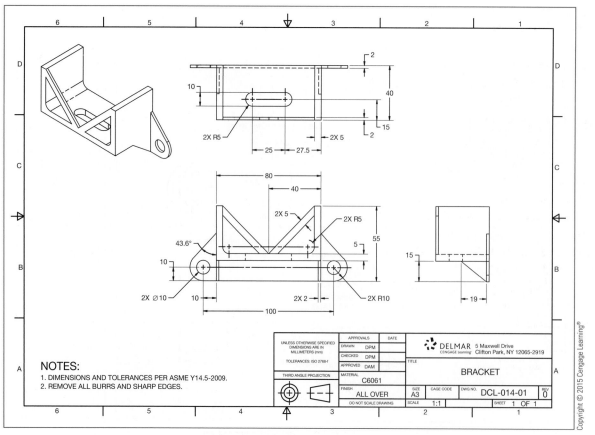

Figure 4-17 *3D CAD drawing.*

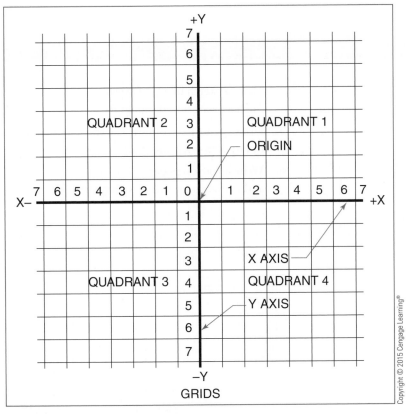

Figure 4-18 *Example of the Cartesian quadrants.*

negative. Note the Cartesian point in quadrant 2 (**Figure 4-19**). The X-coordinate is always given first, followed by a comma, then the Y coordinate. Read the six coordinate points in **Figure 4-19**, and count the positive and negative grid spaces for each.

Conclusion

CAD offers many advantages once the drafter has a knowledge of and background in engineering drawing and has mastered the drawing commands in the software. The time saved, accuracy, and convenience of having the drawings in a computer file will have provide great economic savings in every design project.

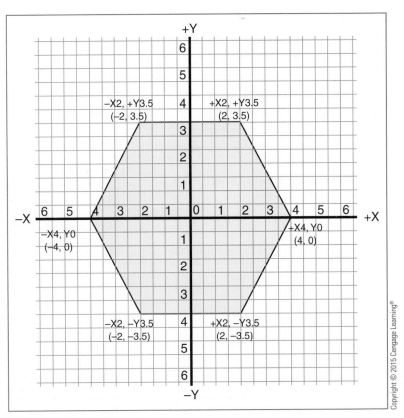

Figure 4-19 *Cartesian point call-outs.*

DRAFTING EXERCISES

1. Take a field trip to a vendor selling CAD systems. Make a list of the various types and the manufactures' addresses, and write for additional information.

2. Take a field trip to an industry or office with a CAD system. Observe its operation and make a list of the software programs observed.

3. Compare the prices of CAD hardware and software.

4. Make a sketch of a CAD system, and label all the hardware and peripherals.

5. Draw the forms from the Cartesian data table in **Figures 4-20** and **4-21** on ¼" grid paper.

6. Write a Cartesian data table for the multiview and isometric drawings in **Figure 4-22**.

DATA TABLE

LN	X	Y	GO TO	X	Y
1	0	0	–	0	7
2	0	7	–	4	3
3	4	3	–	6	3
4	6	3	–	6	0
5	6	0	–	0	0
			NEW START		
1	1	8	–	2.5	9.5
2	2.5	9.5	–	6.5	5.5
3	6.5	5.5	–	5	4
4	5	4	–	1	8
			NEW START		
1	8	7	–	10	7
2	10	7	–	10	0
3	10	0	–	8	0
4	8	0	–	8	7
			NEW START		
5	8	3	–	10	3
			FINISH		

Figure 4-21 *Draw this figure on 1/4" grid paper using the Cartesian data table.*

COORDINATE DATA TABLE

LN	X	Y	go to	X	Y
1	1	3	⟶	4	3
2	4	3	⟶	4	4
3	4	4	⟶	6	3
4	6	3	⟶	4	2
5	4	2	⟶	4	3
			new start		
6	6	4	⟶	6	2
			new start		
7	6	3	⟶	10	3
			FINISH		

Figure 4-20 *Draw this figure on 1/4" grid paper using the Cartesian data table.*

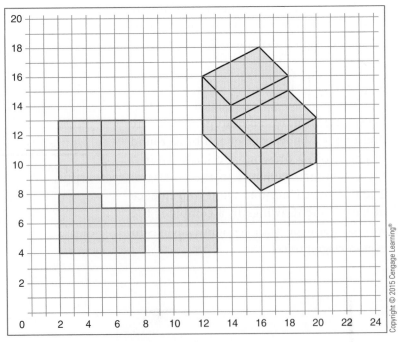

Figure 4-22 *Write a Cartesian data table for the multiview and isometric drawings.*

KEY TERMS

2D

3D

Binary System

Blu-ray disc

Byte

Cartesian coordinate system

Central processing unit (CPU)

Compact disc (CD)

Computer-aided engineering (CAE)

Computer-aided design (CAD)

Computer-aided manufacturing (CAM)

Computer numerical control (CNC)

Cursor

Finite-element analysis

Flash drive

Graphical user interface (GUI)

Handwriting-activation device

Hard disk drive

Hardware

Ink-jet printer

Input

Input device

Integrated circuit (IC)

Internal bus

Keyboard

Laser printer

Light-emitting diode

Magnetic tape

Mainframe system

Microcomputer

Modem

Monitor

Motherboard

Mouse

Networked

Optical disc

Optical image scanner

Output device

Parallel port

Photocopier

Pixels

Power supply unit (PSU)

Printer

Primary storage

Product data management (PDM)

Random-access memory (RAM)

Read-only memory (ROM)

Secondary storage

Serial port

Software

Solid model

Stand-alone

Surface model

Tape drive

Tertiary storage

USB port

Virtual reality system

Voice-activation device

Wireframe

5

Drafting Room Design Teams

The student will be able to:

- Recognize the importance of teamwork
- Understand the design process for mechanical products
- Understand the design process for residential planning
- Understand the problem-solving process
- Understand the importance of creativity
- Work as part of a design team
- Understand the advantages of scientists, engineers, technicians, and nontechnical personnel working together as a team

Introduction

The curiosity, imagination, and creativity of women and men through history have led to the development of tools that make our lives more efficient and more comfortable. Through the design process, scientists, engineers, architects, technicians, and artists have helped to create the high living standards we enjoy today.

The design process involves taking an orderly approach to combining engineering and scientific data with practical and off-the-cuff ideas to reach a final solution. These ideas are processed with a series of actions when solving a problem or creating a new design. An organized design process is critical because it will meet the demand for faster, better, and less expensive products.

Experience shows that the design process works best with a design team. A design team will have accumulated knowledge and will bring diverse views to the design process. Teams should be selected for their compatibility, knowledge, and skills.

Depending on the size of the project, the team size should be from three to five students. The time element will also vary. For a school project, do not select a project that will take more than a full semester to complete. A smaller project taking five to ten weeks would be preferable.

Creativity in the Design Process

An important quality for the team members to have or to develop is **creativity**. Many of the aspects of creativity can be developed with practice. The aspects of creativity are:

- Using new methods in design
- Looking at a problem from several different points of view
- Having a good attitude and confidence in your design abilities
- Creating new ideas and putting them in logical order
- Observing patterns and relationships within the design process
- Being curious and always asking *why*
- Having an open mind when brainstorming within the team
- Having the ability to accept new ideas and reject old ideas
- Having the ability to take existing ideas and combine them in different ways for new purposes
- Continually thinking of alternatives for a design

The Process for an Engineering Product Design

1. *Team selection:* Team members should be compatible and have diverse backgrounds in drafting, CAD, math, science, organizing, and management.

2. *Identifying a need for a product:* Product identification should include research and **survey** data indicating the need for a product.

3. *Preliminary group ideas:* Any ideas, good or bad, for the design of the product should be open to discussion. Interview the involved persons and take notes about their needs.

4. *Market potential:* Ask questions such as What is the market potential? What will production costs be? What can the product be sold for? Will the costs be justified with a market analysis?

5. *Student creativity:* This is a critical ability for the design team.

6. *Decision making:* After the design team has a list of possible solutions, the time has arrived to select the best design.

7. *Delegation of group responsibilities:* Team members are each given responsibilities based on their skills within the overall design process.

8. *Sketching design ideas:* A picture is worth many words. All team members will produce various sketches for the selected design. Design concepts start in the minds of the designers. To verbalize an idea may be sufficient in a few cases, but the ideas should be transferred onto paper as sketches and reviewed for their strengths and weaknesses. Many sketches should be made. The final design may be the product of many **concepts** or of a single idea.

9. *Organization:* Compile and organize all the notes, ideas, concepts, and sketches, then return to the persons interviewed, review all data, and make the necessary revisions.

10. *Design selection:* After all the revisions are completed, the final sketch should be unanimously agreed upon by the designers, engineers, manufacturing, marketing personnel, and consumers.

11. *Material analysis:* A critical step is selecting from the many different types of materials that could be used in the manufacture of the product.

12. *Environmental study:* Before production begins, an **environmental study** should be made. If it is determined that the production or the use of the product will be detrimental to the environment, then the project should be dropped.

13. *Engineering working drawings:* When all parties involved with the design, manufacturing, and marketing agree, then a full set of **working drawings** should be produced.

14. *Building and testing a prototype:* A **prototype** should be produced and tested to be absolutely certain that it functions perfectly. If the project is built in the school shop, all team members must take and pass the safety exams for each shop. If they are working at home, the operating manuals must be read on the operation and safety of all power tools.

15. *Tool design for mass production:* The next step is to create the tooling setup for the manufacturing. This is an expensive phase of the design process, so everything in the design must check out. 3D printing should be considered to develop the prototype.

16. *Manufacturing/production:* The production line should now be developed for efficient machining operations.

17. *Assembly:* If the product has a subassembly with mating parts, tolerance checks must be made. All the mating parts and the integration of the part into the main assembly must be acceptable.

18. *Quality assurance:* It is critical to keeping checking the parts being manufactured for minor or major flaws. Manufacturing flaws can create poor-quality or nonfunctioning products. A final testing of the product should be made before it is released.

19. *Replacement parts:* Extra parts should be produced and stored for the time when replacement parts are required because of worn or failed parts.

20. *Documentation/operating instructions:* A clear and concise instructional manual should accompany each product with details about assembly, installation, usage, and safety.

21. *Advertising and sales promotion:* A budget should be set aside for marketing with brochures and advertising.

22. *Planning upgrade:* Do not wait until the item is outdated before planning the next **upgrade**.

Note: The design of an industrial product may vary with each design process and the personnel making up the design team.

Design Team Procedure for a Simple Product: Design Project 1

1. People have complained about the lack of a good doorstop. Needed is a door stop that will: securely hold a door open, be small in size, not be lost, be low in cost, and not be easily broken.

2. The design team brainstorms ideas:
   - Heavy object on the floor
   - Velcro holders
   - Magnets
   - Wedge block
   - Rubber banding
   - Hook and latch
   - Offset door hinges
   - Rusty hinges

3. The design team discusses the merits of all the ideas, including the unusual ones, for:
   - Ease of use
   - Cost of production
   - Usefulness
   - Sturdy design
   - Simplicity of design
   - Market potential

4. The team must now choose the best idea.

5. Each member of the team is assigned a specific responsibility. Each member is still responsible for having an overview of the full design process.

6. The team produces sketches (**Figure 5-1**).

7. The team selects the best ideas from the sketches.

8. For a design review, the team members should do a survey for further input about the design of the doorstop and make the necessary revisions.

9. An environmental study should cover any detriments to the environment that may occur from the production and use of the doorstop.

10. The team creates a set of working drawings. This may be done with manual drafting or with a CAD system (**Figure 5-2**).

11. The team builds and tests the prototype. This includes:
    - Checking to see if the magnets are too strong or too weak.
    - Adjusting the size or strength of the magnets for the correct amount of holding force.
    - Checking ease of installation

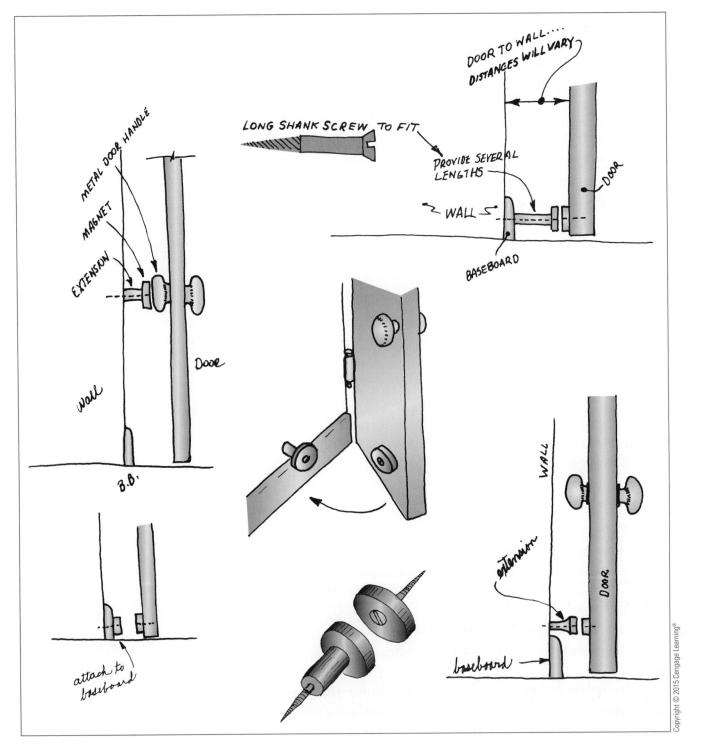

Figure 5-1 *Sketches for a magnetic doorstop.*

- Checking for proper function
- Testing for reliability
- Following all safety rules when using tools

12. When the team is ready to go into production, it must decide:

- What items are to be purchased off the shelf
- Cost factors and budgeting

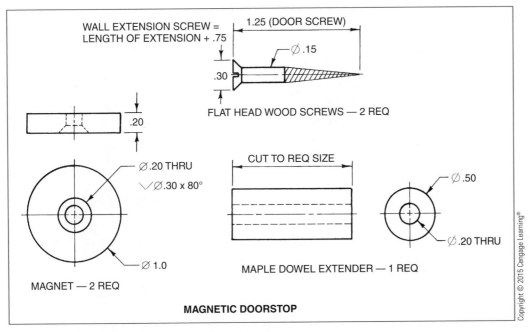

Figure 5-2 *Working drawing for a magnetic doorstop.*

- Where production will take place
- What tools and jigs will be necessary
- The quality control procedure for the production line

13. Write the documentation:
    - How to install the doorstop
    - Operating procedures
    - Safety precautions

14. Decide on marketing and sales promotions.

15. Plan an upgrade.

16. Review the whole process to improve it for the next project.

17. Write a report on the design process and make an oral presentation.

Design Project 2

The design team's second project is to develop a design for a bearing that will securely hold a circular bar supporting a steel strut for a small airplane's front landing gear.

- Airgo Corporation recognizes the need for the product.
- There is no need for a market analysis.
- The design team studies the front landing mechanism of the Airgo airplane and refers to other aircraft front landing gear designs.
- The team sketches possible types of bearings (**Figure 5-3**).
- After conferring with Airgo, the best design is selected.
- Decisions about different types of materials, manufacturing, and costs are made.

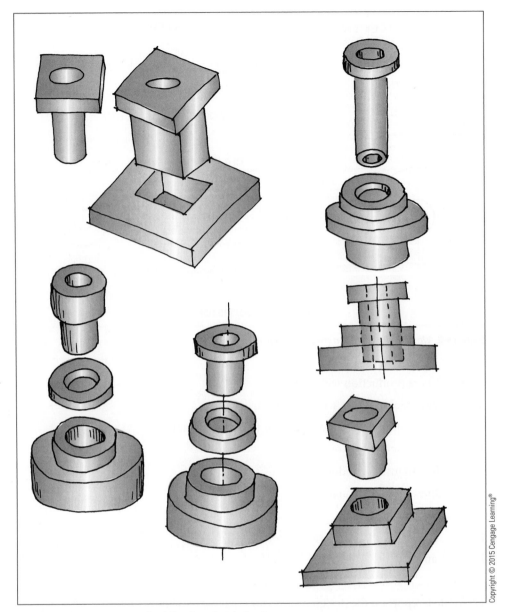

Figure 5-3 *Design concepts should be shared with sketches.*

- An environmental study is undertaken to see if the manufacture or use of the design is detrimental to the environment.
- A design review and analysis is made with Airgo and the manufacturers.
- Engineering working drawings are prepared (**Figure 5-4**).
- A prototype is built and tested for reliability.
- The tool design for the manufacturing is set up.
- Quality assurance for machining and mating parts during manufacturing is closely checked for the final testing.
- A manual with instructions about installation, assembly, and safety is prepared.

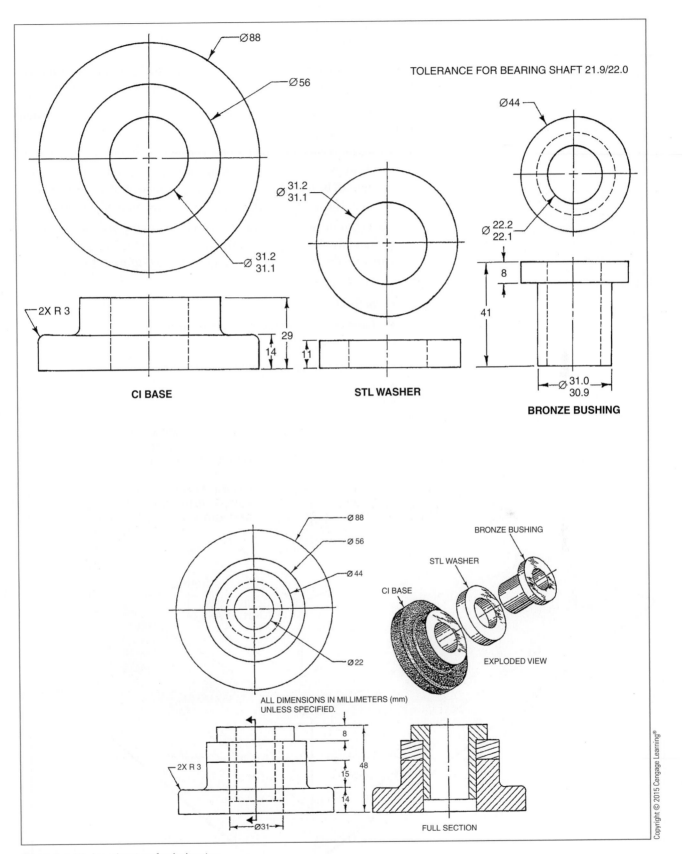

Figure 5-4 *Working drawings for the bearing.*

Architectural Design

Architectural design is another area in which a design team may work. The design process for residential design will vary from that of industrial design.

The following is a sample procedure for a team problem of residential design:

1. *Define the client's needs and wants:* The design team must interview the family for whom they are designing the residence. A lists of *needs* and *wants* must be recorded. The *needs* are design items that must be incorporated. The *wants* are items that may be compromised if necessary.

2. *Zoning ordinances:* It is critical to check the **zoning ordinances** in the community's building department to be certain the team's design concurs with the following zoning requirements (**Figure 5-5**):

   - Square footage of the building site
   - **Setbacks**
   - **Land coverage** of the structure or structures
   - Height limitations
   - **Daylight plane** for a multistory structure
   - Parking

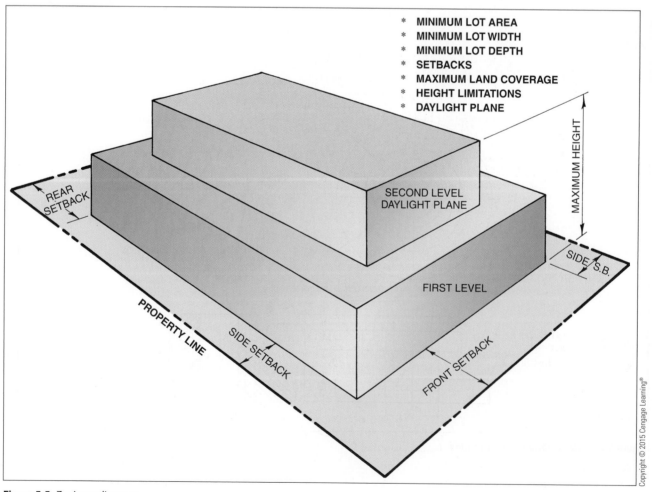

* **MINIMUM LOT AREA**
* **MINIMUM LOT WIDTH**
* **MINIMUM LOT DEPTH**
* **SETBACKS**
* **MAXIMUM LAND COVERAGE**
* **HEIGHT LIMITATIONS**
* **DAYLIGHT PLANE**

Figure 5-5 *Zoning ordinances.*

3. *Orientation:* Two factors affect the placement of rooms on the site, the wants and needs of the clients and the site's features. The site features are (**Figure 5-6**):

- *The sun's direction.* Which side of the house should receive the morning, afternoon, and late afternoon sunlight? Remember the sun rises in the east, goes across the southern sky, and sets in the west. The north side receives no sun.

- *The best view.* Which rooms do the clients want to have the best views?

- *Cool breezes.* Usually the breezes come from the same direction. Which rooms do the clients want to receive these breezes?

- *Noise.* Block off the noise with the service area, garage, fences, and landscaping.

- *The geologic features of the site.* The slope of the site, type of soil, trees, rocks, and water must be considered.

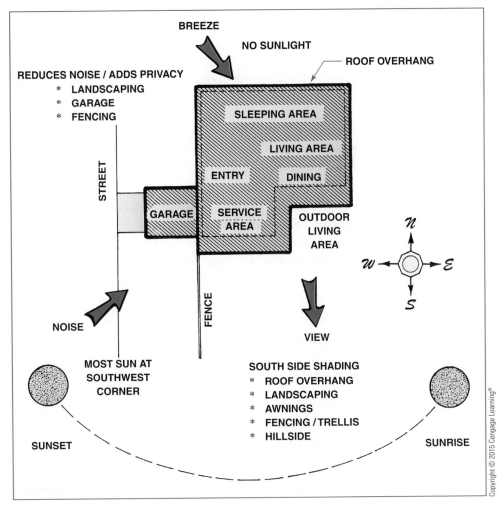

Figure 5-6 *Site orientation.*

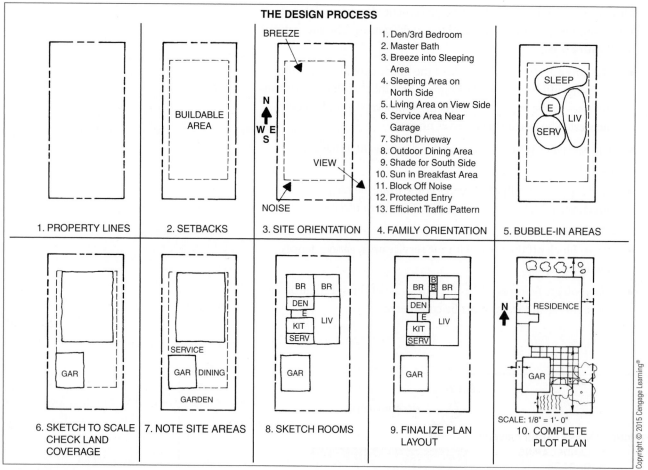

Figure 5-7 *Preliminary sketches.*

4. *Preliminary sketches:* Using the data from the clients, zoning department, and the orientation features, make several preliminary sketches (**Figure 5-7**).

5. *Final sketch selection:* Select the best sketch to discuss with the clients, contractor, community committees, and zoning department. Make the design changes that are recommended. If all parties agree to the final sketch, prepare the set of drawings. To obtain a building permit, you must submit the following architectural working drawings:

- Plot plan
- Floor plan
- Exterior elevations
- Foundation
- Roof plan
- Floor framing plan
- Wall framing plan
- Roof framing plan
- Rendered pictorial

EXERCISES FOR A DESIGN TEAM

1. A curbside mailbox that signals a delivery
2. A tool to turn off a residence's gas line
3. Personalized bookends
4. A tent stake that will not easily be pulled out of the ground
5. A bathroom soap bar holder allowing all sides of the soap bar to dry
6. A school shop lab
7. A school chemistry lab
8. A school gymnasium
9. A doctor or dentist's office
10. Any small manufacturing business (provide flowchart diagrams)
11. New home design
12. Remodel design
13. Home design for the impaired:
    - Wheelchair bound
    - Sight impaired
    - Physically impaired
    - Hearing impaired
18. Interview people with a trailer. Design a safer trailer hitch.
19. Interview mail carriers. Design a better mailbox.
20. Interview typists and drafters. Design an ergonomic chair and workstation.
21. Make a list of products for which you believe people in your community may have a need.
22. How would your design team improve the rudder in **Figure 5-8**?
23. How might your design team develop a better product design for small pieces of usable soap?
24. Design a coffee cup that is comfortable to handle and is also slip-proof.
25. Design a lighting system for reading in bed so that others sleeping in the same room are not disturbed.
26. Feel free to discuss any other ideas with your instructor for your group project.

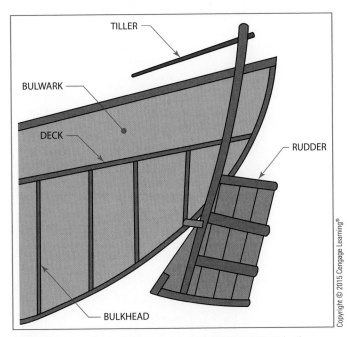

Figure 5-8 *How did the improvement of the rudder in 200BC, by the Chinese, improve seamanship?*

KEY TERMS

Concept

Creativity

Daylight plane

Environmental study

Geologic features

Land coverage

Orientation

Prototype

Quality assurance

Setback

Survey

Upgrade

Working drawing

Zoning ordinance

6

Drafting Conventions and Formats

The student will be able to:

- Draw a basic working drawing using ASME line conventions
- Describe the types of working drawing conventions
- Lay out and draw a school drafting format sheet
- Lay out and draw an industrial company format sheet

Introduction

Different industries and companies use unique processes and procedures for designing and manufacturing products. However, drawings used in the design and manufacture of products are prepared to identical sets of **standards**. Standardization of drawings enables all manufacturers and builders to interpret the drawings of many different designers, drafters, engineers, and architects. Before standardization, working drawings could only be interpreted by the designer or drafter who prepared the drawing. Today's standardized types of engineering drawings are shown in **Figure 6-1**.

National and World Standards

Dictionaries are used to standardize words in a written or spoken language. Likewise, the American Society of Mechanical Engineers (ASME) publications are used as the universal reference work for drafting standards in the United States. The International Organization for Standardization (ISO) publication R-1000 describes metric standards used throughout the world. The consistent use of these two international standards ensures that a product can be

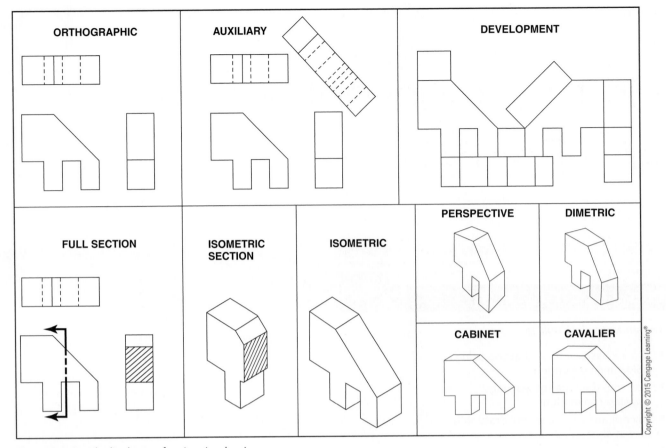

Figure 6-1 *Standardized types of engineering drawings.*

designed anywhere in the world and built or manufactured in any other part of the world. Each specialized standard for drafting is found in ASME's publication with the following identification numbers:

ASME Y14.1	Drawing Sheet Size and Format
ASME Y14.1M	Metric Sheet Size and Format
ASME Y14.100	Engineering drawing and practices
ASME Y14.2	Line Conventions and Lettering
ASME Y14.3	Multiview and Sectional-View Drawings
ASME Y14.4	Pictorial Drawings
ASME Y14.5	Dimensioning and Tolerancing
ASME Y14.24	Types and Applications of Engineering Drawings.
ASME Y14.34	Associated Lists
ASME Y14.35	Drawing Revisions
ASME Y14.38	Abbreviations
ASME Y14.41	Digital Product Definition Drawing Practices
ASME Y14.42	Electronic Approval Systems

The following are Y14 Specialty Standards

ASME Y14.6	Screw-Thread Representation
ASME Y14.7	Gears, Splines, and Serrations
ASME Y14.8	Castings and Forgings
ASME Y14.13	Springs
ASME Y14.18	Drawings for Optical Parts
ASME Y14.31	Undimensioned Drawings
ASME Y14.32	Ground Vehicle Drawing Practices
ASME Y14.36	Surface Texture Symbols
ASME Y14.37	Composite Drawing Parts
ASME Y14.40	Graphic Symbols
ASME Y14.43	Dimensioning and Tolerancing Principles for Gages and Fixtures
ASME Y14.44	Reference Designations

Line Conventions

Technical drawings contain many types of lines, each with a special meaning. Through the consistent use of the ASME line conventions, technical drawings become a language that can be read by anyone who understands the standard. The ASME **line conventions** are shown in **Figure 6-2**.

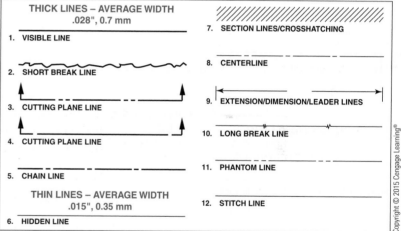

Figure 6-2 *ASME line connections.*

LINE WIDTHS

All lines are either thick (.028" or 0.7 mm average width) or thin (0.15" or 0.35 mm average width). Both thick and thin lines are always dark (**opaque**) to ensure high reproduction quality. Lines used for layout and lettering guide lines (**Figure 6-3**) are drawn very lightly and should be invisible on the finished drawing reproduction. Some drafters use nonreproducible blue pencils to ensure invisibility.

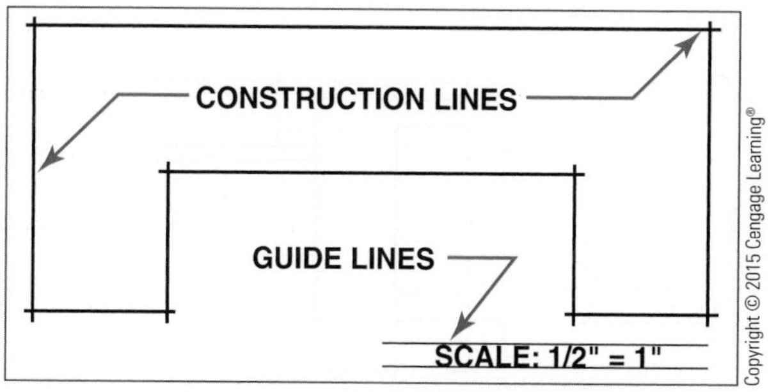

Figure 6-3 *Construction lines and guide lines are drawn very lightly so they will not be seen in the finished working drawing or print.*

LINE TYPES

Visible **object lines** (**Figure 6-4**) are solid, thick lines used to define the outline of objects. **Break lines** are used to interrupt the drawing if the object will not fit on a drawing sheet. Short break lines are thick, wavy lines, and long break lines are thin lines with zigzags spaced

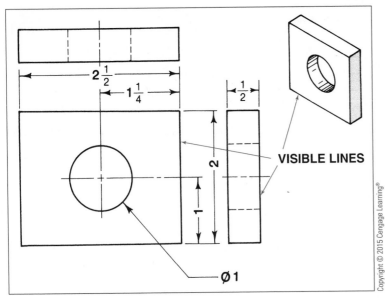

Figure 6-4 *Note the width of the object lines and compare them to the widths of the other lines.*

at approximately 1.5" intervals (**Figure 6-5**). When objects are visually cut (sectioned) to show an interior view, a thick **cutting plane line** is used to show the location of the cut. Thin **section lines** are used to represent the material (**Figure 6-6**) through which the cut is made. **Chain lines** (**Figure 6-7**) are used to indicate specific areas. Object lines that fall behind the front plane of an object are drawn with dashed lines representing **hidden lines** (**Figures 6-8a** and **6-8b**). (Note that when a hidden line appears as the continuation of a solid object line, the drafter must leave a space.) When objects are symmetrical with a common center, the center is defined with a **centerline** constructed with thin long and short dash lines (**Figure 6-9**). The size of objects is described with the use of thin **leaders**, **extension lines**, and **dimension lines**

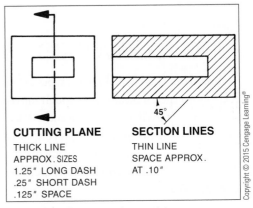

Figure 6-6 *The cutting plane and the section lines (cross-hatching).*

Figure 6-5 *Long and short break lines.*

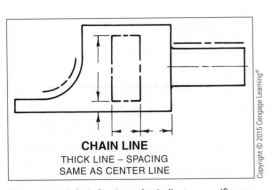

Figure 6-7 *A chain line is used to indicate a specific area.*

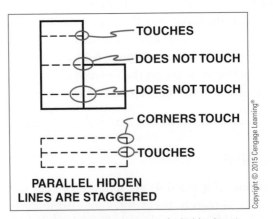

Figure 6-8a *Drafting conventions for hidden lines in a working drawing. The hidden line is approximately 1/8" long with a 1/32" space.*

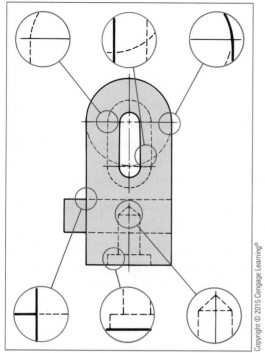

Figure 6-8b *General rules for hidden lines.*

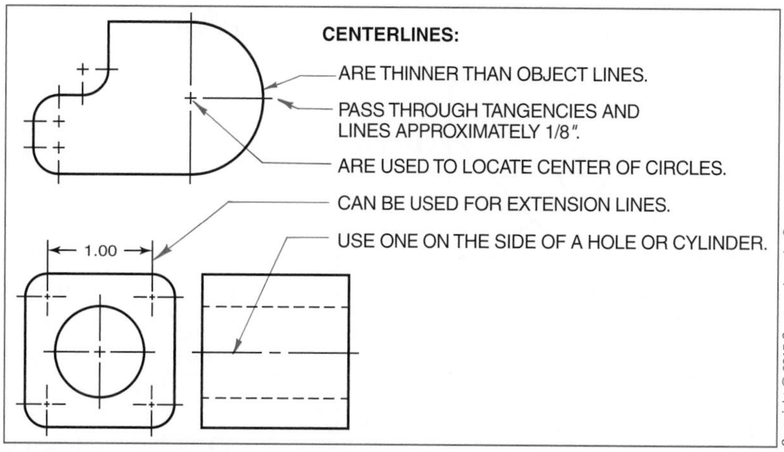

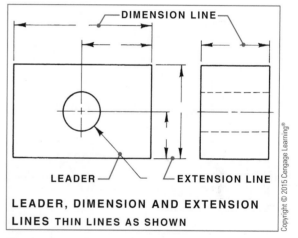

Figure 6-9 *Drafting conventions for centerlines on a working drawing. The approximate dimensions are: long dash 1 1/2″, short dash 1/8″, space 1/16″.*

(**Figure 6-10**). When alternate positions of an object or matching part must be shown without interfering with the main drawing, thin **phantom lines** are used (**Figure 6-11**). To show the location of stitches in a material, **stitch lines** are used (**Figure 6-12**).

Drafting Conventions

The working drawings of a machine part shown in **Figures 6-13** through **6-24** show the application of many drafting conventions. **Figure 6-13** is a **multiview drawing** with inch-fraction dimensions and surface symbols.

Figure 6-10 *Leader, dimension, and extension lines.*

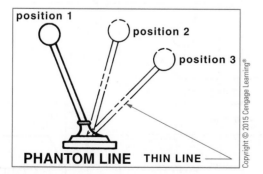

Figure 6-11 *Phantom lines show alternate positions. Line sizes are the same as centerlines.*

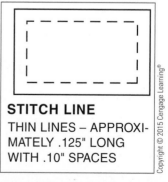

Figure 6-12 *Stitch lines show various stitches sewn into materials.*

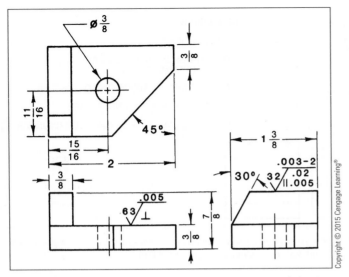

Figure 6-13 *Multiview drawing with third angle projection, inch-fraction dimensions, and surface finish symbols.*

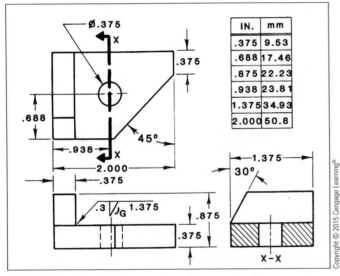

Figure 6-14 *Multiview drawing with inch-decimal dimensions, full section, weld symbol, and dual dimension table.*

Figure 6-14 shows the same multiview drawing with inch-decimal dimensions. In this figure, a cutting plane line is positioned on the top view, while the corresponding full section is shown in the right-side view. Note the standard weld symbol and dual inch-millimeter chart in **Figure 6-14**.

Metric (millimeter) dimensions and positional tolerancing symbols are added to the machine part drawing shown in **Figure 6-15**. Tolerance dimensions are used to dimension the same drawing in **Figure 6-16**.

A pictorial drawing is usually used to show the finished shape of the assembled object.

Isometric drawings (**Figure 6-17**) and isometric **section drawings** (**Figure 6-18**) are drawn with receding sides positioned at 30° and extending to the right and left.

Oblique drawings show the front view of an object in true scale and form, with receding angles projected to only one side at 30° or 45°.

When receding sides are drawn at one-half the actual depth, the drawing is known as a **cabinet drawing** (**Figure 6-19**). When sides are drawn at full-scale depth, the drawing is known as a **cavalier drawing** (**Figure 6-20**). Since cavalier oblique drawings appear more distorted than cabinet drawings, using half-depth dimensions on the receding angle helps reduce the appearance of distortion.

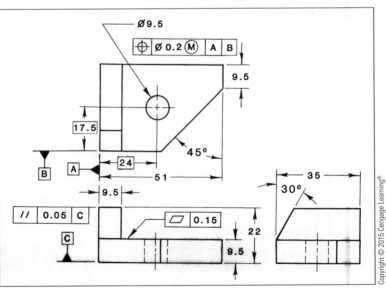

Figure 6-15 *Multiview drawing with metric (millimeters) dimensioning and positional tolerancing symbols.*

Perspective drawings (**Figure 6-21**) are classified into one-, two-, and three-point perspective drawings (see Chapter 15). **Dimetric drawings** (**Figure 6-22**) are similar to isometric drawings, except that two axes have equal angles other than 30°. In **trimetric drawings** (**Figure 6-23**), all three axes' angles are different.

Pattern development drawings show the outline of all connected surfaces of an object when laid flat. Pattern development drawings are used to define the outline of an object when a continuous material is used for all surfaces. The pattern development drawing represents all the surfaces of an object (**Figure 6-24**).

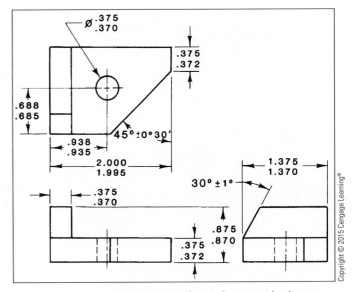

Figure 6-16 *Multiview drawing with tolerance dimensions.*

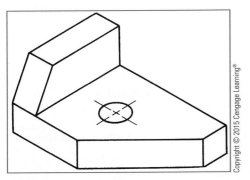

Figure 6-17 *Isometric drawing.*

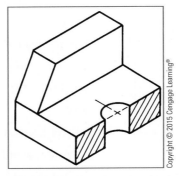

Figure 6-18 *Isometric section drawing.*

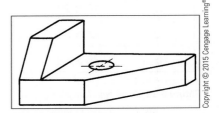

Figure 6-19 *Cabinet drawing.*

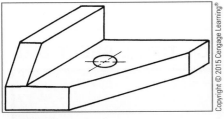

Figure 6-20 *Cavalier drawing.*

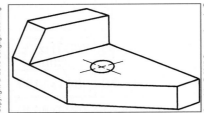

Figure 6-21 *Perspective drawing.*

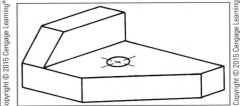

Figure 6-22 *Dimetric drawing.*

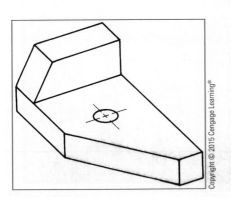

Figure 6-23 *Trimetric drawing.*

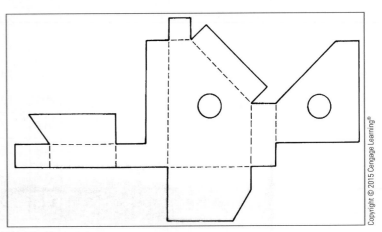

Figure 6-24 *Pattern development drawing.*

E = TOTAL SHEET SIZE

AMERICAN STANDARD SHEET SIZES (INCHES)		
DESG	**WIDTH x LEN**	**WIDTH x LEN**
A	9 x 12	8.5 x 11
B	12 x 18	11 x 17
C	18 x 24	17 x 22
D	24 x 36	22 x 34
E	36 x 48	34 x 44

Figure 6-25 *ASME standard paper sizes.*

Drawing Formats

Drawing sheet sizes are standardized by ASME. Letters A through E are used to specify standard U.S. customary sheet sizes (**Figure 6-25**). In the same manner, metric sheet sizes are standardized by ISO. Designations of A0 through A10 are used to identify standard metric sheet sizes **(Figure 6-26)**. A typical industrial sheet layout and title boxes are shown in **Figure 6-27**. Standard border sizes and layouts are shown in **Figure 6-28**.

Figures 6-29a and **6-29b** are examples of horizontal and vertical A-size industrial formats.

Although the layout of drawing **title blocks** varies greatly among companies, title block information usually contains space for the following:

Company name and division	Drawing title
Drawing number	Part identification number
Date	Tolerance limits
surface quality	Designer's name
Drafter's name	Checker's name
Angle of projection	Material type
Scale	Next assembly
Microfilming marks	Photoreduction marks
Revision record	Material lists

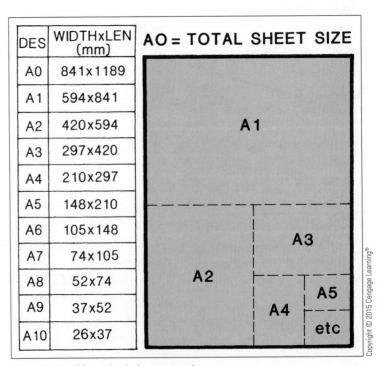

DES	WIDTHxLEN (mm)
A0	841x1189
A1	594x841
A2	420x594
A3	297x420
A4	210x297
A5	148x210
A6	105x148
A7	74x105
A8	52x74
A9	37x52
A10	26x37

A0 = TOTAL SHEET SIZE

Figure 6-26 *ISO metrics A-size paper series.*

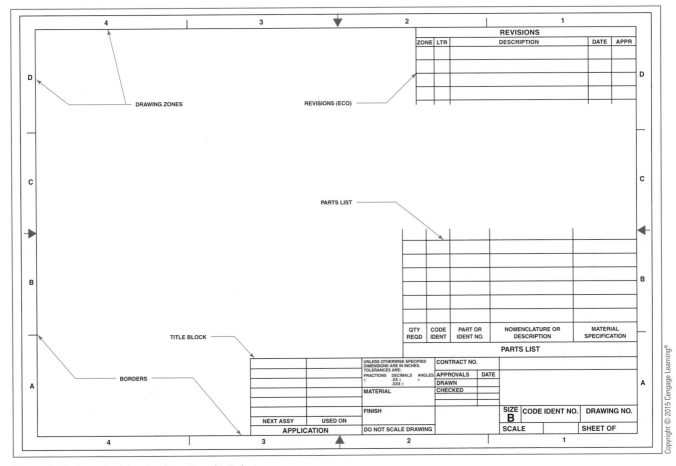

Figure 6-27 *A standard drawing format used in industry.*

Title blocks are abbreviated for educational use. They usually include the following:

- Drawing title
- Scale
- Drawing number date
- Student drafter's name
- School
- Instructor's approval
- Grade

Figure 6-30 shows an A-size school format, and **Figure 6-31** shows a B-size school format.

Folding methods for prints are standardized for consistency and ease of use. **Figure 6-32** shows how standard sheet sizes fold to 8.5" × 11" or 9" × 12". Prints are always folded with the title block and drawing identification number on top for convenient recognition and filing.

Note that in the layering of drawings, a single format only will be needed on the base drawing (**Figure 6-33**).

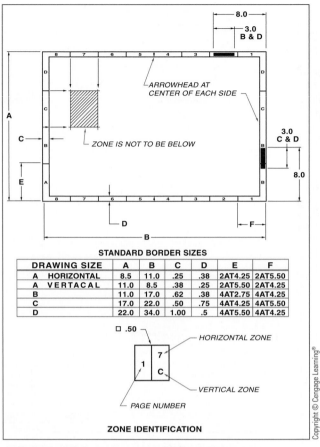

DRAWING SIZE	A	B	C	D	E	F
A HORIZONTAL	8.5	11.0	.25	.38	2AT4.25	2AT5.50
A VERTACAL	11.0	8.5	.38	.25	2AT5.50	2AT4.25
B	11.0	17.0	.62	.38	4AT2.75	4AT4.25
C	17.0	22.0	.50	.75	4AT4.25	4AT5.50
D	22.0	34.0	1.00	.5	4AT5.50	4AT4.25

STANDARD BORDER SIZES

ZONE IDENTIFICATION

Figure 6-28 *Border data and sizes for drawing formats.*
(Reprinted from Technical Drawing and Engineering Communication by Goetsch, Chalk, Nelson, and Rickman, Cengage Learning.)

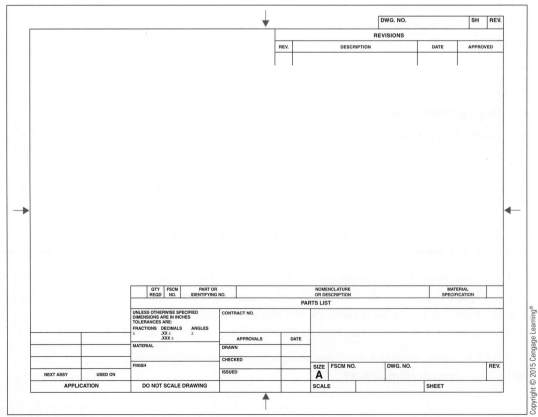

Figure 6-29a *A horizontal layout for an A-size industrial sheet layout.*

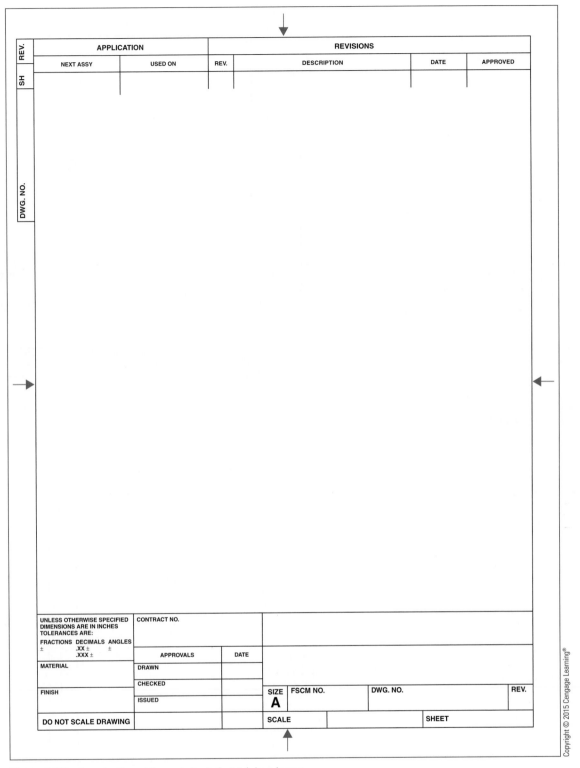

Figure 6-29b *A vertical layout for an A-size industrial sheet layout.*

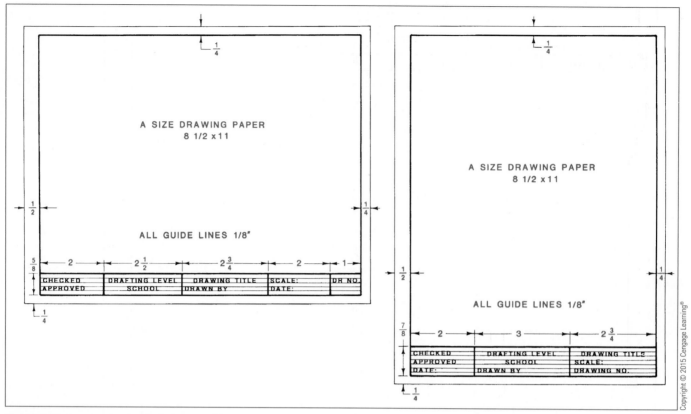

Figure 6-30 *An example of an A-size horizontal format used in drafting classes.*

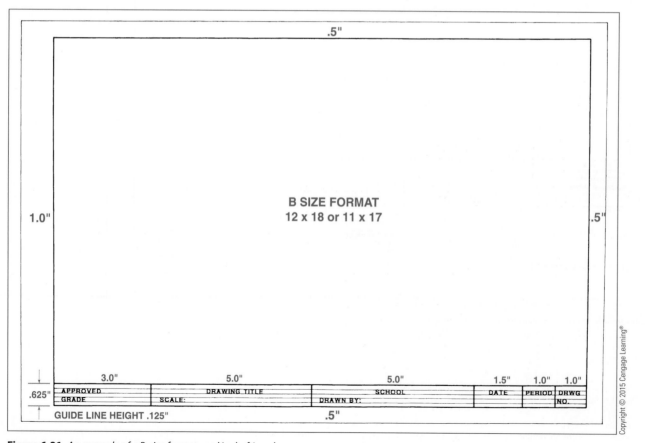

Figure 6-31 *An example of a B-size format used in drafting classes.*

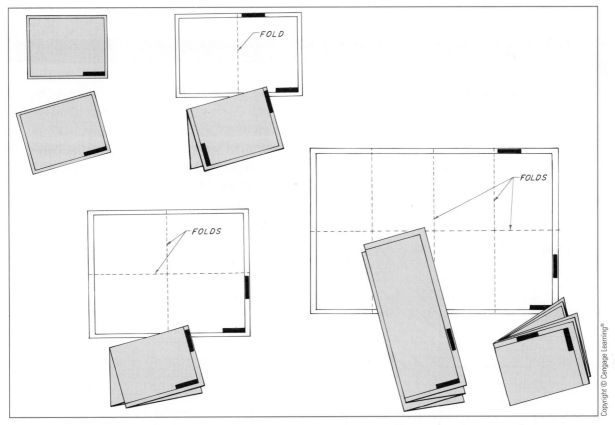

Figure 6-32 *All standard formats will fold down to an A size for easy storage. (Reprinted from Technical Drawing and Engineering Communication by Goetsch, Chalk, Nelson, and Rickman, Cengage Learning.)*

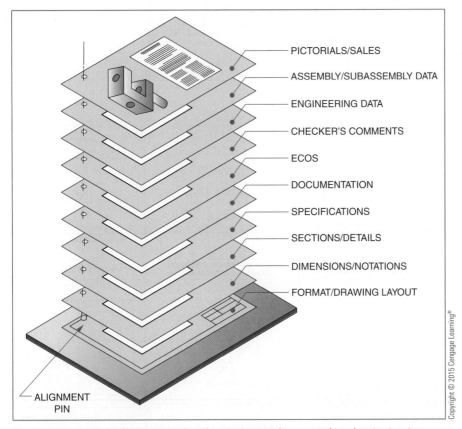

PICTORIALS/SALES

ASSEMBLY/SUBASSEMBLY DATA

ENGINEERING DATA

CHECKER'S COMMENTS

ECOS

DOCUMENTATION

SPECIFICATIONS

SECTIONS/DETAILS

DIMENSIONS/NOTATIONS

FORMAT/DRAWING LAYOUT

ALIGNMENT PIN

Figure 6-33 *Example of how pin graphics (layering) may split-up a working drawing into its separate elements with only the base drawing with a sheet layout format.*

DRAFTING EXERCISES

1. Draw **Figure 6-34** on a CAD system and/or with instruments using the proper line weights or line types.

2. Design a C-size drawing format for school use. Make it a prototype drawing.

3. Design a C-size drawing format for the Ace Tool Design Company.

4. List each ASME drafting standard used in this chapter with the ASME reference page numbers.

5. Set up a standard drawing format for A-, B-, and C-size paper on a CAD system, and create prototype drawings.

6. With a CAD system, draw the figures shown in **Figure 6-35**. Adjust the line widths as needed for the final hard copy plot. Double all dimensions.

7. Name each line convention in **Figure 6-36**.

8. Practice drawing **Figure 6-37** with a CAD system.

Figure 6-34 *Draw these patterns with instruments or CAD.*

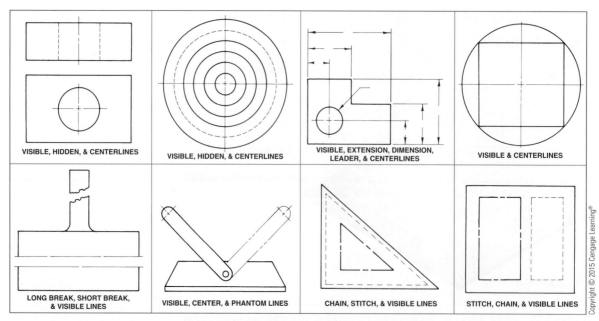

Figure 6-35 *Redraw each figure two times as large using correct line widths.*

LINE SYMBOLS	LINE WIDTH	PENCIL	PEN SIZES	
————————	THICK	H,F	2	0.50 mm
– – – – – – – –	MEDIUM	2H,H	0	0.35 mm
—— · — — · ——	THIN	2H,3H,4H	0	0.35 mm
—— ⌇ —— ⌇ ——	THIN	2H,3H,4H	0	0.35 mm
∿∿∿∿∿	THICK	H,F	2	0.50 mm
—— — — ——	THIN	2H,3H,4H	0	0.35 mm
— — — — — —	THIN	2H,3H,4H	0	0.35 mm
▬▬▬▬▬	VERY THICK	F,HB	3	0.80 mm
⟵————————⟶	THIN	2H,3H,4H	00	0.25 mm
⟍	THIN	2H,3H,4H	00	0.25 mm
⬇⬇⬇	VERY THICK	F,HB	3	0.80 mm
▨▨▨	THIN	2H,3H,4H	00	0.25 mm
≡≡≡	VERY THIN LIGHT	4H		
ARCHITECTURAL	THICK	H,F	1	0.40 mm

Copyright © 2015 Cengage Learning®

Figure 6-36 *Name each line convention.*

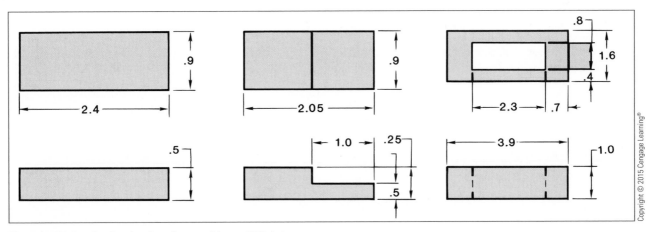

Figure 6-37 *Practice drawing these figures with your CAD system.*

Copyright © 2015 Cengage Learning®

DESIGN EXERCISES

1. Design a B-size format for a small mechanical engineering company named D&P Engineering Inc. Include a full title block, borders, zones, and company logo.

2. Design an A-size format for your classwork.

3. Turn off your monitor then create a 3" × 5" rectangle with a 1" hole in the center.

KEY TERMS

Break line

Cabinet drawing

Cavalier drawing

Centerline

Chain line

Cutting plane line

Dimension line

Dimetric drawing

Extension line

Hidden line

Isometric drawing

Leader

Line conventions

Multiview drawing

Object line

Oblique drawing

Opaque

Pattern development drawing

Perspective drawing

Phantom line

Section drawing

Section line

Standards

Stitch line

Title block

Trimetric drawing

Geometric Construction

The student will be able to:

- Understand geometric form terminology
- Recognize two-dimensional geometric forms
- Recognize three-dimensional geometric forms
- Construct basic geometric forms using drafting instruments
- Construct basic geometric forms using a CAD system

Introduction

- Geometric forms are pervasive in our natural and manmade environment. Where ever you may be looking you will see some type of geometric forms (**Figure 7-1**). The shapes may be of a simple or a very complex pattern (**Figure 7-2**).

- The construction of geometric forms with manual drafting instruments has always been a tedious but necessary drafting task. The use of geometric drafting templates will be a significant time saver for the manual drafter. However, once a CAD operator is familiar with the CAD's software program, drawing geometric forms will be much simpler and faster.

CAD versus Drawing Instruments

Drawing geometric forms on a CAD system is simpler and faster than constructing them manually. A CAD drafter must be familiar with the basics of **geometric construction** in order to fully utilize the geometric drawing capacity of a CAD system.

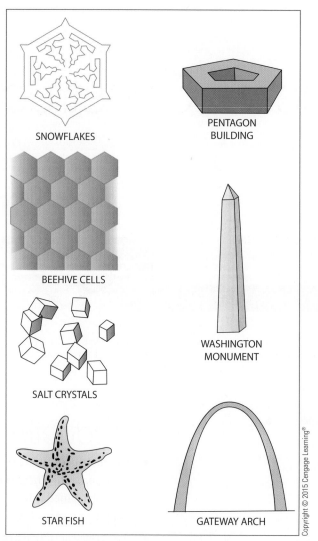

SNOWFLAKES

PENTAGON
BUILDING

BEEHIVE CELLS

WASHINGTON
MONUMENT

SALT CRYSTALS

STAR FISH

GATEWAY ARCH

Figure 7-1 *Natural and manmade geometric shapes.*

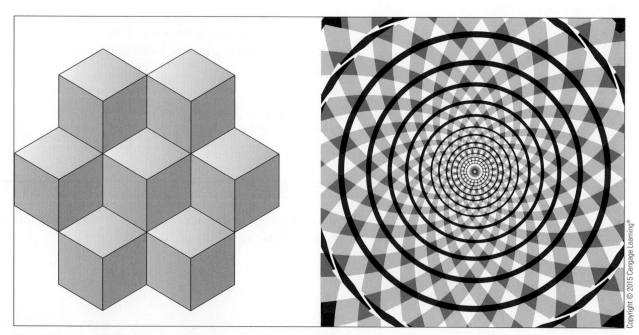

Figure 7-2 *Geometric forms may be simple or very complex.*

Geometric Forms

To communicate effectively, a CAD drafter must be able to use geometric terminology correctly. Geometric terminology commonly used in most fields of drafting and design is shown in **Figure 7-3**. Geometric forms are either **two-dimensional** or **three-dimensional**.

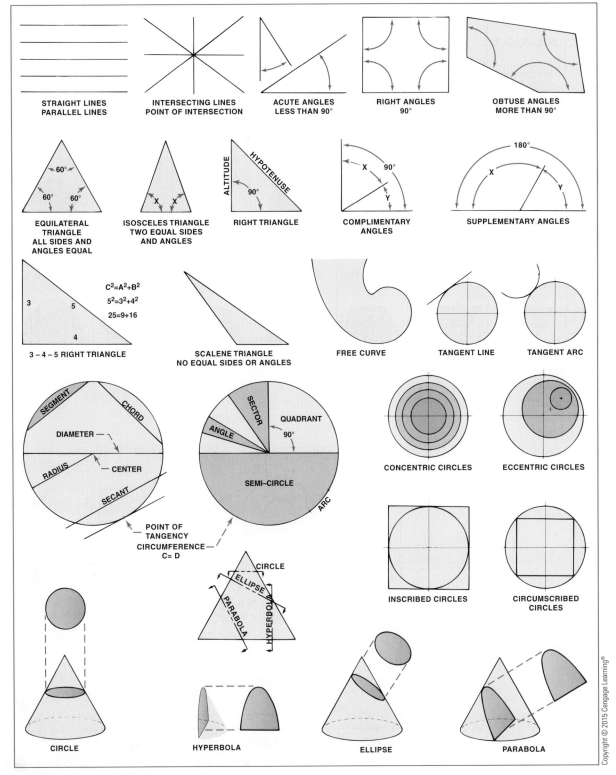

Figure 7-3 *Geometric terminology.*

Figure 7-4 shows basic two-dimensional forms that are variations of circles, triangles, squares, polygons, ellipses, and free forms. **Figure 7-5** shows a two-dimensional engineering drawing constructed with a combination of straight lines, arcs, circles, and curves. Study **Figures 7-4** and **7-5**, and identify the different types of geometric forms.

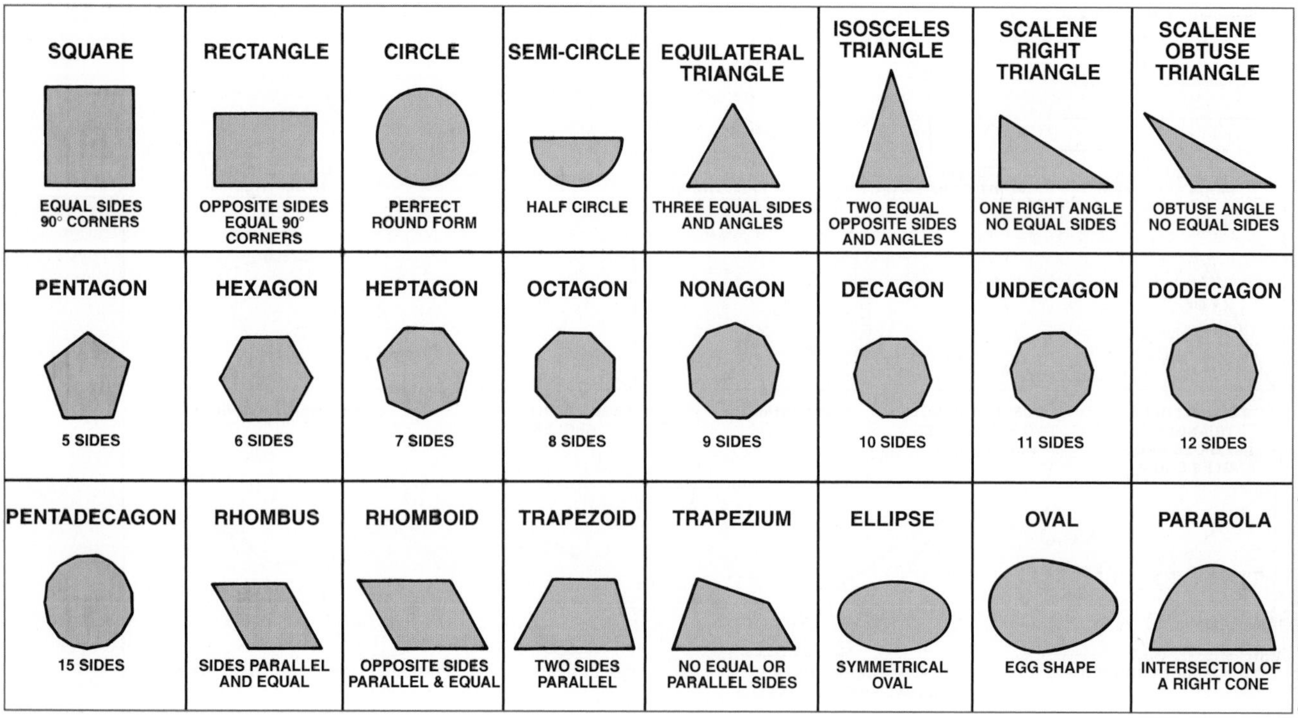

SQUARE	RECTANGLE	CIRCLE	SEMI-CIRCLE	EQUILATERAL TRIANGLE	ISOSCELES TRIANGLE	SCALENE RIGHT TRIANGLE	SCALENE OBTUSE TRIANGLE
EQUAL SIDES 90° CORNERS	OPPOSITE SIDES EQUAL 90° CORNERS	PERFECT ROUND FORM	HALF CIRCLE	THREE EQUAL SIDES AND ANGLES	TWO EQUAL OPPOSITE SIDES AND ANGLES	ONE RIGHT ANGLE NO EQUAL SIDES	OBTUSE ANGLE NO EQUAL SIDES
PENTAGON	HEXAGON	HEPTAGON	OCTAGON	NONAGON	DECAGON	UNDECAGON	DODECAGON
5 SIDES	6 SIDES	7 SIDES	8 SIDES	9 SIDES	10 SIDES	11 SIDES	12 SIDES
PENTADECAGON	RHOMBUS	RHOMBOID	TRAPEZOID	TRAPEZIUM	ELLIPSE	OVAL	PARABOLA
15 SIDES	SIDES PARALLEL AND EQUAL	OPPOSITE SIDES PARALLEL & EQUAL	TWO SIDES PARALLEL	NO EQUAL OR PARALLEL SIDES	SYMMETRICAL OVAL	EGG SHAPE	INTERSECTION OF A RIGHT CONE

Copyright © 2015 Cengage Learning®

Figure 7-4 *Two-dimensional geometric forms.*

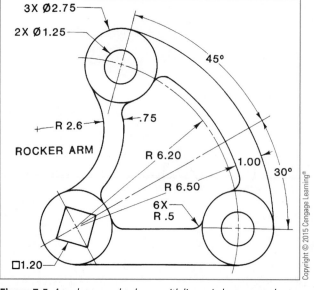

Figure 7-5 *Any shape can be drawn with lines, circles, arcs, and curves.*

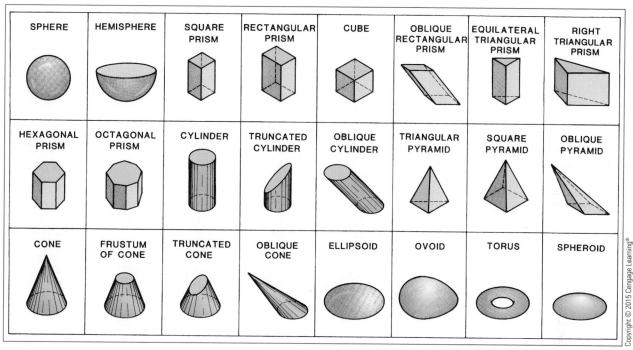

Figure 7-6 *Three-dimensional geometric forms.*

Three-dimensional geometric forms possess width, depth, and height. Although there are hundreds of different forms, all are variations of the basic forms found in **Figure 7-6**.

Conventional Geometric Construction

The procedures for manually constructing the most common geometric forms are shown in **Figures 7-7** through **7-27**.

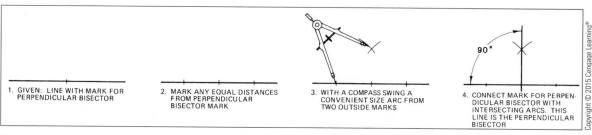

Figure 7-7 *Perpendicular bisector.*

Figure 7-8 *Parallel lines.* Copyright © 2015 Cengage Learning®

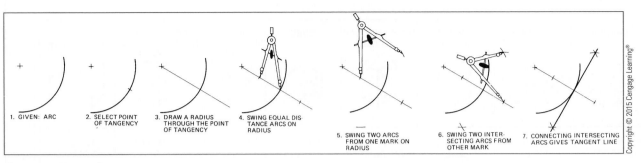

Figure 7-9 *Tangent line.*

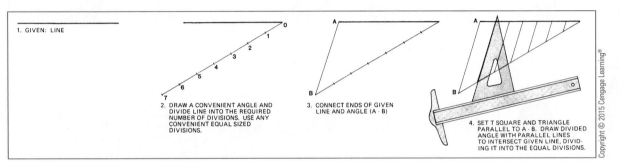

Figure 7-10 *Divide a line into equal parts.*

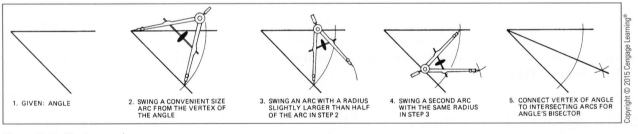

Figure 7-11 *Bisect an angle.*

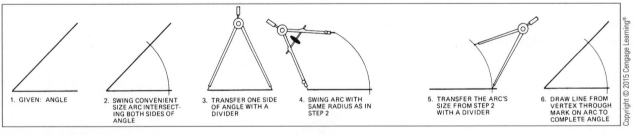

Figure 7-12 *Duplicate an angle.*

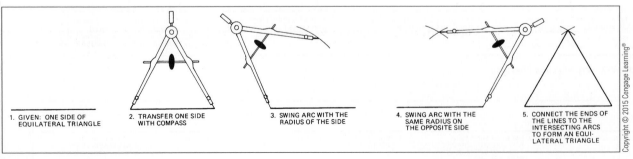

Figure 7-13 *Constructing an equilateral triangle.*

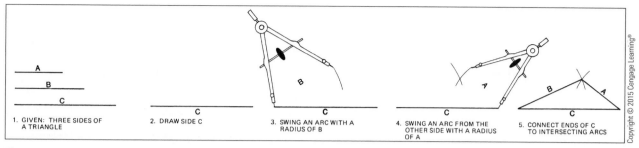

Figure 7-14 *Constructing a triangle.*

1. GIVEN: THREE SIDES OF A TRIANGLE
2. DRAW SIDE C
3. SWING AN ARC WITH A RADIUS OF B
4. SWING AN ARC FROM THE OTHER SIDE WITH A RADIUS OF A
5. CONNECT ENDS OF C TO INTERSECTING ARCS

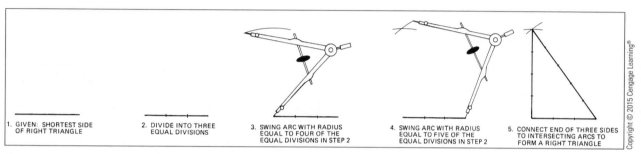

Figure 7-15 *Constructing a right triangle.*

1. GIVEN: SHORTEST SIDE OF RIGHT TRIANGLE
2. DIVIDE INTO THREE EQUAL DIVISIONS
3. SWING ARC WITH RADIUS EQUAL TO FOUR OF THE EQUAL DIVISIONS IN STEP 2
4. SWING ARC WITH RADIUS EQUAL TO FIVE OF THE EQUAL DIVISIONS IN STEP 2
5. CONNECT END OF THREE SIDES TO INTERSECTING ARCS TO FORM A RIGHT TRIANGLE

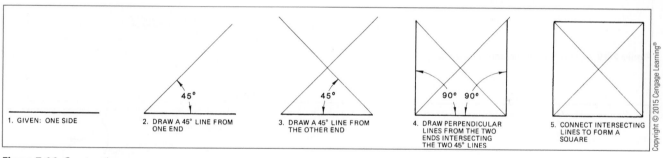

Figure 7-16 *Constructing a square.*

1. GIVEN: ONE SIDE
2. DRAW A 45° LINE FROM ONE END
3. DRAW A 45° LINE FROM THE OTHER END
4. DRAW PERPENDICULAR LINES FROM THE TWO ENDS INTERSECTING THE TWO 45° LINES
5. CONNECT INTERSECTING LINES TO FORM A SQUARE

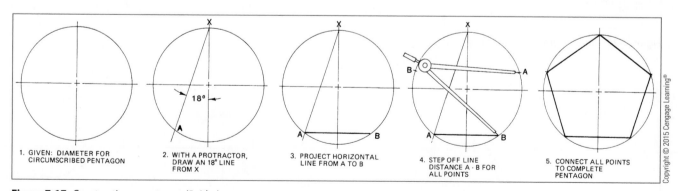

Figure 7-17 *Constructing a pentagon (5 sides).*

1. GIVEN: DIAMETER FOR CIRCUMSCRIBED PENTAGON
2. WITH A PROTRACTOR, DRAW AN 18° LINE FROM X
3. PROJECT HORIZONTAL LINE FROM A TO B
4. STEP OFF LINE DISTANCE A - B FOR ALL POINTS
5. CONNECT ALL POINTS TO COMPLETE PENTAGON

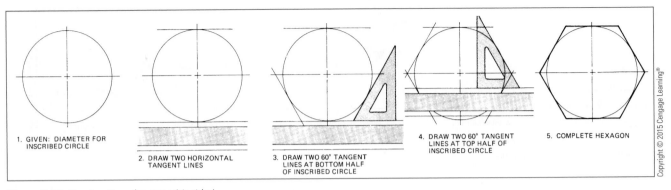

Figure 7-18 *Constructing a hexagon (six sides).*

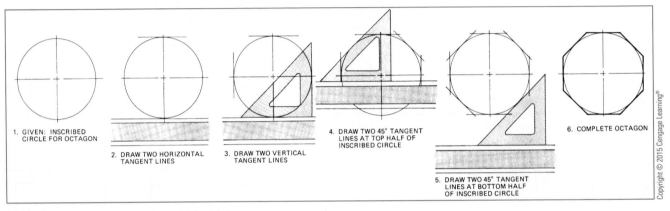

Figure 7-19 *Constructing an octagon (eight sides).*

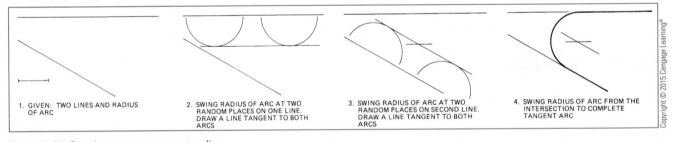

Figure 7-20 *Drawing an arc tangent to two lines.*

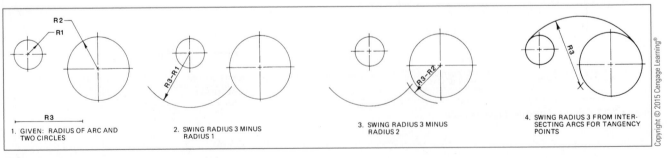

Figure 7-21 *Drawing an arc tangent to two circles.*

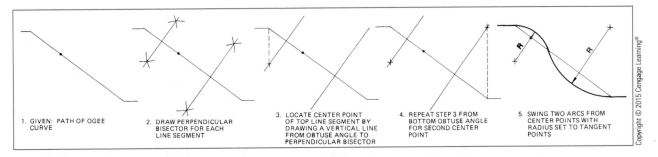

Figure 7-22 *Drawing an ogee curve (also called a reverse curve). An ogee curve occurs in situations where a smooth contour is needed between two offset features.*

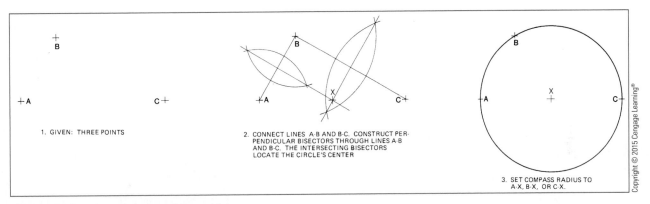

Figure 7-23 *Drawing a circle through three points.*

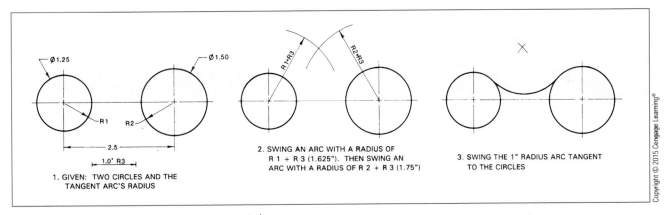

Figure 7-24 *Drawing a concave arc tangent to two circles.*

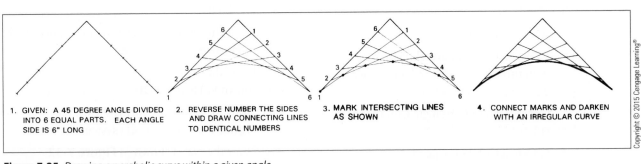

Figure 7-25 *Drawing a parabolic curve within a given angle.*

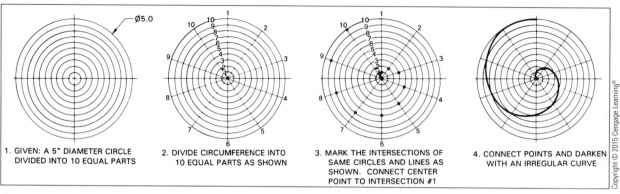

1. GIVEN: A 5" DIAMETER CIRCLE DIVIDED INTO 10 EQUAL PARTS

2. DIVIDE CIRCUMFERENCE INTO 10 EQUAL PARTS AS SHOWN

3. MARK THE INTERSECTIONS OF SAME CIRCLES AND LINES AS SHOWN. CONNECT CENTER POINT TO INTERSECTION #1

4. CONNECT POINTS AND DARKEN WITH AN IRREGULAR CURVE

Figure 7-26 *Drawing a spiral of Archimedes.*

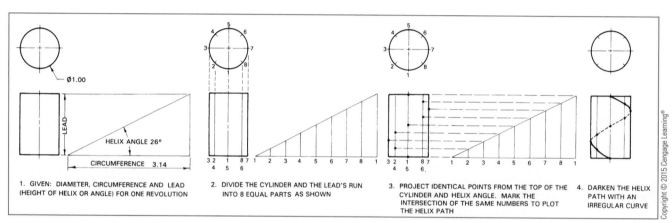

1. GIVEN: DIAMETER, CIRCUMFERENCE AND LEAD (HEIGHT OF HELIX OR ANGLE) FOR ONE REVOLUTION

2. DIVIDE THE CYLINDER AND THE LEAD'S RUN INTO 8 EQUAL PARTS AS SHOWN

3. PROJECT IDENTICAL POINTS FROM THE TOP OF THE CYLINDER AND HELIX ANGLE. MARK THE INTERSECTION OF THE SAME NUMBERS TO PLOT THE HELIX PATH

4. DARKEN THE HELIX PATH WITH AN IRREGULAR CURVE

Figure 7-27 *Drawing a helix.*

DRAFTING EXERCISES

Do these exercises using drawing instruments or CAD.

1. Construct a perpendicular bisector for a 4.75" line.

2. Construct a parallel line to a 3.5" line set at 30°.

3. Construct a tangent line to a 1.75" diameter circle.

4. Divide a 5" line into eight equal parts.

5. Bisect a 42° angle.

6. Construct an **equilateral triangle** (a triangle with three equal sides and angles) with 2.25" sides.

7. Construct a right triangle with the shortest side 1.5".

8. Construct a 1.375" square.

9. Construct a **pentagon** (five-sided figure) in a circumscribed circle with a 1.5" radius.

10. Construct a **hexagon** (six-sided figure) with a 2.0" radius inscribed circle.

11. Construct an **octagon** (eight-sided figure) with a 3.75" diameter inscribed circle.

12. Construct a tangent arc within a 30° angle with a .5" radius.

13. Construct a right triangle with the short side 2.0" long.

14. Draw the basic geometric constructions shown in **Figures 7-28** and **7-29**.

15. Draw the two-dimensional geometric forms shown in **Figure 7-30**.

16. Draw the geometric forms shown in **Figure 7-31** with a CAD system.

17. Draw the plates shown in **Figure 7-32**.

18. Draw the four geometrics starting with a 3" circle (**Figure 7-33**).

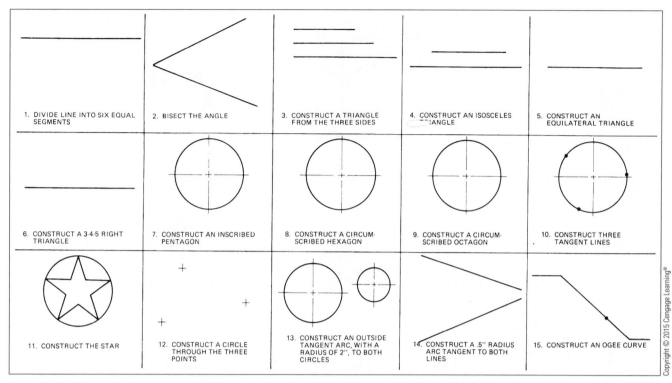

Figure 7-28 *With drawing instruments or a CAD system, draw the geometric forms. Double the size of each exercise.*

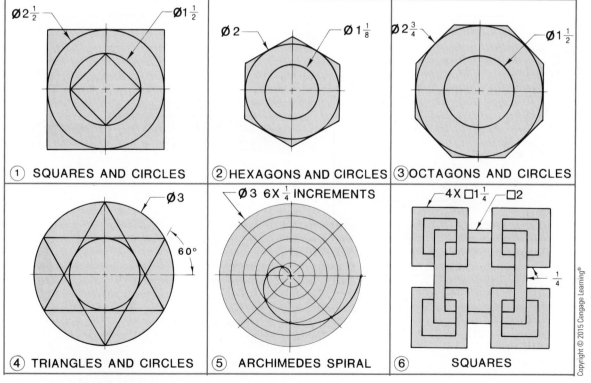

Figure 7-29 *With drawing instruments or a CAD system, draw the geometric exercises.* *(continued)*

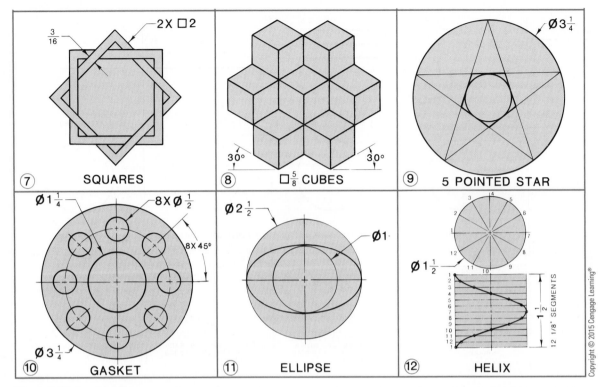

Figure 7-29 *(continued).*

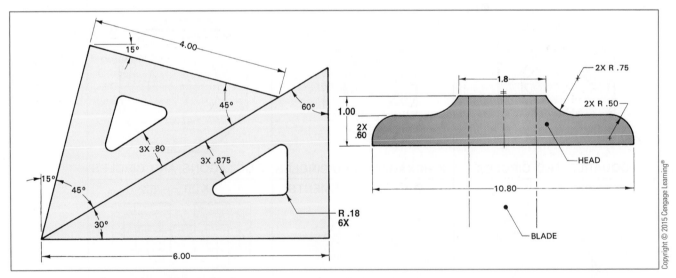

Figure 7-30 *Draw the triangles and T square head.*

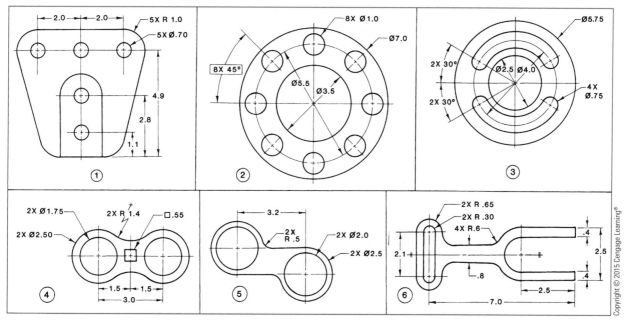

Figure 7-31 *With drawing instruments or a CAD system, draw the geometric forms.*

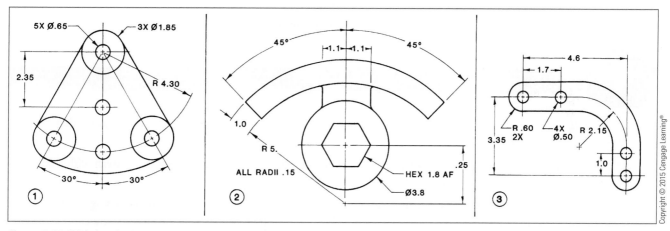

Figure 7-32 *With drawing instruments or a CAD system, draw three plates.*

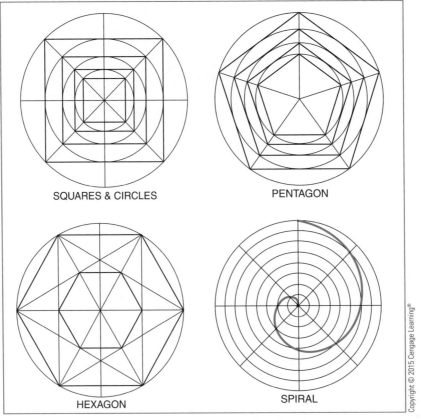

SQUARES & CIRCLES

PENTAGON

HEXAGON

SPIRAL

Figure 7-33 *Draw the four geometric shapes starting with a 3" circle.*

DESIGN EXERCISES

1. Design a family crest for yourself and/or your family.

2. Design a company logo for a business called Space & Aeronautics Inc.

3. Design a new school logo using the school's colors.

4. Design a business card for your mother and/or father.

5. Design a new geometric pattern with a compass and straightedge.

6. Design and draw an object using as many basic lines and curved surfaces as you can.

KEY TERMS

Equilateral triangle

Geometric construction

Hexagon

Octagon

Pentagon

Three-dimensional form

Two-dimensional form

Multiview Drawings

Multiview Orthographic Projections

Orthographic projection is a projection of a single view of an object on a drawing surface that is perpendicular to both the view and the lines of projection. Horizontal orthographic projection lines are used to align views with the same height dimensions. Vertical orthographic projection lines are used to align views with the same width dimensions (**Figure 8-1**). The technique of orthographic projection is used to create **multiview drawings** of a single object. The related views of the object appear as if they were all in the same plane. Multiview drawings describe the exact size and shape of an object. **Figure 8-2** shows a CAD-generated multiview drawing of a machine part.

Selection of Views

Describing the shape of an object with a multiview drawing should always be accomplished with the minimum number of views. The first view chosen is the front view because it best shows the form of an object. **Figure 8-3** shows the selection of the front view of several objects. In each case, the front view reveals the most unique and distinguishing shape feature of the object.

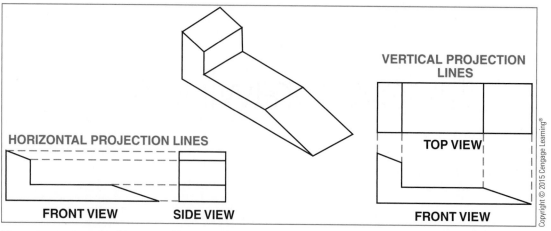

Figure 8-1 *An example of the orthographic projection of the side and top views from the front view.*

Simple flat objects, such as gaskets (**Figure 8-4**), require only one view, since the thickness is uniform and can be described in a note.

Symmetrical objects may need only one or two views. It is not necessary to draw another view if it is identical to an existing view. In a multiview drawing of a solid sphere (ball), all views are identical so only one view is needed. Only two views are required to describe the object in **Figure 8-5**, since the front and side views are the same.

Three views are used to describe a clock radio. The front view (**Figure 8-6**) shows the height and width of the clock radio, the **top view** (**Figure 8-7**) shows the depth and width, and the right-side view (**Figure 8-8**) shows the height and depth. The size and shape of most objects can be described in three views (**Figure 8-9**) when the bottom, left-side, and rear views would not help to further clarify any aspect of the object.

Other objects may require up to six views to show the exact size and shape of each side (**Figure 8-10**). **Figure 8-11** shows a six-view drawing of the same clock radio shown in the three views in **Figures 8-6** through **8-8**. Although the three views have identical dimensions, the details of each of the six views shown in **Figure 8-11** are different. Six views, therefore, are needed to show all the details that may be necessary for manufacturing the clock radio case. Many additional multiview working drawings are needed for the manufacture of the interior of the clock radio.

Planes of Projection

Each view in a multiview drawing is drawn as it appears on an imaginary **plane of projection**. Imagine a sheet of glass (plane of projection) placed between a viewer and an object (**Figure 8-12**). Now imagine the viewer's lines of sight as parallel and connected through the glass to each line and corner of the object. The shape of the object can be visualized where the lines of sight intersect the plane of projection.

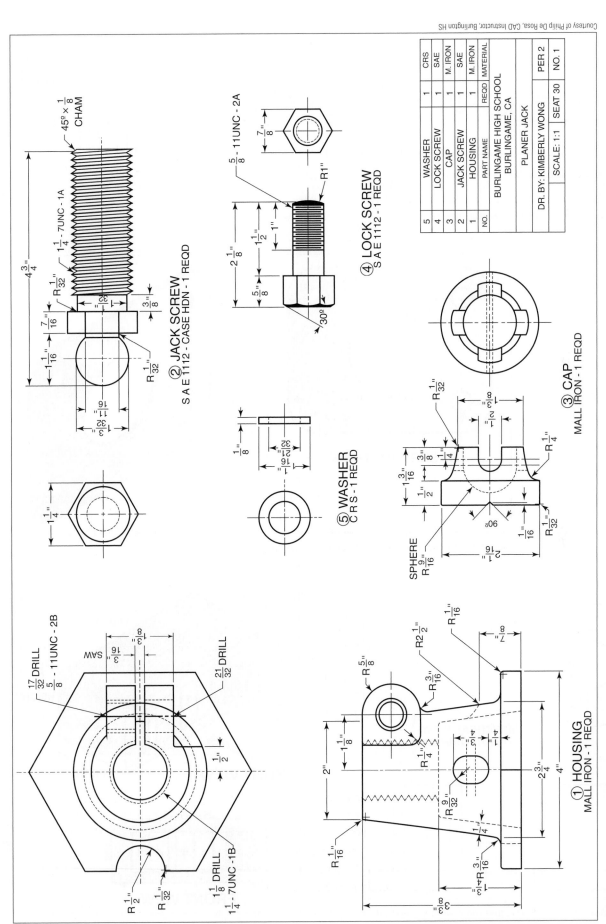

Figure 8-2 *A CAD generated multiview drawing.*

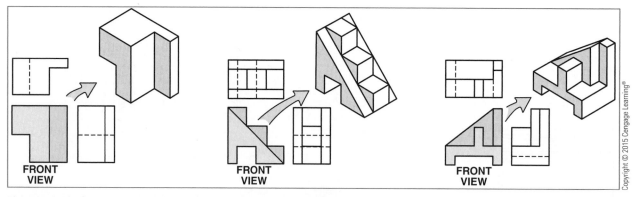

Figure 8-3 *The front view selection depends on which view best describes the shape of the item being drawn.*

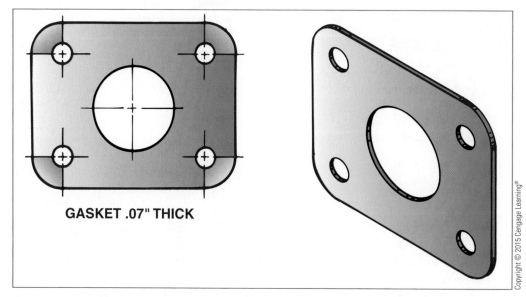

GASKET .07" THICK

Figure 8-4 *Simple flat objects can be defined with a single view.*

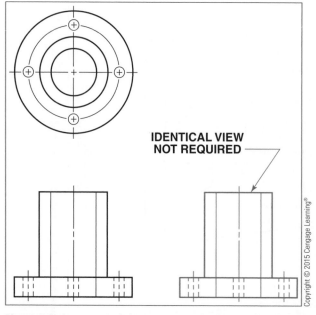

IDENTICAL VIEW
NOT REQUIRED

Figure 8-5 *A symmetrical object requires only two views for a full description.*

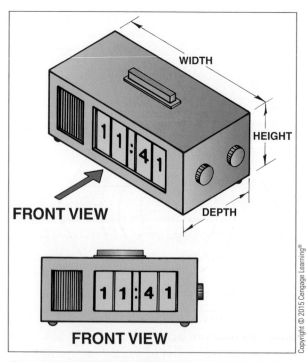

WIDTH

HEIGHT

FRONT VIEW

DEPTH

1 1 : 4 1

FRONT VIEW

Figure 8-6 *A multiview drawing's front view of a digital clock.*

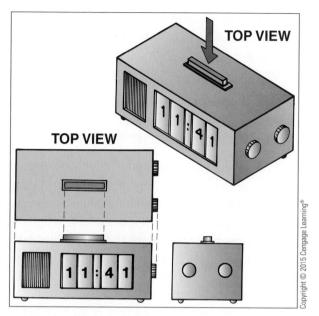

Figure 8-7 *The digital clock's top view.*

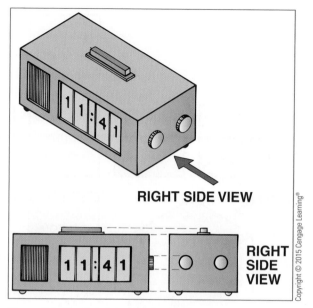

Figure 8-8 *The digital clock's right-side view.*

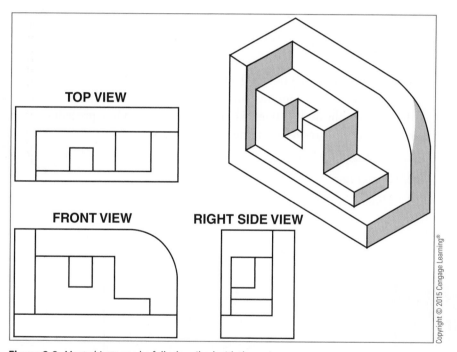

Figure 8-9 *Most objects can be fully described with three views.*

To visualize all six planes of projection, imagine a glass box surrounding an object (**Figure 8-13**) with each view of the object drawn on each glass plane as viewed from the outside. When the sides of the glass box are hinged open (**Figure 8-14**), the relative position of each view in a multiview drawing is revealed. The hinge lines between the planes are also called **fold lines** or **reference lines** and are used to locate the position of each view (**Figure 8-15**).

Views can be located also by drawing a 45° line that is used to project a third view once two base views are established. **Figure 8-16** shows the use of this method in projecting a side view from a front and top view, or a top view from a front and side view.

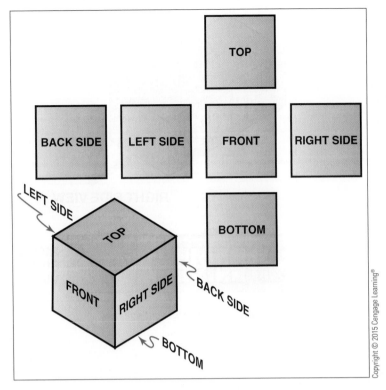

Figure 8-10 *Orthographic projection of six views using third-angle projection for the placement of views.*

Another method of drawing a third view from two established views is the use of dividers to transfer distances from completed views. For example, depth dimensions from the top view can be transferred to the side view with dividers.

Angles of Projection

The three basic views used in multiview drawings in the United States are the front view, right-side view, and top view. This is known as **third-angle projection** (**Figure 8-17**). In Europe, the three basic views are the front view, left-side view, and bottom view. This is known as **first-angle projection** (**Figure 8-18**). **Figure 8-19** shows a comparison of a multiview drawing of the same object using first-angle and third-angle projection.

LINE PRECEDENCE

When objects are viewed through a plane of projection, some lines fall on an identical perpendicular plane, although one line may be further from the viewer than the other. When lines overlap in this manner, the darkest line takes precedence (called **line precedence**) on the drawing (**Figure 8-20**) in the following order:

1. Cutting plane line
2. Object line
3. Hidden line
4. Centerline

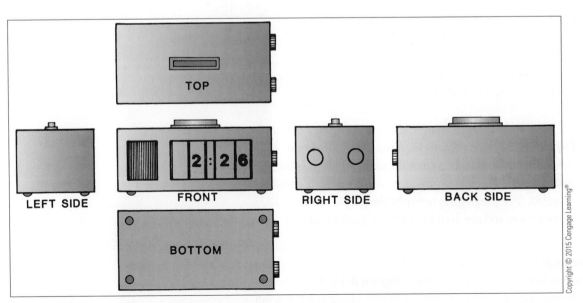

Figure 8-11 *A multiview drawing may have a maximum of six orthographic views.*

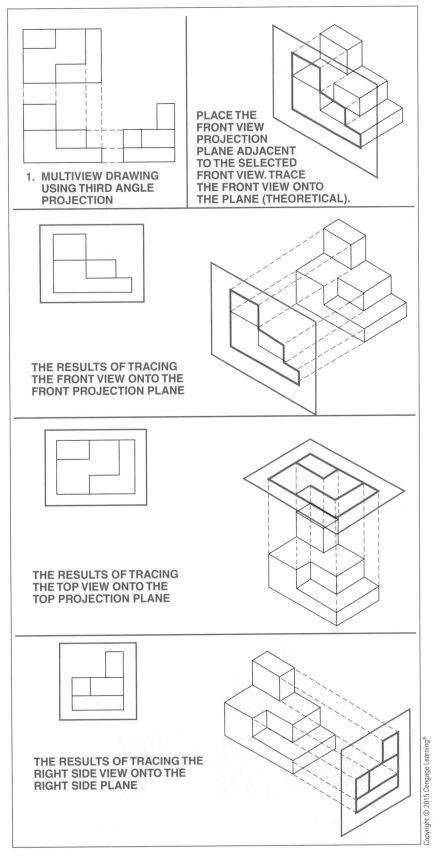

1. MULTIVIEW DRAWING USING THIRD ANGLE PROJECTION

PLACE THE FRONT VIEW PROJECTION PLANE ADJACENT TO THE SELECTED FRONT VIEW. TRACE THE FRONT VIEW ONTO THE PLANE (THEORETICAL).

THE RESULTS OF TRACING THE FRONT VIEW ONTO THE FRONT PROJECTION PLANE

THE RESULTS OF TRACING THE TOP VIEW ONTO THE TOP PROJECTION PLANE

THE RESULTS OF TRACING THE RIGHT SIDE VIEW ONTO THE RIGHT SIDE PLANE

Figure 8-12 *Planes of projection.*

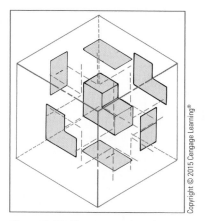

Figure 8-13 *Views are projected perpendicular to the six planes.*

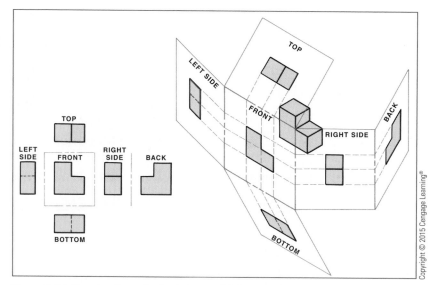

Figure 8-14 *Opening the projection box on its fold lines (reference lines).*

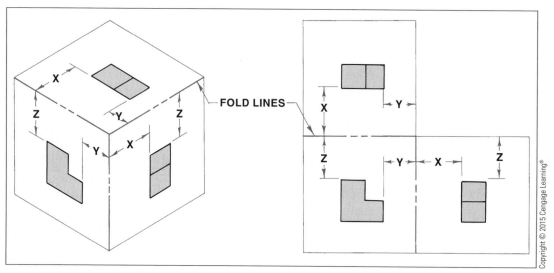

Figure 8-15 *Locating views from the fold lines (reference lines).*

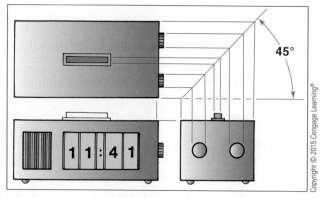

Figure 8-16 *A third view can be projected from two established views.*

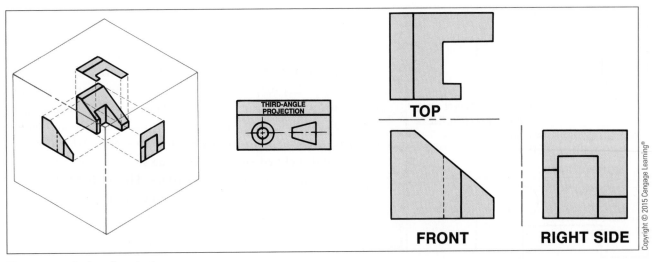

Figure 8-17 *Third-angle projection.*

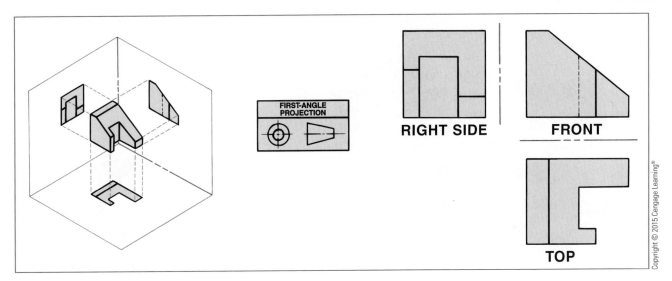

Figure 8-18 *First-angle projection.*

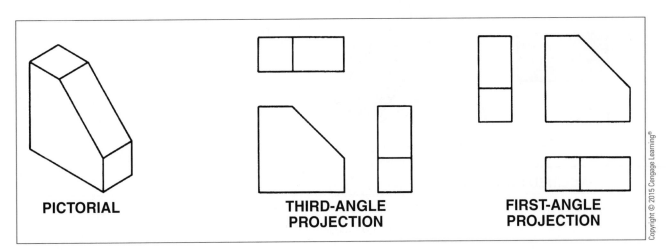

Figure 8-19 *A comparison of first-angle and third-angle projection.*

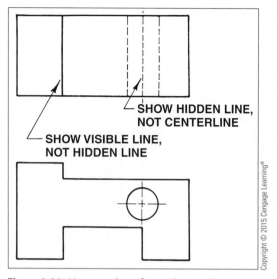

Figure 8-20 *Line precedence for overlapping lines.*

Visualization

All flat surfaces in multiview drawings are either normal, inclined, or oblique (**Figure 8-21**). **Normal surfaces** are **true size surfaces**, and are parallel or perpendicular to the plane of projection. **Inclined surfaces** are angular; their planes are not parallel or perpendicular to the plane of projection. Inclined surfaces are slanted to normal surfaces. **Figure 8-22** shows the relationship between normal and inclined surfaces. Oblique surfaces are not parallel or perpendicular to any normal surface. **Figure 8-23** shows the relationship between normal and oblique surfaces. Numbers and letters are used on **Figures 8-22** and **8-23** to identify identical surfaces on different views.

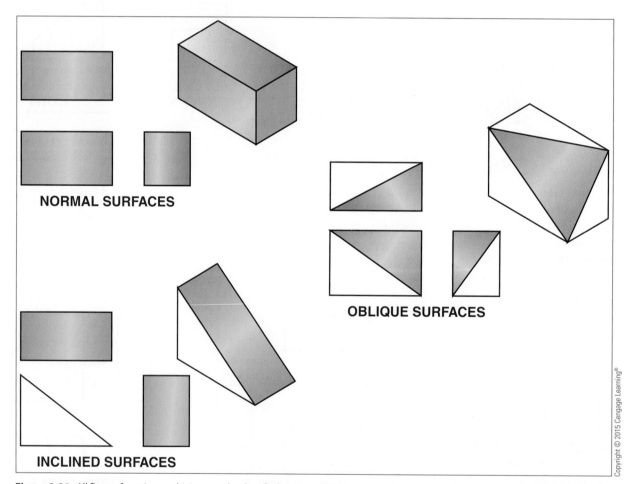

Figure 8-21 *All flat surfaces in a multiview may be classified as normal, inclined, or oblique.*

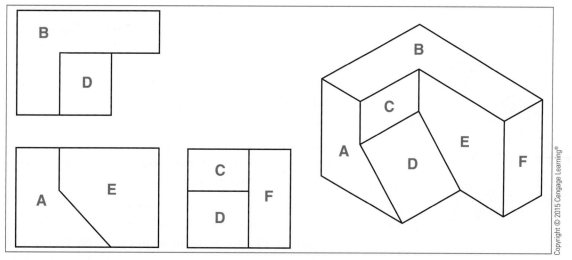

Figure 8-22 *Identifying normal and inclined surfaces.*

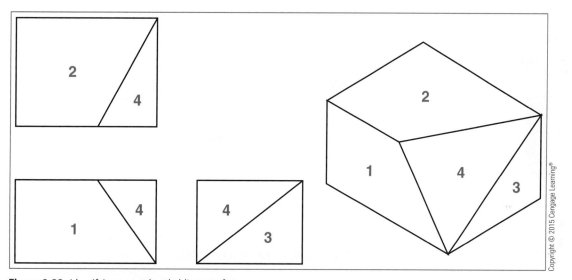

Figure 8-23 *Identifying normal and oblique surfaces.*

Inclined and Oblique Surface Projections

Surfaces perpendicular or parallel to the plane of projection always appear as true size surface or edge on the orthographic plane. Surfaces that lie on any angle to the plane of projection appear foreshortened (not true size surfaces) on the orthographic plane. **Figure 8-24** shows a comparison of true size surfaces and foreshortened surfaces on an orthographic view.

Holes and cylinders appear as true circles on normal surfaces but appear as **ellipses** on inclined surfaces (**Figure 8-25**). Since inclined surfaces recede from the plane of projection at right angles, the ellipse's major diameter parallel to the plane of projection is true size. The ellipse dimension perpendicular to the plane of projection is foreshortened.

The amount of foreshortening is related to the angle of incline. Small angles produce narrow ellipses and large angles produce wide ellipses (**Figure 8-26**). **Figure 8-27** shows ellipse widths derived from common inclined angles.

The use of standard ellipse templates eliminates having to plot ellipse layouts on inclined surfaces. The steps shown in **Figure 8-28** must be followed to ensure the accuracy of the ellipse.

Since oblique surfaces (**Figure 8-29**) are not parallel to any normal plane, oblique surfaces, including circles, must be plotted from known true distances in other views.

RUNOUTS

Surfaces with round or smooth contours do not intersect with straight lines. When surfaces blend together in this manner, the merging surface is called a **runout**. Square runouts and round runouts are shown in **Figure 8-30**.

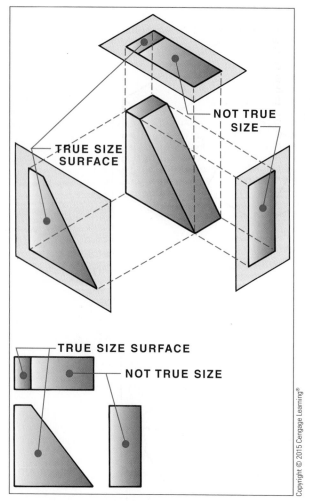

Figure 8-24 *Projection of perpendicular and inclined surfaces.*

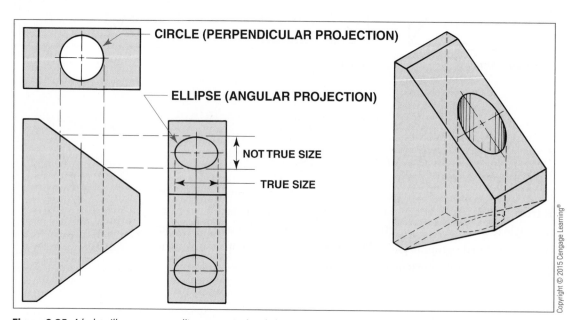

Figure 8-25 *A hole will appear as an ellipse on an inclined plane.*

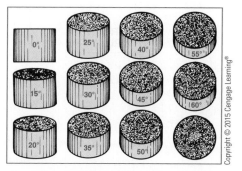

Figure 8-26 *Example of ellipses.*

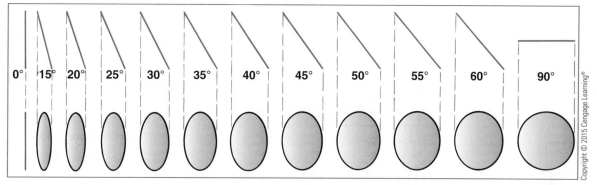

Figure 8-27 *The angle of projection determines the minor diameter of the ellipse.*

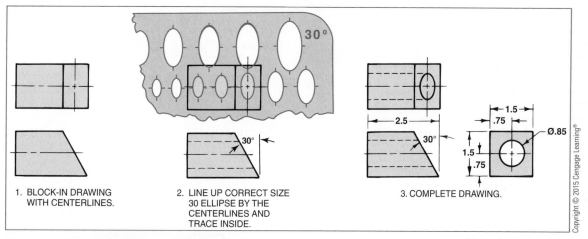

1. BLOCK-IN DRAWING WITH CENTERLINES.

2. LINE UP CORRECT SIZE 30 ELLIPSE BY THE CENTERLINES AND TRACE INSIDE.

3. COMPLETE DRAWING.

Figure 8-28 *Drawing an ellipse with an ellipse template.*

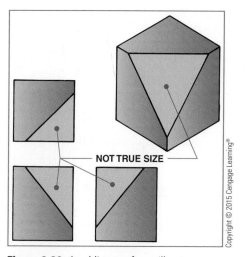

NOT TRUE SIZE

Figure 8-29 *An oblique surface will not appear true size in any view.*

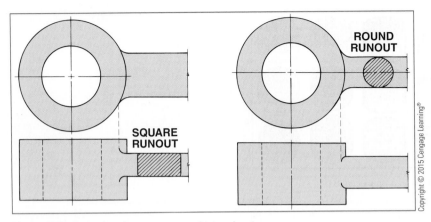

Figure 8-30 *Examples of runouts on a multiview drawing.*

Dimensioning

Conventions for dimensioning are covered in Chapter 9. Some consideration of dimensioning needs is necessary, however, when planning and preparing multiview drawings. Use the following guidelines in planning for the dimensioning of multiview drawings:

1. Spacing between views depends on the number of dimensions to be placed between the views (**Figure 8-31**).

2. Place the closest dimension to the view at .375" clearance. Every additional dimension is spaced .25" apart (**Figure 8-32**).

3. Place as many dimensions as possible between views (**Figure 8-33**), rather than on the outside.

4. Place dimensions on the most descriptive view of that feature (**Figure 8-34**).

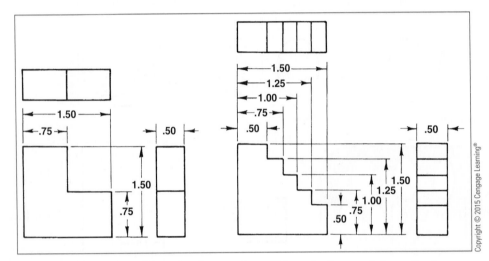

Figure 8-31 *The numbers of dimensions between views will determine the space between the views.*

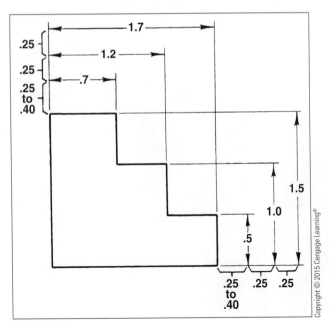

Figure 8-32 *Typical dimensioning spacing.*

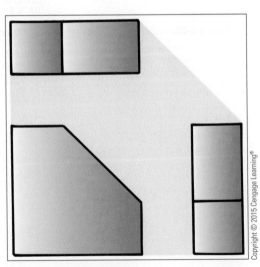

Figure 8-33 *Place as many dimensions as possible between the views.*

5. A working drawing must be completely dimensioned with enough information to manufacture the item without guesswork or fabrication redesign.

6. Do not duplicate the same dimensions on a different view.

7. The overall dimensions must always be given.

8. One dimension in a full chain of dimensions must be omitted.

9. Holes are always dimensioned on the circular view.

10. Shafts are dimensioned on their side view.

Most companies position the first drawing on a sheet in a corner to allow space for additional details. Nevertheless, a drafter must know how to center a drawing in a given space to be certain the finished drawing will fit into the available space. To center a drawing, follow the steps in **Figure 8-35**.

Drawing Procedures

When drawing an object, first determine the number of views needed, and then decide which surface will become the front view. Next, establish the space needed between views for dimensioning and calculate the center of the drawing. Once these preliminary decisions have been made, follow the steps in **Figure 8-36**.

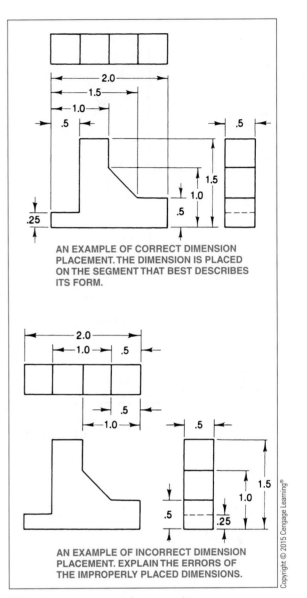

AN EXAMPLE OF CORRECT DIMENSION PLACEMENT. THE DIMENSION IS PLACED ON THE SEGMENT THAT BEST DESCRIBES ITS FORM.

AN EXAMPLE OF INCORRECT DIMENSION PLACEMENT. EXPLAIN THE ERRORS OF THE IMPROPERLY PLACED DIMENSIONS.

Figure 8-34 *Comparison of correct and incorrect dimension placements on a multiview drawing.*

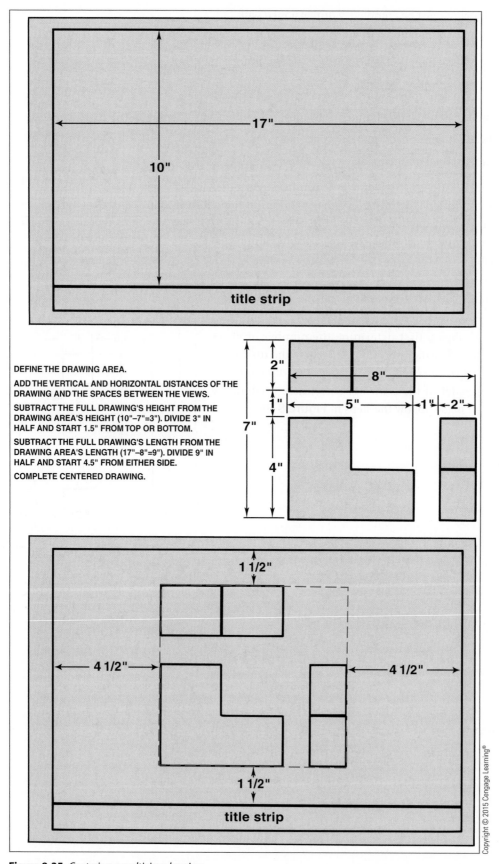

DEFINE THE DRAWING AREA.

ADD THE VERTICAL AND HORIZONTAL DISTANCES OF THE DRAWING AND THE SPACES BETWEEN THE VIEWS.

SUBTRACT THE FULL DRAWING'S HEIGHT FROM THE DRAWING AREA'S HEIGHT (10"–7"=3"). DIVIDE 3" IN HALF AND START 1.5" FROM TOP OR BOTTOM.

SUBTRACT THE FULL DRAWING'S LENGTH FROM THE DRAWING AREA'S LENGTH (17"–8"=9"). DIVIDE 9" IN HALF AND START 4.5" FROM EITHER SIDE.

COMPLETE CENTERED DRAWING.

Figure 8-35 *Centering a multiview drawing.*

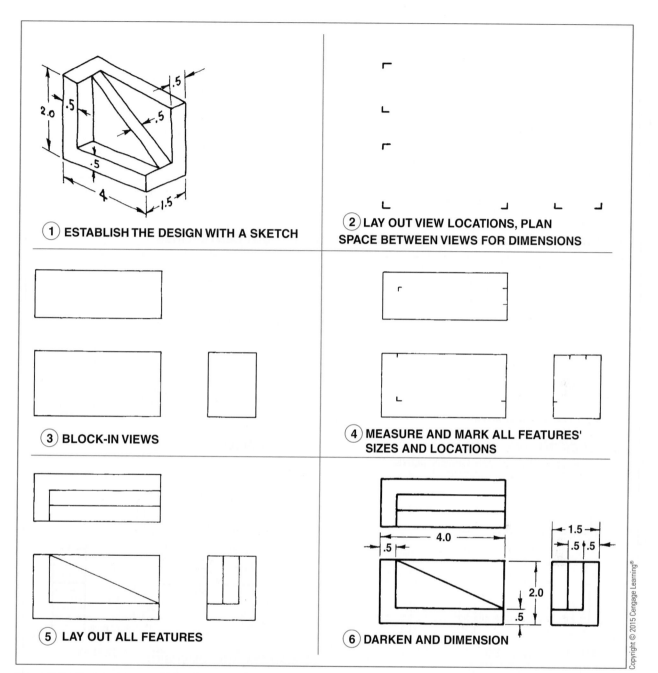

Figure 8-36 *Steps to draw a multiview drawing with an inclined surface.*

Follow the steps in **Figure 8-37** to draw arcs and circles in a multiview drawing. Refer to **Figure 8-38** to prepare multiview drawings that contain oblique surfaces.

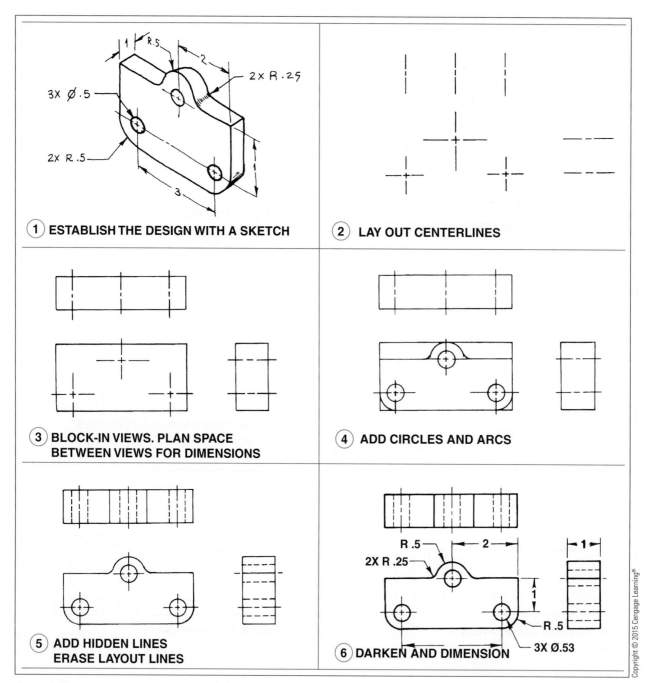

Figure 8-37 *Steps to draw a multiview drawing with circles and arcs.*

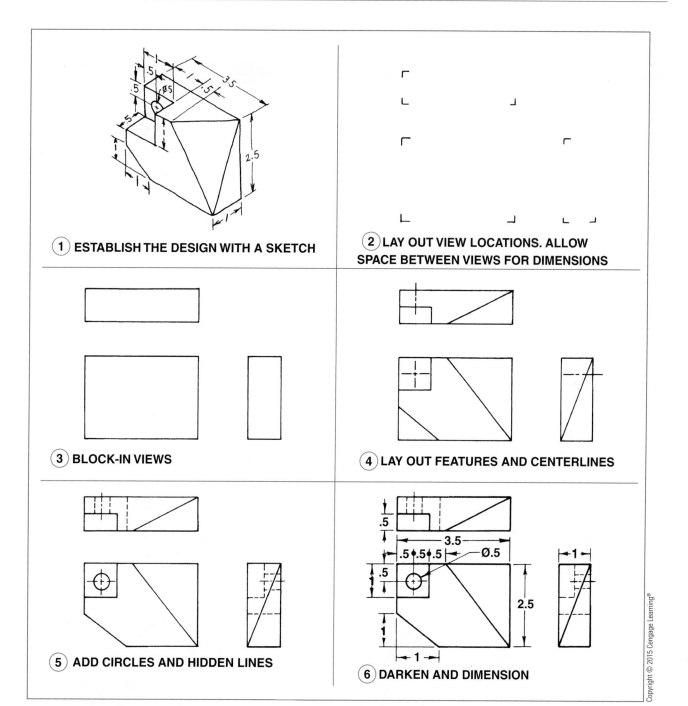

Figure 8-38 *Steps to draw a multiview drawing with an oblique surface.*

Multiview drawings are combinations of lines, circles, and arcs to create different views of the same object. The use of grid marks, orthographic projection lines, and different layers for object, hidden, center, and construction lines simplifies and increases the speed of preparing multiview drawings.

DRAFTING EXERCISES

1. Complete the nondrawing exercises in **Figures 8-39** through **8-43**. Follow the instructions for each exercise.

2. Complete the freehand sketching exercises in **Figures 8-44** through **8-49**.

    All of the following in this chapter may be done by sketching, instrument drawing, and CAD.

3. Complete the two- or three-view multiview drawings in **Figures 8-50** through **8-68**.

4. Complete the two- or three-multiview drawings and the isometric in **Figures 8-69** through **8-78**.

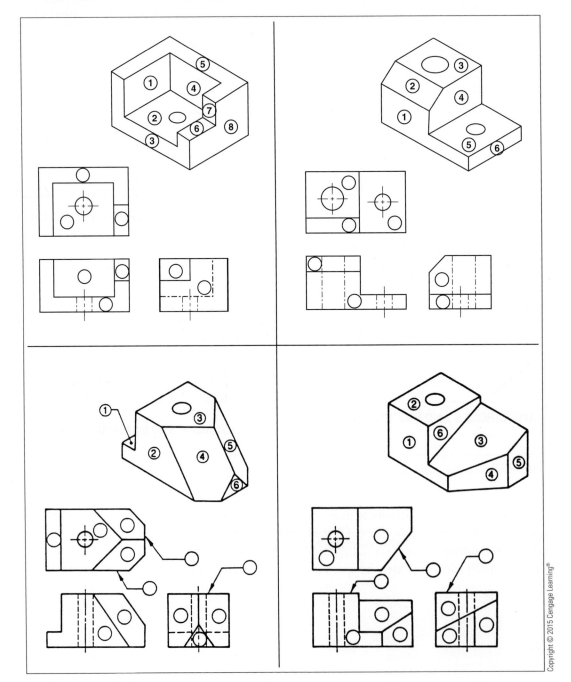

Figure 8-39 *Identify each surface with the correct number.*

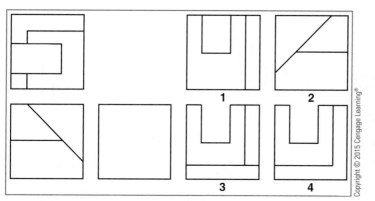

Figure 8-40 *Identify the correct right-side view(s).*

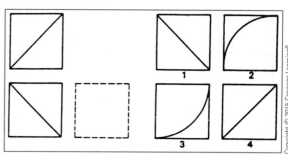

Figure 8-43 *Identify the correct right-side view(s).*

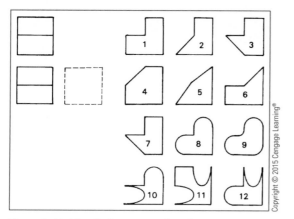

Figure 8-41 *Identify the correct right-side view(s).*

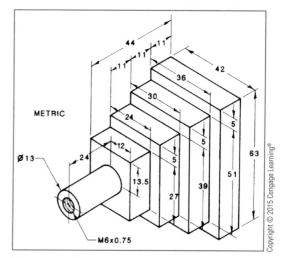

Figure 8-44 *Sketch a two-view multiview drawing of the centering guide.*

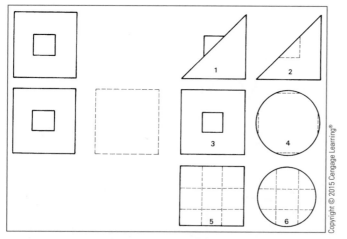

Figure 8-42 *Identify the correct right-side view(s).*

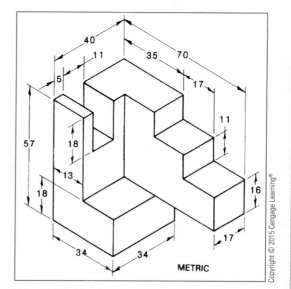

Figure 8-45 *Sketch a two-view multiview drawing of the metric step block.*

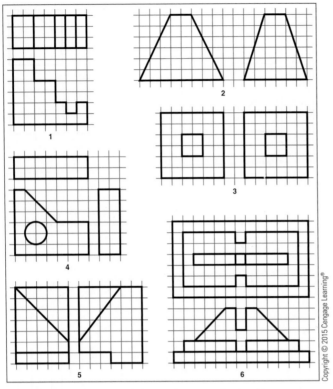

Figure 8-46 *Sketch the multiview and isometric for the objects.*

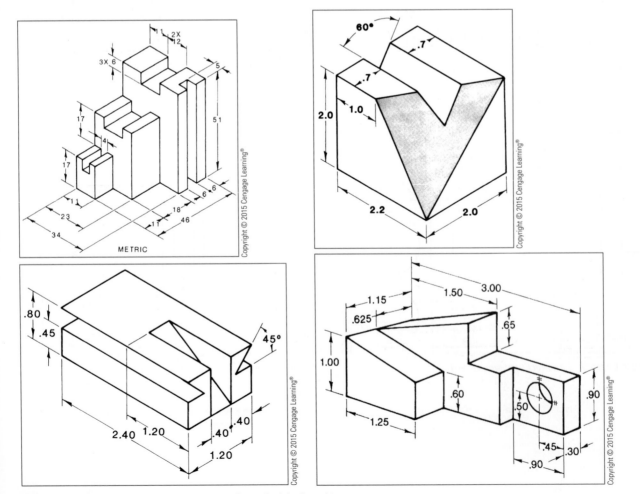

Figure 8-47 *Sketch the multiview and isometric for each of the four objects.*

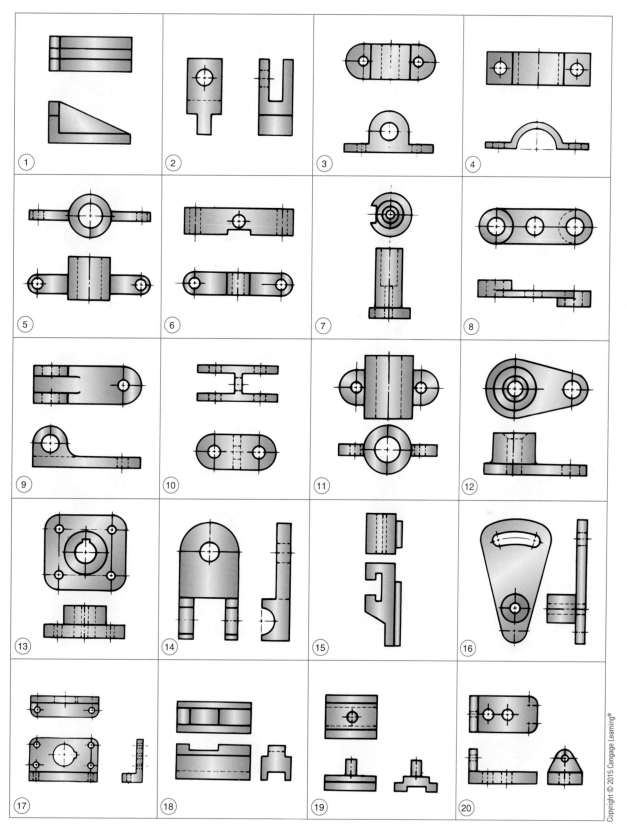

Figure 8-48 *Sketch the multiview drawings.*

Figure 8-49 *Sketch the multiview drawings for each isometric drawing.*

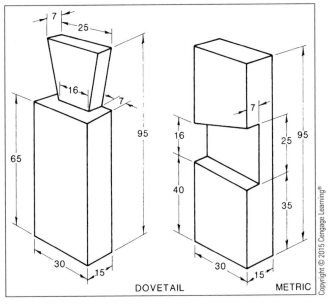

Figure 8-50 *Draw the multiview drawing (metric) with instruments or a CAD system.*

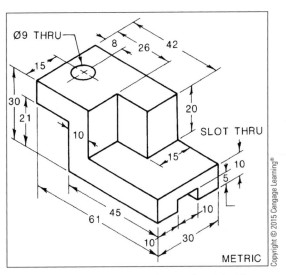

Figure 8-51 *Draw the multiview drawing (metric) with instruments or a CAD system.*

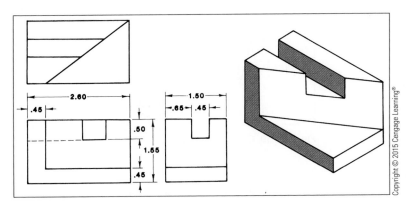

Figure 8-52 *Draw the multiview drawing of the base guide.*

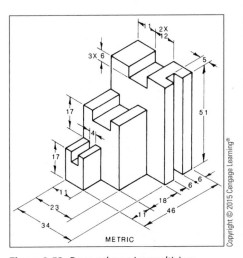

Figure 8-53 *Draw a three-view multiview drawing of the slotted-guide jig.*

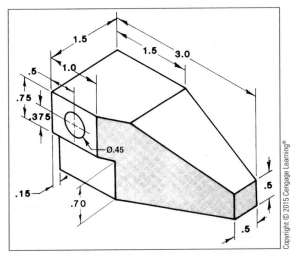

Figure 8-54 *Complete the multiview drawing.*

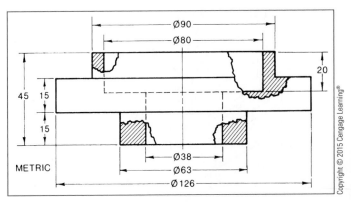

Figure 8-55 *Complete the multiview drawing (metric) with instruments or a CAD system.*

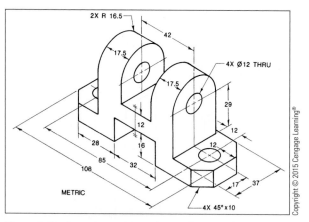

Figure 8-56 *Draw a two-view multiview drawing of the metric bearing support.*

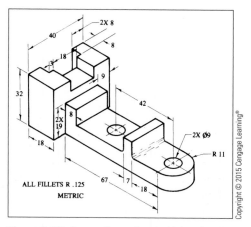

Figure 8-57 *Draw a three-view multiview drawing of the location jig.*

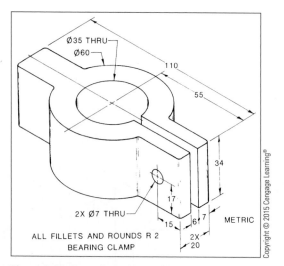

Figure 8-58 *Draw the multiview drawing (metric) with instruments or a CAD system.*

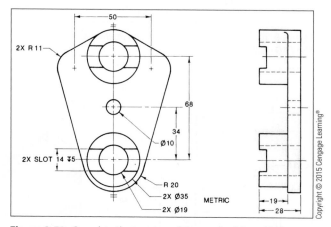

Figure 8-59 *Complete the top view of the metric slide-guide jig.*

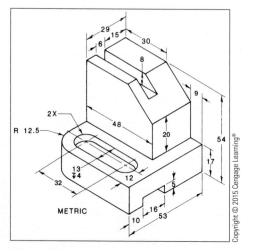

Figure 8-60 *Draw a three-view multiview drawing of the metric slip guide.*

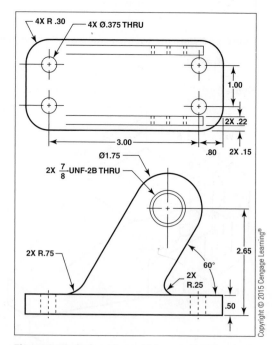

Figure 8-61 *Redraw the multiview working drawing of the angle bracket. Scale 3/4″ = 1′- 0″.*

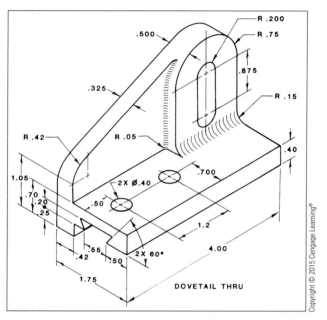

Figure 8-62 *Draw the multiview drawing of the dovetail slide guide.*

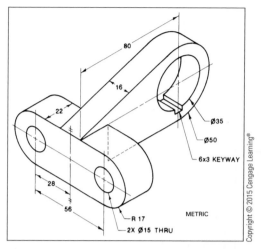

Figure 8-63 *Draw a three-view multiview drawing of the metric single-bearing support.*

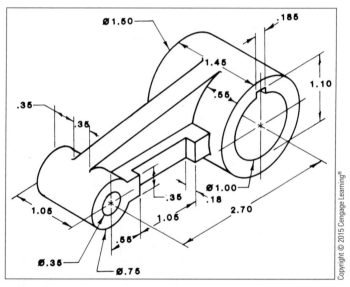

Figure 8-64 *Draw the multiview drawing of the bearing bracket.*

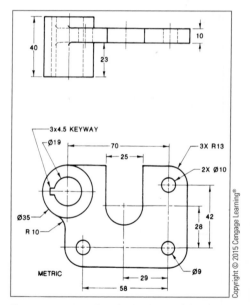

Figure 8-65 *Complete the third view of the metric bearing fixture.*

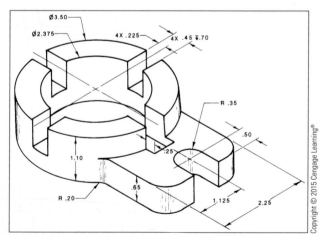

Figure 8-66 *Draw a two-view multiview drawing of the ring alignment jig.*

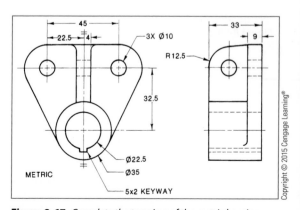

Figure 8-67 *Complete the top view of the metric bearing locator fixture.*

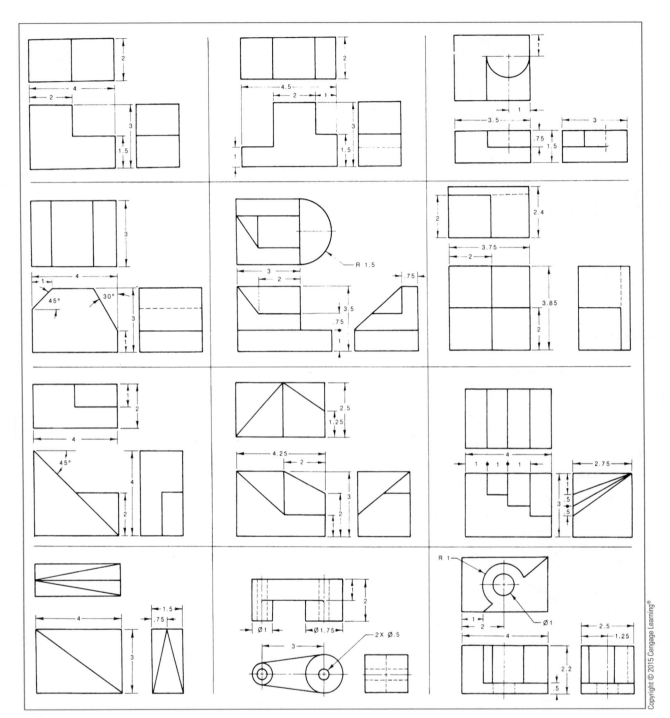

Figure 8-68 *Draw the exercise with drawing instruments or a CAD system.*

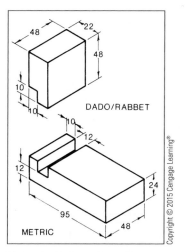

Figure 8-69 *Draw the multiview drawing (metric) with instruments or a CAD system.*

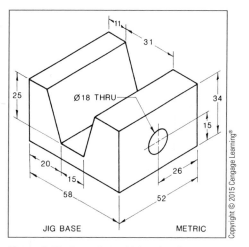

Figure 8-70 *Draw the multiview drawing (metric) with instruments or a CAD system.*

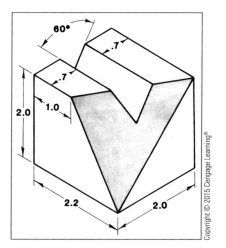

Figure 8-71 *Draw the multiview drawing of the V-groove corner block.*

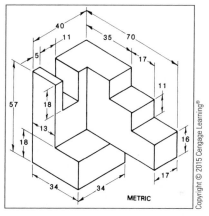

Figure 8-72 *Draw a two-view multiview drawing of the metric step block.*

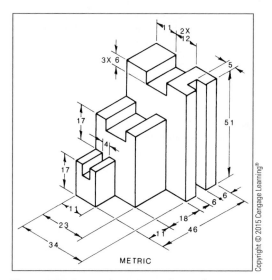

Figure 8-73 *Draw a three-view multiview drawing of the slotted-guide jig.*

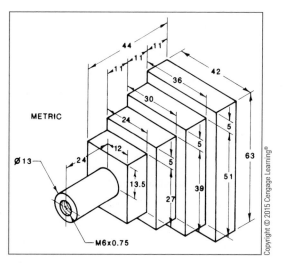

Figure 8-74 *Draw a two-view multiview drawing of the centering guide.*

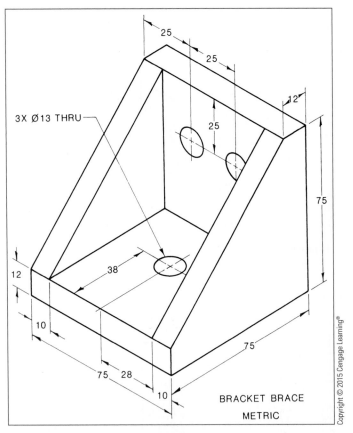

3X Ø13 THRU

BRACKET BRACE

METRIC

Figure 8-75 *Draw the multiview drawing (metric) with instruments or a CAD system.*

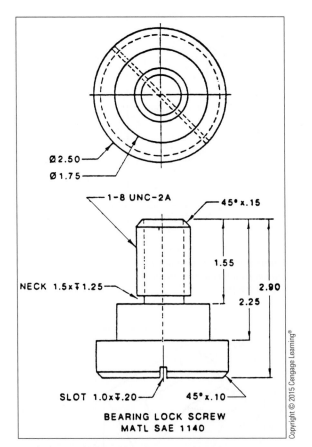

Ø2.50
Ø1.75

1-8 UNC-2A   45° x.15

1.55

2.90

NECK 1.5x↧1.25   2.25

SLOT 1.0x↧.20   45° x.10

BEARING LOCK SCREW
MATL SAE 1140

Figure 8-76 *Redraw the multiview working drawing of the lock screw.*

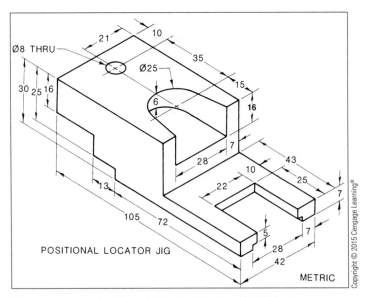

Ø8 THRU

Ø25

POSITIONAL LOCATOR JIG

METRIC

Figure 8-77 *Complete the multiview drawing (metric) with the instruments or a CAD system.*

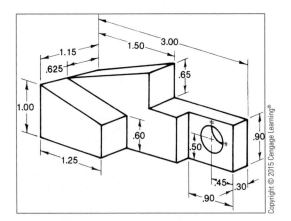

Figure 8-78 *Draw a two-view multiview drawing of the double incline guide.*

Copyright © 2015 Cengage Learning®

DESIGN EXERCISES

1. Draw the Multiview of **Figures 8-79** through **8-81**. Feel free to redesign each figure.

2. Design and draw a working drawing for a small storage cabinet that will store music cassettes, VCR tapes, and CD-ROM disks.

3. Design and draw a multiview drawing for a doorstop for a factory's 10′ × 8′ steel door. The space between the bottom of the door and the concrete floor is .5″. The stop should not project more than 2″ from the door into the passageway.

4. A heavy wood 10′ × 3′ worktable, with four 3″ × 3″ legs, is 30″ in height. Design and draw a multiview drawing of an adjustable leg lift capable of changing the height to 34″. Changes in the table's height will be made several times a week.

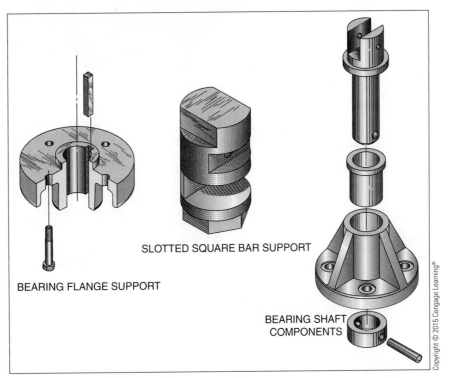

SLOTTED SQUARE BAR SUPPORT

BEARING FLANGE SUPPORT

BEARING SHAFT COMPONENTS

Figure 8-79 *Obtain dimensions with dividers directly from the page. Double the size of your drawing if necessary.*

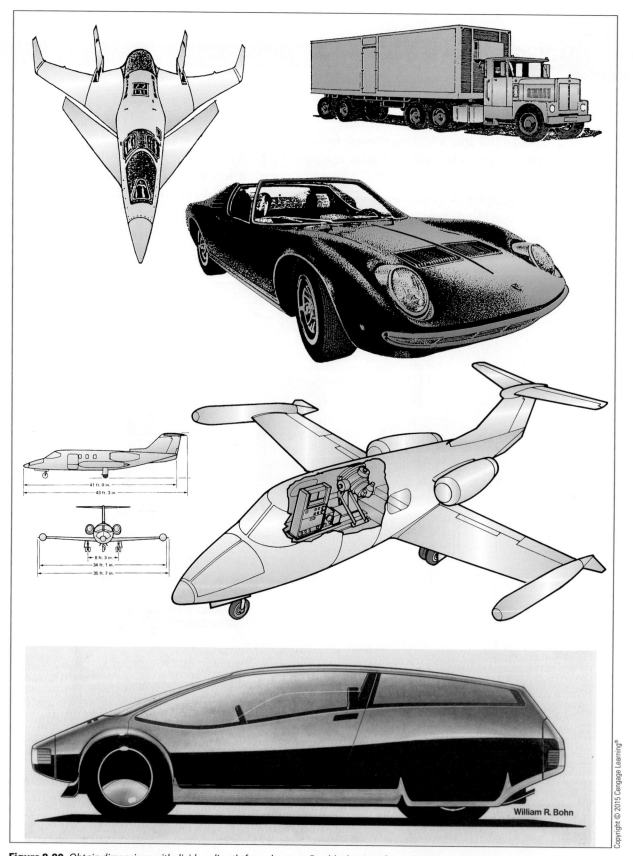

Figure 8-80 *Obtain dimensions with dividers directly from the page. Double the size of your drawing if necessary.*

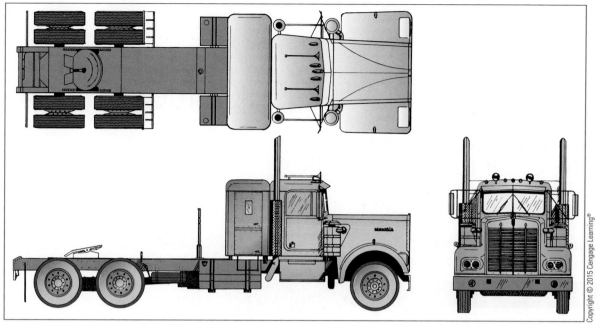

Figure 8-81 *Obtain dimensions with dividers directly from the page. Double the size of your drawing if necessary.*

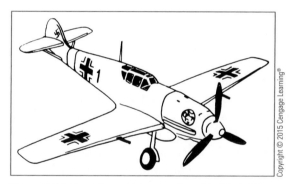

Figure 8-82 *Draw the multiview drawing of the World War II German plane.*

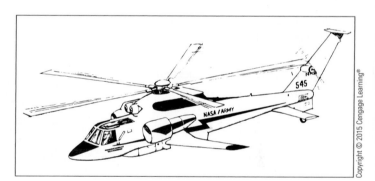

Figure 8-83 *Draw the multiview drawing of a helicopter. (Courtesy of NASA.)*

 KEY TERMS

Ellipses	Multiview drawing	Runout
First-angle projection	Normal surface	Symmetrical
Fold line	Orthographic projection	Top view
Inclined surface	Plane of projection	Third-angle projection
Line precedence	Reference line	True size surface

9

Dimensioning Conventions and Surface Finishes

OBJECTIVES

The student will be able to:

- Recognize standardized dimensioning symbols
- Apply correct dimensioning techniques to a working drawing
- Dimension all geometric forms
- Dimension machine operations
- Dimension an isometric drawing
- Add surface control symbols to a working drawing

Introduction

Engineering working drawings must contain all information necessary for manufacturing. This includes exact descriptions of the size, shape, and material specifications of every part of a product. These descriptions are accomplished through standardized **dimensioning** conventions (**Figure 9-1**). The machine part in **Figure 9-1** can be manufactured in any plant because the drawing contains universally accepted ASME (American Society of Mechanical Engineers) dimensional symbols and **notations**. Without these symbols and notations, the drawing (**Figure 9-2**) is meaningless and the part could not be manufactured without extensive guesswork. Product quality and accuracy would depend on the interpretation of the manufacturer. Adhering to established standards is important in preparing working drawings manually or with a CAD system.

Working drawing dimensions are of two types: **size dimensions** and **location dimensions**. Size dimensions describe the size of each geometric form (**Figure 9-3**). Location dimensions provide the exact location of each geometric part of a drawing (**Figure 9-4**). In addition to size and location dimensions, manufacturing notes with leaders and general notations are required to specify materials and other elements (**Figure 9-5**). If any dimension or notation is omitted, costly production errors and delays could result.

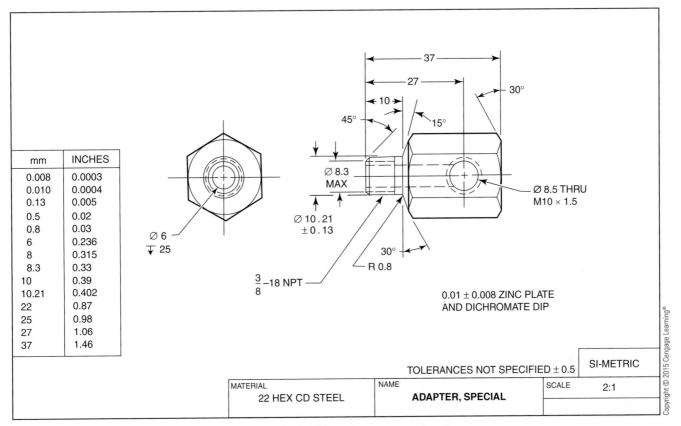

mm	INCHES
0.008	0.0003
0.010	0.0004
0.13	0.005
0.5	0.02
0.8	0.03
6	0.236
8	0.315
8.3	0.33
10	0.39
10.21	0.402
22	0.87
25	0.98
27	1.06
37	1.46

Figure 9-1 *Using the latest dimensioning and notation standards will assure accurate manufacturing.*

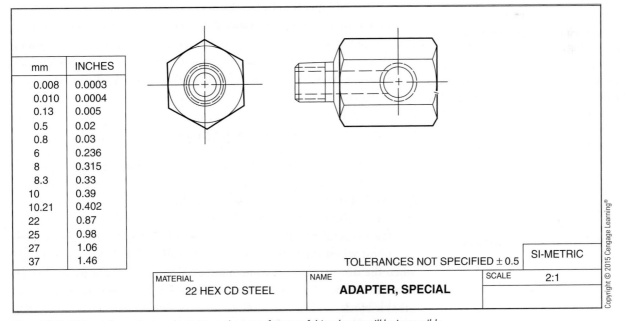

mm	INCHES
0.008	0.0003
0.010	0.0004
0.13	0.005
0.5	0.02
0.8	0.03
6	0.236
8	0.315
8.3	0.33
10	0.39
10.21	0.402
22	0.87
25	0.98
27	1.06
37	1.46

Figure 9-2 *Without dimensions and notations, the manufacture of this adapter will be impossible.*

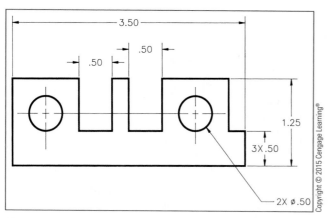

Figure 9-3 *Size dimensions.*

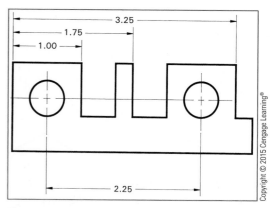

Figure 9-4 *Location dimensions.*

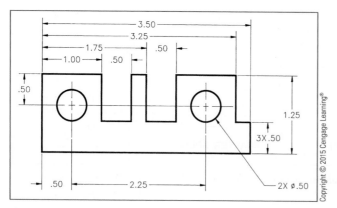

Figure 9-5 *A fully dimensioned drawing.*

1. Use the decimal inch for dimension values (5.75).
2. Do not use a zero before the decimal point for values less than one inch (.50).
3. The dimension value is expressed to the same number of decimal places as its tolerance (5.500 ± .002).
4. The number of decimal places may represent the tolerance:
   • 5.5 = large tolerance
   • 5.50 = average tolerance
   • 5.500 = small tolerance

Figure 9-6 *ASME conventions for U.S. customary dimensions.*

Systems of Dimensioning

The principles and rules for dimensioning, including dimensioning systems, are standardized by the ASME.

U.S. CUSTOMARY DIMENSIONING

ASME standards for **U.S. customary dimensioning**, with the use of decimal-inch values, are shown in **Figure 9-6**. When using decimal-inch dimensions, a zero is not used to the left of the decimal point for values less than one inch. The same number of decimal places should always be used for both dimensions in a tolerance dimension range. **Tolerance** is the total acceptable variation within a dimension (see Chapter 10). Establishing the number of decimal places represents the level of tolerance desired. Using three decimal places implies a higher level of tolerance than using two.

METRIC DIMENSIONING

ASME standards for **metric dimensioning** (**Figure 9-7**) require all dimensions to be expressed in millimeters. Therefore, the millimeter symbol (mm) is not needed on each dimension, but it is used when a dimension is used in a notation. In this case, a single space separates the numeral and the millimeter symbol. Zeros precede the decimal point when the value is less than 1 millimeter; zeros are not used when the dimension is a whole number. Commas are not used between thousand units. A metric dimensional note should be displayed in or near the title block.

DUAL DIMENSIONING

Working drawings are normally prepared with all U.S. customary or all metric dimensions. However, when the object is to be manufactured using both metric and U.S. customary measuring systems, **dual dimensions** may be necessary. Dual-dimensioning may also be necessary when converting a metric drawing to a U.S. customary drawing and vice versa. When dual dimensions are used, the metric dimensions are placed under the dimension line in parentheses. Sometimes a metric conversion chart is used (**Figure 9-8**).

Dimensioning Elements

ASME has also standardized the use of lines, numerical values, symbols, notations, and their placement on working drawings. **Figures 9-9** and **9-10** show the application of basic ASME dimensioning symbols. They illustrate the positioning of dimensioning lines, leaders, and notes to define the size and shape of common geometric forms and angles.

ASME dimensioning standards use dimension lines with arrowheads, dimension numerals, and **extension lines**. The extension line projects from the drawing but does not touch it. The dimension line touches the extension lines and is drawn parallel to the part being dimensioned. Leave a space for the numbers and add arrowheads. **Leaders** can also be used to connect a specific area of a drawing to a dimension or note. **Figure 9-11** shows the application of these standards to a typical view. **Figure 9-12** illustrates the correct method of drawing and spacing these lines. **Figure 9-13** shows the accepted proportion of a standard dimension arrowhead.

Dimensioning Guidelines

Figure 9-14 illustrates basic dimensioning types. **Figure 9-15** shows the two standard dimension positioning systems: unidirectional and aligned. **Unidirectional dimensions** all read from the bottom and are easier to read. **Aligned dimensions** read from the bottom or right of the drawing. Aligned dimensions conserve space on a drawing since they are aligned with dimension lines. Either system is acceptable if used consistently throughout a drawing.

1. All dimensions will be in millimeters.
2. Do not use millimeter symbol mm with dimensions (25 not 25 mm).
3. When a dimension value is less than one millimeter, a zero will precede the decimal point (0.55 not .55).
4. When the dimension is a whole number, do not use a zero or decimal point (150 not 150.0).
5. When the dimension uses a decimal value, do not end it with a zero (75.5 not 75.50).
6. An exception to convention No. 5 is: if the dimension value is followed by a tolerance, the number of decimal places must be equal (75.50 ± 0.05 not 75.5 ±0.05).
7. Do not use commas or spaces (1575 not 1,575 or 1 575).
8. When specifying millimeters of the drawing, as in a document, leave a space between the number and the symbol (2585 mm not 2585mm).
9. Display the word "METRIC" with large letters near the title block.
10. Display note:"UNLESS OTHERWISE SPECIFIED,ALL DIMENSIONS ARE MILLIMETERS."
11. The abbreviation for system international (metric system) is SI.

Figure 9-7 *ASME conventions for metric dimensions.*

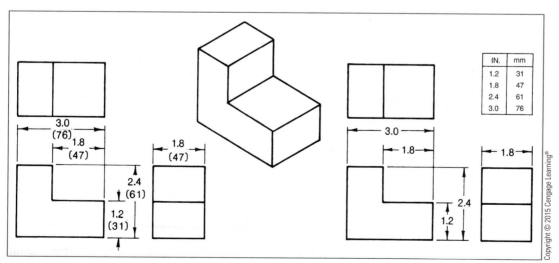

Figure 9-8 *Two procedures used for dual dimensioning.*

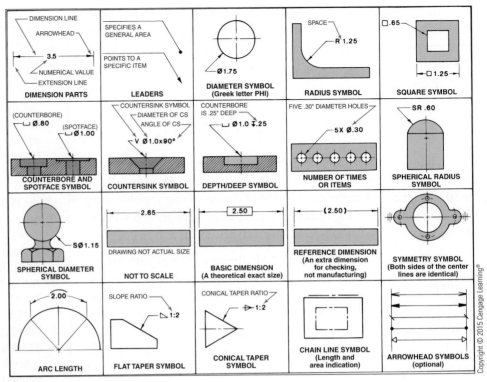

Figure 9-9 *ASME dimensioning conventions.*

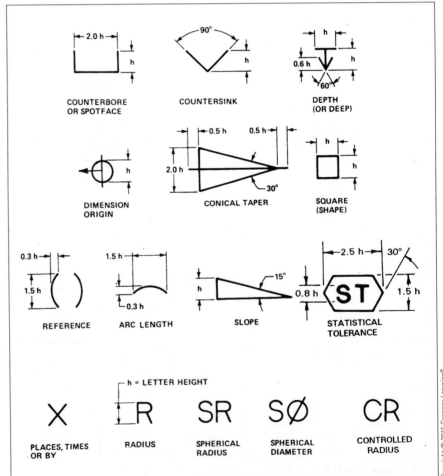

Figure 9-10 *Dimensioning Symbols*

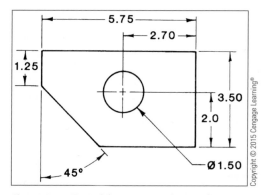

Figure 9-11 *Typical dimension application for ASME conventions.*

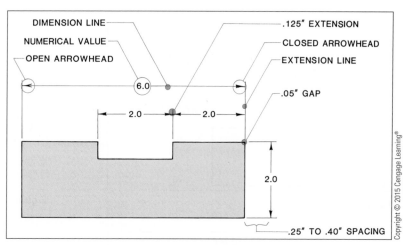

Figure 9-12 *General dimensioning conventions.*

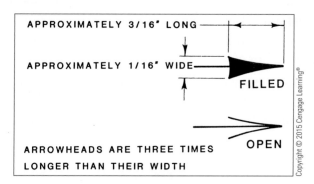

Figure 9-13 *The freehand arrowhead and its proportions.*

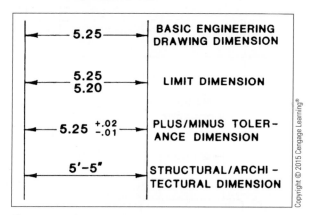

Figure 9-14 *Dimension types.*

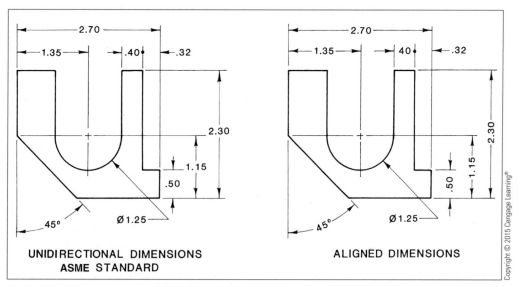

Figure 9-15 *Positioning systems for reading dimensions.*

If dimensions are not spaced properly, the drawing surface area will be wasted or dimensions may be too close to be read without confusion. Closely spaced dimensions can also inhibit the easy reading of object lines. For these reasons, spacing of dimensions (**Figure 9-16**) is recommended.

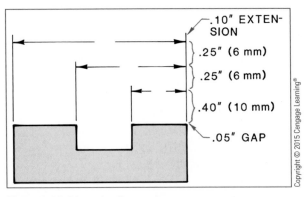

Figure 9-16 *Dimension line spacing.*

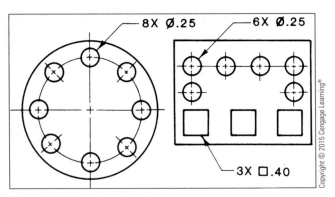

Figure 9-17 *Repetitive dimensions.*

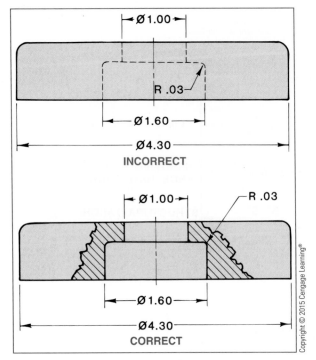

Figure 9-18 *Always dimension to visible lines.*

Many manufactured objects contain geometric forms or parts that are identical. When this occurs, only one dimension is needed (**Figure 9-17**). A dimension that covers many identical (repetitive) parts is always accompanied by a label indicating the number of parts of the same size.

Regardless of the object, always dimension to visible lines, not to hidden lines (**Figure 9-18**). If it is necessary to dimension to a hidden line, another view or sectional drawing is required.

POLAR COORDINATE DIMENSIONING

When parts of a drawing align on a common radial centerline, **polar coordinate dimensions** are used (**Figure 9-19**). In this case, all radial dimension lines originate from a baseline with the largest dimension on the outside. All other dimensions progress inward to the smallest dimension.

RECTANGULAR COORDINATE DIMENSIONING

When a large number of parts, projections, holes, or complex contours are contained in a small area, **rectangular coordinate dimensions** may be used in lieu of conventional dimension practices. Rectangular coordinate systems eliminate the need for dimension lines by establishing distances from base intersections, usually a corner (**Figure 9-20**). The baseline is assigned a value of zero. The dimension to each extension line represents the distance from zero. To eliminate size dimensions, each standard hole or projection is assigned a letter that relates to the size specified in an accompanying chart.

Rectangular coordinate grid systems are used extensively to define and locate irregular contour lines (**Figure 9-21**). In this application, the grid lines are spaced evenly so each square can be increased (or decreased) in size to duplicate the outline of the design. Some rectangular coordinate systems completely eliminate all detail dimensions

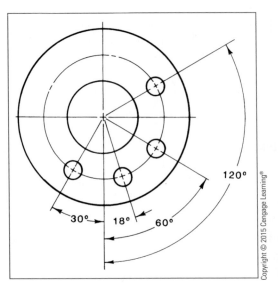

Figure 9-19 *Example of polar coordinate dimensioning.*

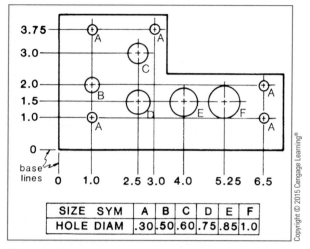

Figure 9-20 *Rectangular coordinate dimensioning without dimension lines.*

SIZE SYM	A	B	C	D	E	F
HOLE DIAM	.30	.50	.60	.75	.85	1.0

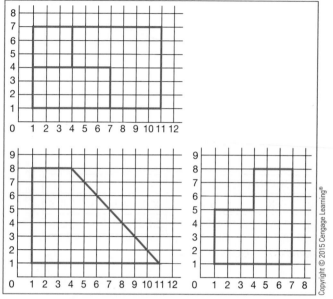

Figure 9-21 *Rectangular coordinate dimensioning system using a grid layout.*

with a chart that contains distances from established baselines. In **Figure 9-22**, the location and depth of holes on X, Y, and Z surfaces are shown in the top chart. The bottom chart shows the number and diameter of each hole.

TABULATED OUTLINE DIMENSIONING

Another type of rectangular coordinate dimensioning is the **tabulated outline** method (**Figure 9-23**). In this example, the X-Y distance of each plotting point on a contour line is tabulated on an X-Y coordinate chart. The conventional method of dimensioning a curve of this type is shown in **Figure 9-24**. Notice all dimensions originate from baselines.

Special Dimensioning Procedures

Since some object configurations or areas are difficult to dimension, special dimensioning procedures are standardized.

DIMENSIONING CROWDED AREAS

Figure 9-25 shows examples of dimensioning practices for crowded areas. **Figure 9-26** shows how dimensions should be staggered for easier reading. In rare cases, oblique dimensioning (**Figure 9-27**) is used to remove a dimension from a crowded area. Often, a corner or area to be dimensioned is outside the object outline or may be eliminated during manufacturing. In this case, extension lines are drawn to imaginary intersections. Extension and dimension lines are then added (**Figure 9-28**).

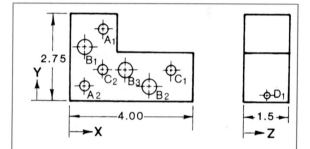

HOLE	FROM SURF	X	Y	Z
A_1	X,Y	1.0	2.25	THRU
A_2	X,Y	.5	.5	THRU
B_1	X,Y	.5	1.7	.50
B_2	X,Y	2.6	.5	.50
B_3	X,Y	1.8	1.0	.50
C_2	X,Y	1.0	1.0	THRU
C_1	X,Y	3.3	1.0	THRU
D_1	Z,Y	THRU	.1	.75

HOLE	DESCRIPTION	QUANTITY
A	Ø.30	2
B	Ø.45	3
C	Ø.28	2
D	Ø.20	1

Figure 9-22 *Rectangular coordinate dimensioning in tabular form.*

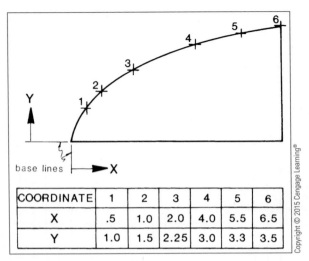

Figure 9-23 *Tabulated outline.*

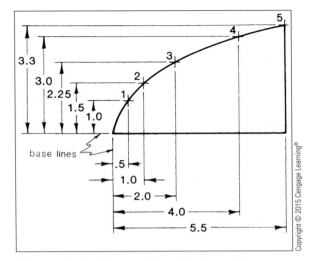

Figure 9-24 *Coordinate or offset outline dimensioning.*

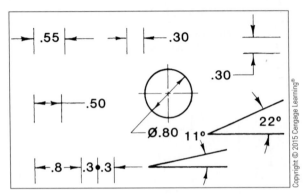

Figure 9-25 *Dimensioning conventions for crowded areas.*

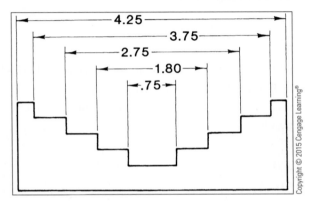

Figure 9-26 *Examples of staggered dimensions.*

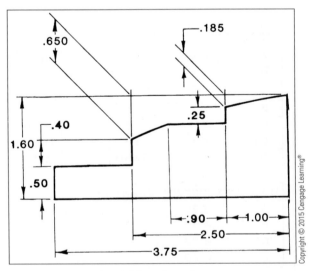

Figure 9-27 *Examples of oblique dimensions.*

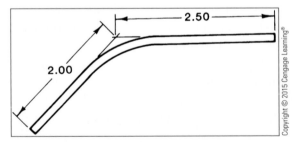

Figure 9-28 *Locating a point for dimensioning.*

Since a wide variety of machined holes are used in manufactured products, dimensioning standards include ASME symbols for holes (**Figure 9-29**). From these notation symbols, the print reader can visualize the information shown in the accompanying sectional view.

DIMENSIONING SLOTS, RADII, FILLETS, ROUNDS, AND CHAMFERS

Often, the dimensioning of rounded ends (**Figure 9-30**), slots, radii, fillets, and chamfers offers an acceptable opportunity to shortcut normal dimension procedures. **Figure 9-31** shows three methods of dimensioning **slots**.

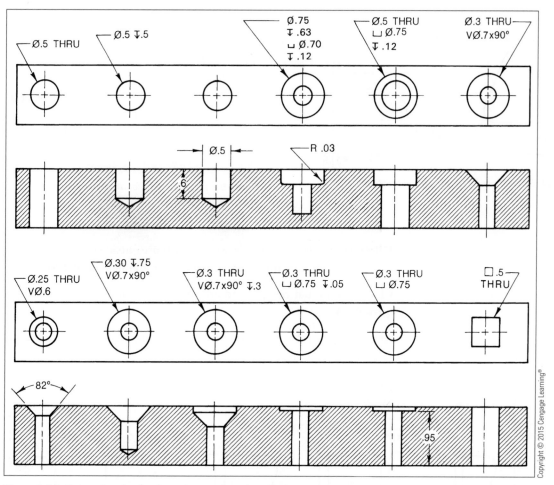

Figure 9-29 *Conventions for the dimensioning of machined holes.*

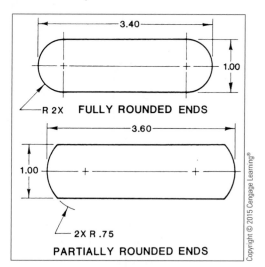

Figure 9-30 *Conventions for the dimensioning of rounded ends.*

Radii dimensioning shows the distance from apex to arc. If the apex of the radii is not located near the drawing, an offset dimension line can be used. Very small radii are dimensioned without the use of an apex (**Figure 9-32**).

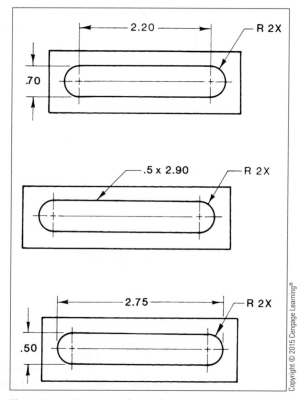

Figure 9-31 *Conventions for the dimensioning of slotted holes.*

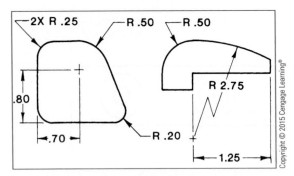

Figure 9-32 *Conventions for the dimensioning of radii.*

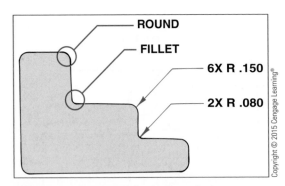

Figure 9-33 *Conventions for the dimensioning of fillets and rounds.*

Fillets are inside rounded corners. **Rounds** are outside rounded corners. Fillets and rounds are dimensioned in a similar manner (**Figure 9-33**).

Chamfers are flat corners and are either external, internal, or oblique (**Figure 9-34**).

Knurling is another common manufacturing process that is dimensioned with a symbolic notation (**Figure 9-35**).

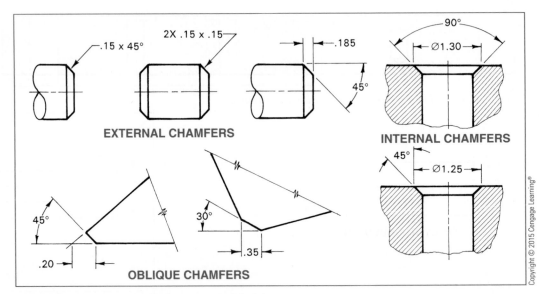

Figure 9-34 *Conventions for the dimensioning of chamfers.*

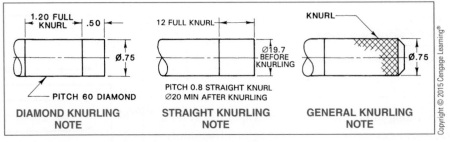

Figure 9-35 *Conventions for knurling dimensions and notations.*

DIMENSIONING CHORDS, ARCS, AND ANGLES

Chords, arcs, and angles may appear similar in many drawings but should be dimensioned differently (**Figure 9-36**). Chords are dimensioned numerically with a straight line, and arcs are dimensioned numerically with a curved line. In dimensioning angles, extend straight extension lines and connect them with curved dimension lines (**Figure 9-37**).

DIMENSIONING PRISMS, CYLINDERS, CONES, PYRAMIDS, AND SPHERES

Other basic forms, such as prisms, cylinders, cones, pyramids, and spheres, are dimensioned by ASME standards (**Figure 9-38**).

PICTORIAL DIMENSIONING

Pictorial drawings are rarely used as working drawings and are usually accompanied by a fully dimensioned multiview drawing. If a pictorial drawing must be used as a working drawing, it should be dimensioned as shown on the isometric drawing in **Figure 9-39**. (See Chapter 15 for pictorial drawing details.)

Surface Control

The **surface texture** for manufacturing processes is specified on working drawings through **finish symbols** (**Figure 9-40**). The surface texture value is controlled with a symbol or notation on the working drawing (**Figure 9-41**). The point of the symbol is placed on the surface for which

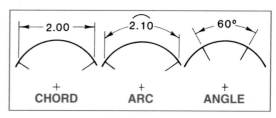

Figure 9-36 *Conventions for the dimensioning of chords, arcs, and angles.* Copyright © 2015 Cengage Learning®

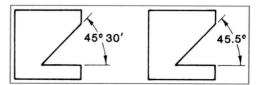

Figure 9-37 *Conventions for dimensioning angles.* Copyright © 2015 Cengage Learning®

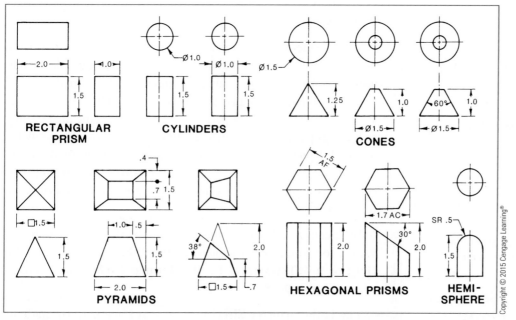

Figure 9-38 *Examples of dimension placement on basic forms.*

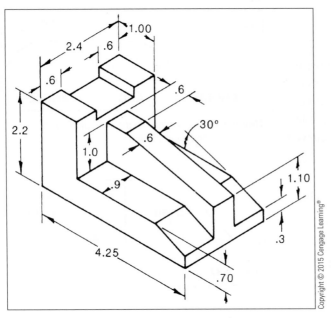

Figure 9-39 *Example of a dimensional isometric drawing.*

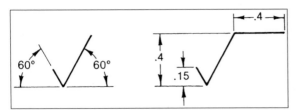

Figure 9-40 *Typical size dimensions for the surface finish symbol.* Copyright © 2015 Cengage Learning®

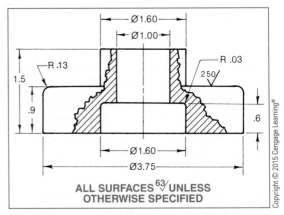

Figure 9-41 *The surface texture value may be controlled with symbols or a notation placed on the drawing.*

ALL SURFACES $^{63}\!\!\sqrt{}$ UNLESS
OTHERWISE SPECIFIED

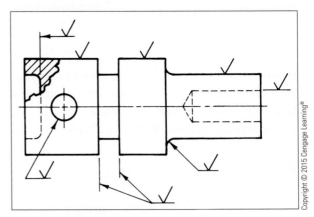

Figure 9-42 *Examples for various positions of the surface finish symbols.*

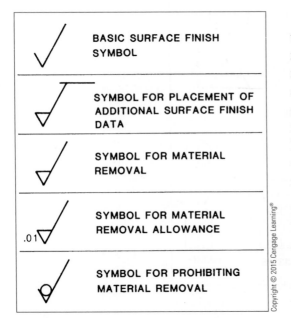

	BASIC SURFACE FINISH SYMBOL
	SYMBOL FOR PLACEMENT OF ADDITIONAL SURFACE FINISH DATA
	SYMBOL FOR MATERIAL REMOVAL
	SYMBOL FOR MATERIAL REMOVAL ALLOWANCE
	SYMBOL FOR PROHIBITING MATERIAL REMOVAL

Figure 9-43 *Various types of surface finish symbols.*

a surface finish is specified (**Figure 9-42**). Different forms of the surface finish symbol are used to designate different surface finishes (**Figure 9-43**).

Surface smoothness is a relative term. No surface appears smooth under a microscope. Some surfaces require a higher degree of smoothness than others. Bearings, seals, gears, pistons, and most matching and moving parts require a high degree of smoothness. Other surfaces, such as outside walls of machine parts, can be as rough as the final castings or forgings permit. It is, therefore, important to correctly specify the required degree of smoothness or roughness on working drawings. Roughness is the opposite of smoothness. **Figure 9-44** shows the basic characteristics of surface textures that are used to define degrees of roughness.

In specifying degrees of roughness, the drafter must specify values for roughness height (**Figure 9-45**), roughness width (**Figure 9-46**), waviness height (**Figure 9-47**), and waviness width (**Figure 9-48**).

Lay is the primary direction of the surface pattern made by machine tool marks. Seven symbols are given for the lay patterns produced by different manufacturing processes (**Figure 9-49**).

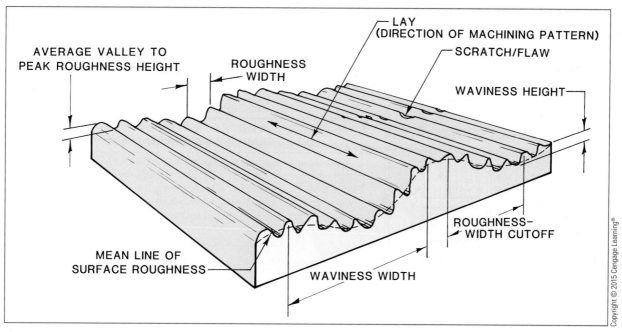

Figure 9-44 *Characteristics of surface texture.*

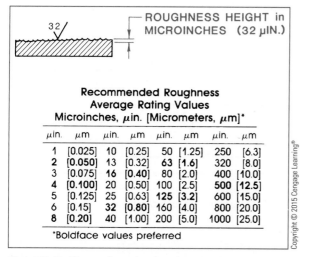

Figure 9-45 *The roughness height values noted on the finish symbol. The prefix micro is one-millionth (.000001 or 10⁻⁶)*

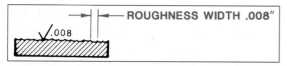

Figure 9-46 *The callout for the roughness width noted on the finish symbol.* Copyright © 2015 Cengage Learning®

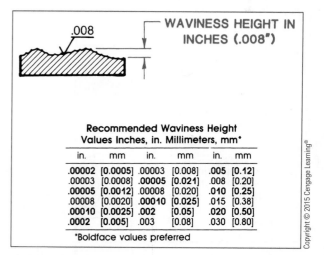

Figure 9-47 *The waviness height values noted on the finish symbol.*

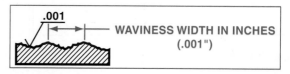

Figure 9-48 *The waviness width noted on the finish symbol.*
Copyright © 2015 Cengage Learning®

Roughness-width cutoff is the greatest spacing of repetitive surface irregularities included in the average roughness height. It is specified in inches on working drawings (**Figure 9-50**).

Figure 9-51 shows a surface control symbol with all its controlling features. **Figure 9-52** shows the method used for specifying minimum **contact area** for mating parts.

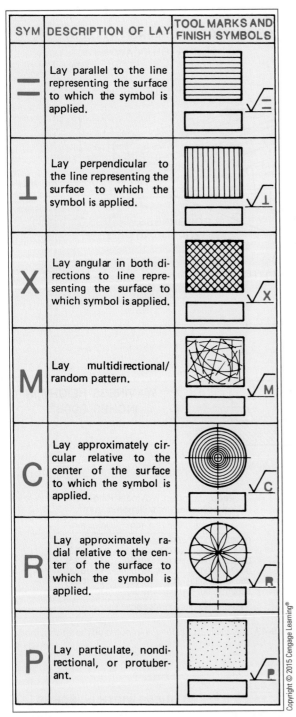

Figure 9-49 *The lay symbols and their characteristics.*

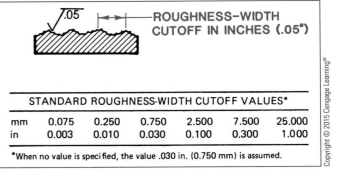

Figure 9-50 *The roughness-width cutoff noted on the finish symbol. (The greatest spacing of repetitive surface irregularities, due to machining, that is included in the average roughness height.)*

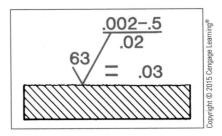

Figure 9-51 *Surface control symbol in full use.*

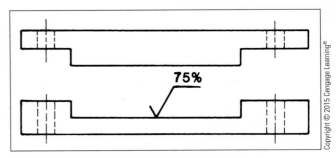

Figure 9-52 *Specifying the minimum contact area.*

ROUGHNESS TOLERANCE

Different manufacturing processes, such as depth of cut, amount of feed, and machining repetition, result in varying degrees of surface smoothness. Engineers must use design processes that produce the proper degree of smoothness with the least amount of manufacturing processing. Producing surfaces that are smoother than necessary may drastically increase production time and costs. Producing surfaces that are too rough may result in an inferior-quality product. In specifying degrees of smoothness—that is, tolerance for roughness—refer to the charts shown in **Figures 9-53** and **9-54**. These charts show the degree of roughness height produced by various manufacturing and machining processes.

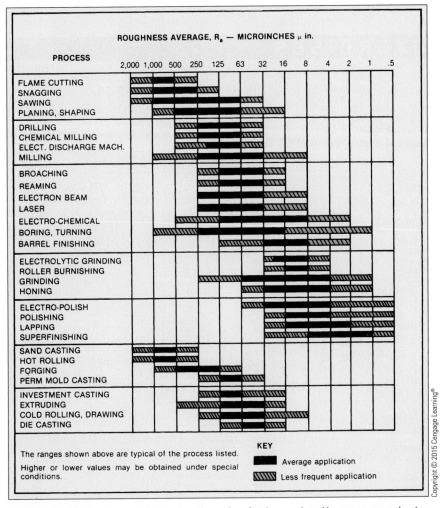

Figure 9-53 *The roughness height ratings for surface finishes produced by common production procedures. (Reprinted from Blueprint Reading for Manufacturing by Hoffman and Wallach.)*

ROUGHNESS HEIGHT MICROINCHES (μ INCH)	SURFACE DESCRIPTION	MACHINING PROCESSES
1000	Very rough	Saw or torch cutting, forging, or sand casting
5000	Rough machining	Coarse feeds and heavy cuts in machining
250	Coarse	Coarse surface grind, medium feeds, and average cuts in machining
125	Medium	Sharp tools, light cuts, fine feeds, high speeds with machining
63	Good finish	Sharp tools, light cuts, extra fine feeds, light cuts with machining
32	High grade finish	Very sharp tools, very fine feeds and cuts
16	Higher grade finish	Surface grinding, coarse honing, coarse lapping
8	Very fine machine finish	Fine honing and fine lapping
3	Extremely smooth machine finish	Extra fine honing and lapping

Figure 9-54 *Typical roughness height values.*

DRAFTING EXERCISES

1. Sketch the single-view drawings in **Figures 9-55** and **9-56** on grid paper and add dimensions.

2. List the dimensioning errors in **Figure 9-57**. Redraw and dimension it correctly.

3. Double the size of **Figure 9-58** and add the dimensions as specified in the table.

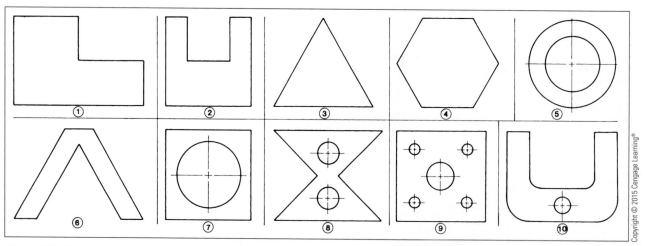

Figure 9-55 *Sketch and dimension each example. Double the size of each drawing with a scale or dividers.*

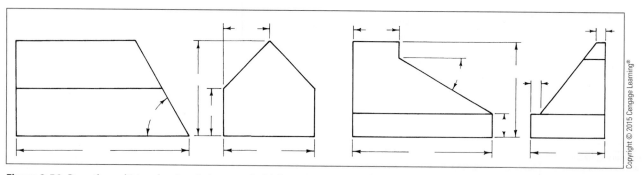

Figure 9-56 *Draw the multiview drawings 3x larger and add dimensions.*

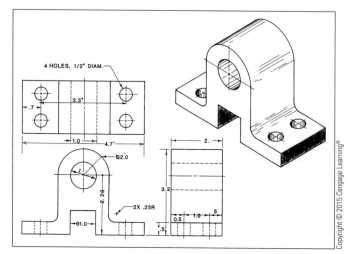

Figure 9-57 *Make a list of the dimensioning errors on this drawing. Redraw and dimension with ASME dimensioning conventions.*

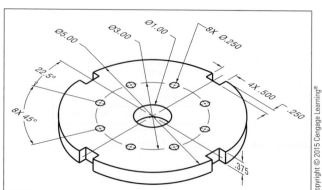

Figure 9-58 *Double the size and add the dimensions using the standards for this figure.*

DESIGN EXERCISES

1. Design a simple toy car made from wood. The parts will consist of the body, two wood dowel axles, and four wood wheels. Draw an assembly drawing and a detailed drawing of each part fully assembled.

2. Design and draw a caddy to hold all your manual instruments in order while you are drafting.

3. Design and draw a CAD manual holder to hold manuals while operating a CAD system.

KEY TERMS

Aligned dimensions

Chamfer

Contact area

Dimensioning

Dual dimensions

Extension line

Fillet

Finish symbol

Knurling

Lay

Leader

Location dimensions

Metric dimensioning

Notations

Polar coordinate dimension

Radius

Rectangular coordinate dimension

Roughness-width cutoff

Rounds

Size dimensions

Surface texture

Tabulated outline dimensioning

Tolerance

Unidirectional dimensions

U.S. customary dimensioning

10

Tolerancing and Geometric Tolerancing

OBJECTIVES

The student will be able to:

- Place limit dimensions on a working drawing
- Place plus and minus tolerance dimensions on a working drawing
- Recognize geometric tolerancing symbols
- Design various fits for mating parts
- Dimension a working drawing with form tolerance symbols
- Understand the concepts for each form tolerancing symbol

Introduction

In early manufacturing, parts produced were never identical. Therefore, parts had to be hand selected to ensure proper assembly. Today's manufacturing does not have the luxury of time to hand-fit product assemblies. Geometric tolerancing ensures the interchangeability of individual manufactured parts.

Absolute accuracy is not completely attainable in any field. An expert marksman cannot hit the bull's-eye dead center every time. But is hitting the edge of the bull's-eye tolerable? Or is hitting the second ring acceptable occasionally? And is hitting the third ring never acceptable? Whichever goal is tolerated becomes the limit of accuracy. There are limits of accuracy in the manufacture of products as well. In fact, each dimension must contain a working limit of accuracy. This limit is called a **tolerance**. A tolerance is the permissible amount a dimension may vary in size during manufacturing or construction. For example, the width of a highway can vary ¼" from the dimension shown on the drawing without serious consequences. However, if a watch part varies only a thousandth of an inch, the part will not fit or function. To control these variances (tolerances), manufacturers are assigned degrees of accuracy required in each product. This is done by assigning a tolerance range to each dimension or by means of a general tolerance notation. Close-fitting mating parts will operate more efficiently but will cause more friction

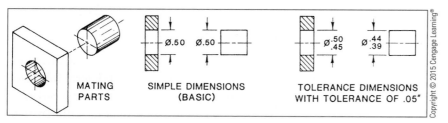

Figure 10-1 *Comparison of basic dimensions and tolerance dimensions.*

and wear. To improve the efficiency of mating parts, the study of *tribology* is used. Tribology is the interaction of sliding surfaces and lubrication, and it is used to improve the life span of modern machinery.

Tolerance Dimensions

The use of **basic (simple) dimensions** that do not include tolerance limits allows manufacturers to establish their own acceptable limits that may not produce the desired result. **Figure 10-1** shows a comparison of a drawing dimensioned with basic dimensions and with tolerance dimensions of .05". The tolerance dimensions, shown on the right, ensure that the shaft will fit into the hole with the clearance desired. Basic dimensions (**Figure 10-2**) show only the general size of a part without any tolerance limitations.

Tolerance is the difference between two dimensions. If a 190'-6" sidewalk is acceptable when constructed at 190'-5" or 190'-7", that is 1 inch over and 1 inch under the basic dimension. However, the tolerance range is actually two inches (190'-7" minus 190'-5" = 2").

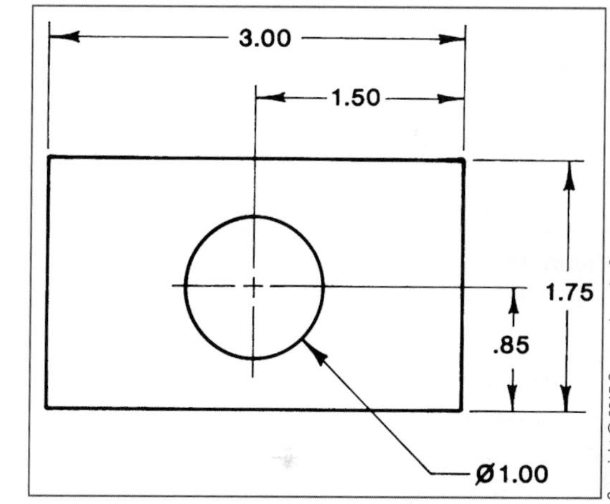

Tolerance Practices

A **limit tolerance** is a statement of the variations that can be permitted from a given dimension. When tolerance limits are aligned vertically, the larger dimension is placed on top. When aligned horizontally, the smaller dimension is placed to the left of the larger one. The number of decimal places in the basic dimension and in the tolerancing dimension must be equal (**Figure 10-3**).

Plus and minus tolerance dimensions (**Figure 10-4**) are used to indicate the tolerance range above and below the basic dimension. The tolerance level higher than the basic dimension is marked with a plus sign (+). The tolerance level less than the basic dimension is marked with a minus sign (−). Plus and minus tolerance dimensions are placed to the right of the basic dimension, with the plus dimension on top of the minus dimension. If the plus tolerance dimension is the same as the minus tolerance dimension, a single number may be used preceded by a combined ± sign. The tolerance of 3.00" ± .03 (**Figure 10-4**) indicates the actual dimension may be as large as 3.03" or as small as 2.97". This represents a tolerance range of .06".

Figure 10-2 *The basic dimension is the theoretical exact size.*

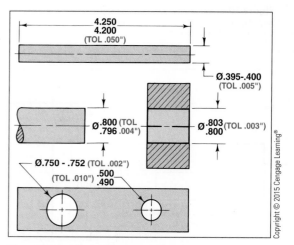

Figure 10-3 *A limit tolerance is a precise statement of the variations that can be permitted within a dimension. Note the large dimension is always on the top of the vertical tolerances, and to the right of the horizontal tolerances.*

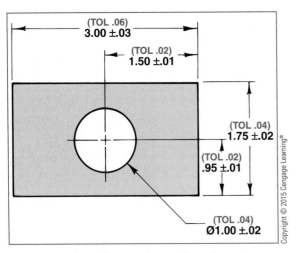

Figure 10-4 *Plus and minus tolerance dimensions.*

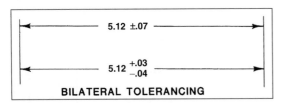

Figure 10-5 *A bilateral tolerance has variations of size of larger and smaller (+ and −).* Copyright © 2015 Cengage Learning®

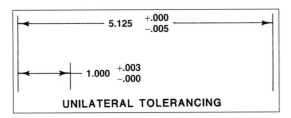

Figure 10-6 *A unilateral tolerance only has a variation in one direction – larger or smaller from the basic dimension.* Copyright © 2015 Cengage Learning®

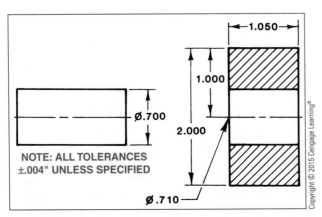

Figure 10-7a *A general tolerance notation may be placed near the drawing or in the title block.*

BILATERAL AND UNILATERAL TOLERANCING

Tolerance dimensions that allow variation in both directions from a basic dimension (+ and −) are **bilateral tolerance dimensions** (**Figure 10-5**). Tolerances that allow variances in only one direction are **unilateral tolerance dimensions** (**Figure 10-6**). In the top of **Figure 10-5**, the dimension may fall between 5.19" and 5.05". In the bottom of **Figure 10-6**, the dimension may fall between 1.000" and 1.003".

GENERAL TOLERANCING

When all tolerances are identical, a **general tolerance note** (**Figure 10-7a**) may be used. This saves repeating tolerance limits on each dimension. A note can apply to an entire product or an individual part. Another type of general tolerancing note is shown in **Figure 10-7b**. The dimensions and tolerances may vary with the design of different items. The tolerance levels in **Figure 10-7b** are expressed as follows:

- All dimensions below one inch may have a tolerance of ±.001".

- All dimensions between 1.01 and 6 inches may have a tolerance of .004".

- All dimensions above six inches may have a tolerance of .008".

CHAIN DIMENSIONING

Tolerance dimensions on **chain dimensions** are not recommended. Chain dimensioning accumulates tolerances. For example, the cumulative tolerance for the length of the part in **Figure 10-8** can be plus .05" (.02 + .01 + .01 + .01) or minus .05", a total tolerance range of .10", or anywhere in between.

DATUM DIMENSIONING

In assigning tolerances to chain dimensions as in **Figure 10-8**, a dimensioned 8" length could be as short as 7.95" or as long as 8.05". For this reason,

TOLERANCES UNLESS OTHERWISE SPECIFIED:
0 TO 1.00 ARE ± .001
1.01 TO 6.00 ARE ± .004
OVER 6.00 ARE ± .008

Figure 10-7b *General tolerance note.* Copyright © 2015 Cengage Learning®

datum dimensioning is recommended when tolerance limits are critical. In datum dimensioning, all dimensions originate from a single point and tolerances are not allowed to accumulate (**Figure 10-9**). The total length in **Figure 10-9** cannot deviate more than ±.03".

When the distance between any two points is critical, a **direct dimension** should be used to insure the least tolerance buildup. Points A and B in **Figure 10-10** should be within .04" of 6.00". Therefore, a tolerance dimension is used to connect these two points even though the remainder of the drawing is datum dimensioned.

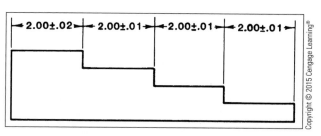

Figure 10-8 *Chain dimensions accumulate tolerances. Accumulation in this drawing can be plus or minus .05". This method is not recommended.*

ANGLE TOLERANCING

While most tolerances relate to linear measurements, angle tolerances are also needed on many working drawings. **Figure 10-11** shows the correct method of specifying angle tolerances in degrees, minutes, and seconds, and with degrees and a decimal part of a degree.

Mating Parts

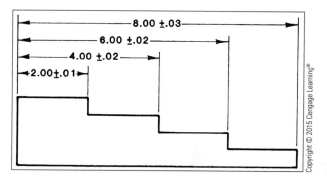

Figure 10-9 *Datum dimensioning will not accumulate tolerances. The maximum tolerance in this drawing is plus or minus .03".*

Appropriate tolerance levels for **mating parts** are very critical. If too much space is allowed between parts, the fit may wobble or fall apart. If too little space is allowed, parts may not fit or excessive friction may occur when they are moved. Very close tolerances also require more precise manufacturing methods and are expensive to produce. For these reasons, it is important that the appropriate fit be designed, dimensioned, and manufactured.

Figure 10-12 is a table showing the appropriate tolerances for various industrial applications at different sizes. The tolerance grade range extends from Grade 4 to Grade 13, with the lower number representing the closest tolerance limits.

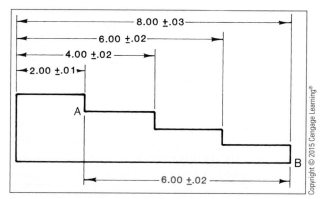

Figure 10-10 *Direct dimensioning for a specific dimension will have the most accuracy.*

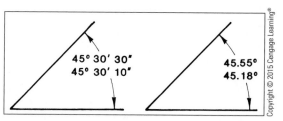

Figure 10-11 *Specifying angle tolerances with degrees, minutes, and seconds, and with degrees and a decimal part of a degree.*

	STANDARD TOLERANCES									
Nominal Size Range in Inches	Slip Gauges, Production and Wear Tolerance of Gauges and Measuring Instruments		Parts Subject to Very Tight Tolerances, Precision Bearings, Precision Assemblies		Precision Engineered Designs (General)		Engineering Work in General, Giving Scope for Wider Tolerances		Rough Work, Steel Structures, Castings, Agricultural Machinery	
Over–To	Grade 4	Grade 5	Grade 6	Grade 7	Grade 8	Grade 9	Grade 10	Grade 11	Grade 12	Grade 13
0.04–0.12	.00015	.00020	.00025	.0004	.0006	.0010	.0016	.0025	.004	.006
0.12–0.24	.00015	.00020	.0003	.0005	.0007	.0012	.0018	.0030	.005	.007
0.24–0.40	.00015	.00025	.0004	.0006	.0009	.0014	.0022	.0035	.006	.009
0.40–0.71	.0002	.0003	.0004	.0007	.0010	.0016	.0028	.0040	.007	.010
0.71–1.19	.00025	.0004	.0005	.0008	.0012	.0020	.0035	.0050	.008	.012
1.19–1.97	.0003	.0004	.0006	.0010	.0016	.0025	.0040	.006	.010	.016
1.97–3.15	.0003	.0005	.0007	.0012	.0018	.0030	.0045	.007	.012	.018
3.15–4.73	.0004	.0006	.0009	.0014	.0022	.0035	.005	.009	.014	.022
4.73–7.09	.0005	.0007	.0010	.0016	.0025	.0040	.006	.010	.016	.025
7.09–9.85	.0006	.0008	.0012	.0018	.0028	.0045	.007	.012	.018	.028
9.85–12.41	.0006	.0009	.0012	.0020	.0030	.0050	.008	.012	.020	.030
12.41–15.75	.0007	.0010	.0014	.0022	.0035	.006	.009	.014	.022	.035
15.75–19.69	.0008	.0010	.0016	.0025	.004	.006	.010	.016	.025	.040
19.69–30.09	.0009	.0012	.0020	.003	.005	.008	.012	.020	.030	.050
30.09–41.49	.0010	.0016	.0025	.004	.006	.010	.016	.025	.040	.060
41.49–56.19	.0012	.0020	.003	.005	.008	.012	.020	.030	.050	.080
56.19–76.39	.0016	.0025	.004	.006	.010	.016	.025	.040	.060	.100
76.39–100.9	.0020	.003	.005	.008	.012	.020	.030	.050	.080	.125
100.9–131.9	.0025	.004	.006	.010	.016	.025	.040	.060	.100	.160
131–171.9	.003	.005	.008	.012	.020	.030	.050	.080	.125	.200
171.9–200	.004	.006	.010	.016	.025	.040	.060	.100	.160	.250

Figure 10-12 *Standardized grades of tolerances.*

Gauge instruments .04" to .12", Grade 4, have the smallest acceptable tolerance, .00015". Grade 13 with nominal sizes for castings, 171.9–200, has the largest acceptable tolerance of .250.

CLEARANCE FITS

Tolerancing of mating parts relies on the control of clearance desired between parts. Clearance (allowance) is the **tolerance zone** between mating parts. **Clearance fits** require separate tolerancing limits for each part.

The shaft's dimensions in **Figure 10-13** are .50"/.45" with a tolerance of .05". The hole's dimensions are .53"/.51" with a tolerance of .02". When the largest shaft diameter of .50" mates with the smallest hole with a diameter of .51", the difference or allowance will be .01". When the smallest shaft (.45' diameter) mates with the largest hole (.53" diameter), the allowance is .08". With mass production, the mating of parts may have any allowance in the range of .01" to .08".

INTERFERENCE FITS

When mating parts are designed to be permanently attached and not movable, an **interference fit** may be specified. Interference fits (force fits) have negative clearances so that the parts must be forced together (**Figure 10-14**). A positive allowance permits a clearance fit. A negative allowance permits an interference fit. The amount of interference tolerance depends on the flexibility and hardness of the material. Soft materials force together more easily than hard materials. A positive allowance is a clearance fit. A negative allowance is an interference fit.

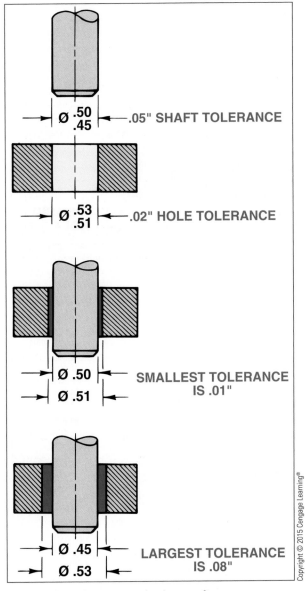

Figure 10-13 *Tolerance zones for clearance fits.*

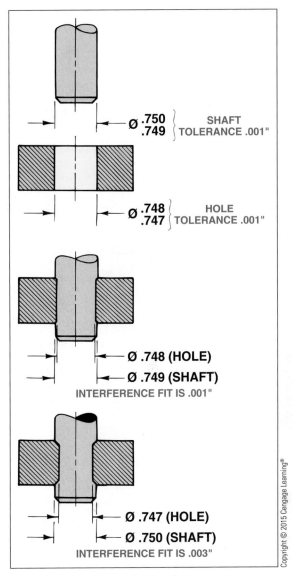

Figure 10-14 *Interference fits (force fits).*

TRANSITION FITS

A **transition fit** is one in which the mating of parts may have a clearance fit or an interference, depending upon which parts are selected to mate (**Figure 10-15**).

BASIC SIZES

Basic size is the theoretical exact size. The appropriate matching of mating parts can be accomplished with custom machining; however, standard-size parts should be specified when possible. For example, in the mating of a hole and shaft, the shaft size can be adjusted to match a standard hole size or vice versa. Matching a shaft size to a basic hole size is shown in **Figure 10-16**. **Basic hole size** refers to the minimum diameter of a hole that was made by a drill or boring tool. The **basic shaft diameter** refers to its standard manufacturing size (**Figure 10-17**).

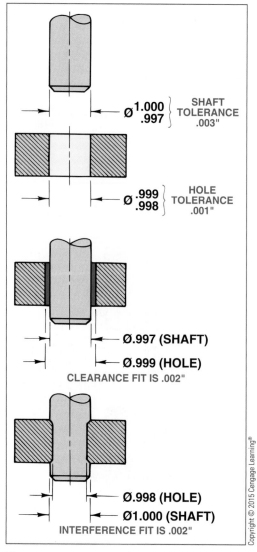

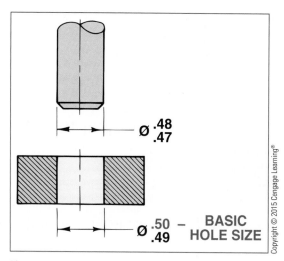

Figure 10-16 *System of basic hole tolerances.*

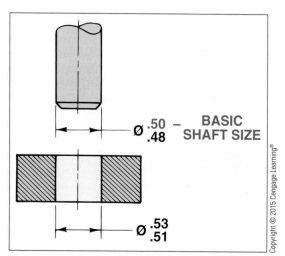

Figure 10-17 *System of basic shaft tolerances.*

Figure 10-15 *Transitional fits may be interference and/or clearance fits.*

MATERIAL CONDITION

When establishing hole tolerance sizes of mating parts, material conditions such as **maximum material condition (MMC)** and **least material condition (LMC)** must be considered. The MMC for a hole is the smallest allowable diameter that leaves the maximum amount of material. The MMC for a shaft is the largest allowable diameter that leaves the maximum amount of material. The LMC for a shaft is the smallest allowable diameter that leaves the minimum amount of material. The MMC and LMC for a shaft are shown in **Figure 10-18a**, and the MMC and LMC for a hole are shown in **Figure 10-18b**. **Regardless of feature size (RFS)** is

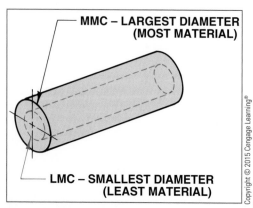

Figure 10-18a *Maximum and minimum material conditions for a shaft.*

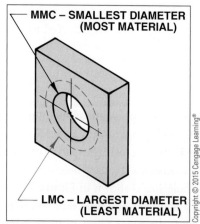

Figure 10-18b *Maximum and minimum material conditions for a hole.*

a geometric tolerance specification that refers to the condition that all geometric tolerance must be met regardless of the size of the item or any specific locations such as holes or protrusions. The symbol for RFS is an "S" within a circle.

Basic dimensioning controls only the shape and size of an item on a technical drawing. **Geometric tolerancing** is an extension of the dimensioning process that describes more precisely how an item is to be manufactured. Geometric tolerancing includes form tolerancing and positional tolerancing. **Form tolerancing** controls the geometric shape of objects (**Figure 10-19**). **Positional tolerancing** controls the exact location of product features such as holes, depressions, and projections (**Figure 10-20**).

Geometric characteristic symbols are used to control the tolerance limits of form, profile, orientation, location, and runouts (**Figure 10-21**).

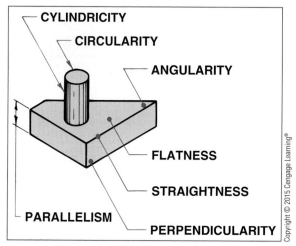

Figure 10-19 *Form tolerancing controls.*

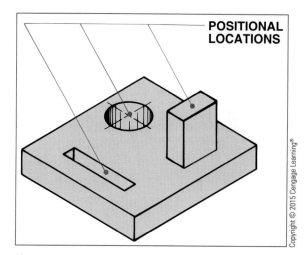

Figure 10-20 *Positional tolerancing controls.*

	TYPE OF TOLERANCE	CHARACTERISTIC	SYMBOL
FOR INDIVIDUAL FEATURES	FORM	STRAIGHTNESS	—
		FLATNESS	▱
		CIRCULARITY (ROUNDNESS)	○
		CYLINDRICITY	⌀
FOR INDIVIDUAL OR RELATED FEATURES	PROFILE	PROFILE OF A LINE	⌒
		PROFILE OF A SURFACE	⌓
FOR RELATED FEATURES	ORIENTATION	ANGULARITY	∠
		PERPENDICULARITY	⊥
		PARALLELISM	//
	LOCATION	POSITION	⊕
		CONCENTRICITY	◎
		SYMMETRY	⹀
	RUNOUT	CIRCULAR RUNOUT	↗ *
		TOTAL RUNOUT	↗↗ *

*Arrowhead(s) may be filled in.

Figure 10-21 *Geometric characteristic symbols.*

TERM	SYMBOL
AT MAXIMUM MATERIAL CONDITION	Ⓜ
AT LEAST MATERIAL CONDITION	Ⓛ
PROJECTED TOLERANCE ZONE	Ⓟ
FREE STATE	Ⓕ
TANGENT PLANE	Ⓣ
DIAMETER	Ø
SPHERICAL DIAMETER	SØ
RADIUS	R
SPHERICAL RADIUS	SR
CONTROLLED RADIUS	CR
REFERENCE	( )
ARC LENGTH	⌒
STATISTICAL TOLERANCE	⟨ST⟩
BETWEEN	↔

Figure 10-22 *Modifying symbols specifying geometric tolerancing.*

These symbols are a form of dimensioning shorthand, each representing a geometric feature to be controlled. Other modifying symbols used for geometric tolerancing are summarized in **Figure 10-22**.

DATUMS

All geometric tolerancing begins with the establishment of a datum. A **datum** is a reference location easy to locate and measure. Theoretically, it is a perfect plane, point, line, or axis. The location of all geometric characteristics originate from a datum. On a technical drawing, the **datum identification box** or **datum target** is used to locate specific surfaces for reference (**Figure 10-23**).

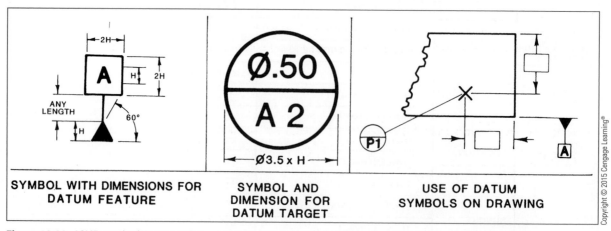

Figure 10-23 *ASME standard Y14.5M, Reference datum symbols and their drawing sizes (Note: H equals the letter height).*

FEATURE CONTROL FRAME

The **feature control frame** contains data for each controlled specification and includes the characteristic symbol, tolerance value, and datum reference letter. In **Figure 10-24**, the marked surface is to be parallel within .003" of datum A. The feature control frames are usually referenced to a datum surface reference.

GEOMETRIC CHARACTERISTIC SYMBOLS

As previously mentioned, each geometric characteristic symbol represents a geometric feature that must be controlled to the level specified. Most symbols are referenced to a datum; however, **straightness**, **flatness**, **circularity (roundness)**, and **cylindricity** may stand alone. **Figures 10-25** through **10-37** show characteristic symbols accompanied by an explanation and a working drawing example.

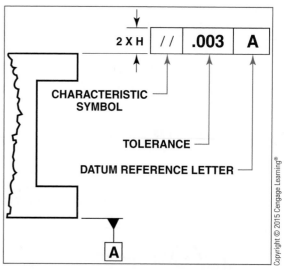

Figure 10-24 *The feature control frame's drawing size. It states "The surface is to be parallel within .003" of surface -A-.*

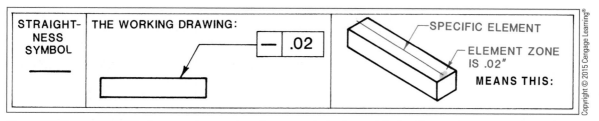

Figure 10-25 *STRAIGHTNESS refers to an element of a surface or an axis that is a straight line. The straightness tolerance specifies a tolerance zone where the element or axis lies.*

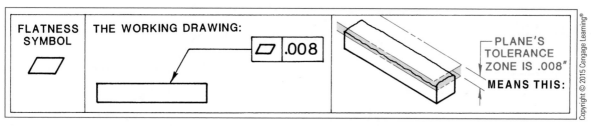

Figure 10-26 *FLATNESS is the condition of a plane. The flatness tolerance specifies a tolerance zone where the surface must lie.*

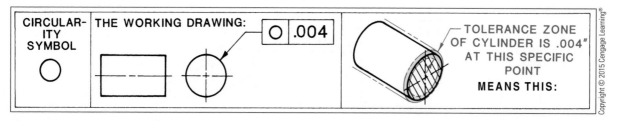

Figure 10-27 *CIRCULARITY (ROUNDNESS) specifies a tolerance zone bounded by two concentric circles at a single specific point.*

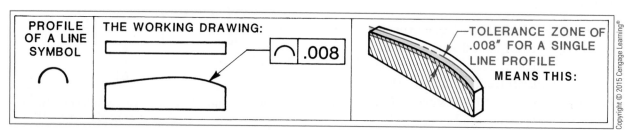

Figure 10-28 *CYLINDRICITY is the combination of roundness and straightness for the whole cylinder. The tolerance zone is bounded by two concentric circles.*

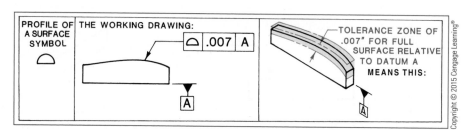

Figure 10-29 *LINE PROFILE controls a single line profile (section) of an irregular surface. The tolerance zone extends on both sides of the profile along its full length.*

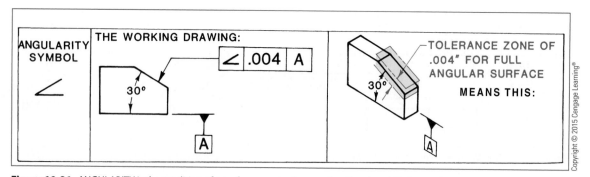

Figure 10-30 *SURFACE PROFILE controls the entire surface for an irregular area. The tolerance zone extends the length and width of the surface.*

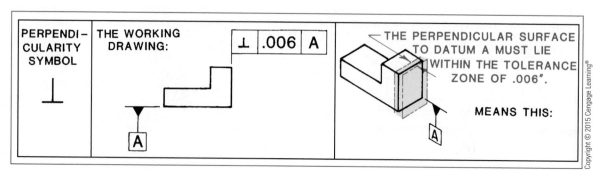

Figure 10-31 *ANGULARITY is the condition of a surface or axis at an angle, other than 90°, from a datum or axis. The tolerance zone is defined by two parallel lines at the specific angle from the datum.*

Figure 10-32 *PERPENDICULARITY is the condition of a surface, plane, or axis, at a right angle to a datum plane or axis.*

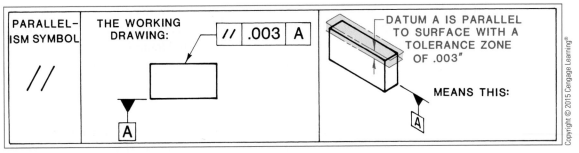

Figure 10-33 *PARALLELISM is the condition of a surface equidistant at all points from a datum plane or axis. The tolerance zone is defined by two planes or lines parallel to the datum.*

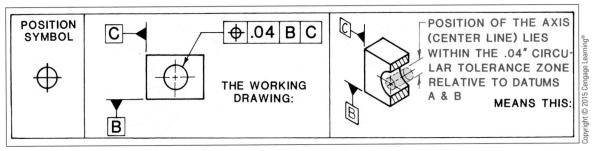

Figure 10-34 *FEATURE POSITION is the theoretical exact location of its axis or center plane from the features from which it is dimensioned. The positional tolerance is the permissible error for the location of a feature relative to its other features.*

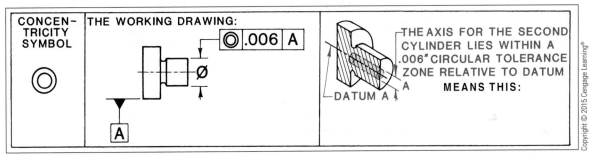

Figure 10-35 *CONCENTRICITY is the condition where the axis of a cylinder aligns to a common axis datum feature. The tolerance zone forms a cylinder where the axis must lie.*

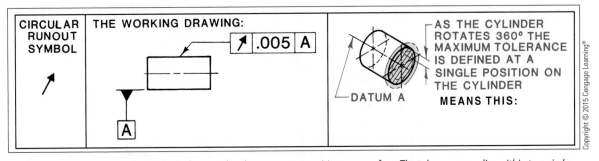

Figure 10-36 *CIRCULAR RUNOUT controls a circular element at one position on a surface. The tolerance zone lies within two circles as the surface rotates one revolution.*

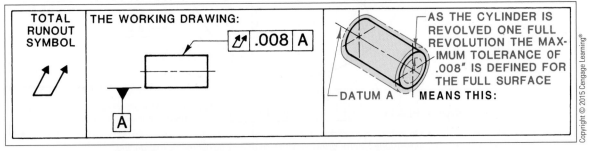

Figure 10-37 *TOTAL RUNOUT controls the entire circular surface as the part is rotated 360°.*

An example of a fully dimensioned metric working drawing, complete with dimensional and geometric tolerancing notations, is shown in **Figure 10-38**. Interpret each dimension and feature control frame in this illustration.

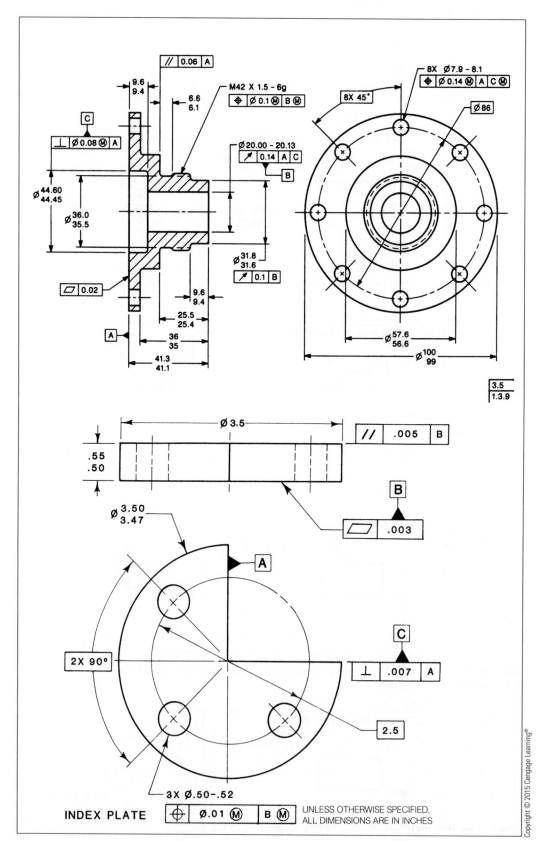

Figure 10-38 *Full-dimensioned metric drawing with tolerances and geometric tolerancing.*

DRAFTING EXERCISES

1. With drawing instruments or a CAD system, draw the dimensioning and tolerancing symbols in **Figure 10-39**.

2. Calculate the maximum and minimum tolerances and clearances for the mating parts in **Figures 10-40** through **10-42**.

3. With instruments or a CAD system, draw the working drawings and complete the dimensioning, with tolerances and geometric tolerancing, as instructed in the captions of **Figures 10-43** through **10-45**.

4. Write an explanation of all the symbols in **Figure 10-46**.

5. Draw **Figure 10-47**. Follow the instructions in the caption.

6. Draw **Figure 10-48**. How many datums are in **Figure 10-48**? Write a description of the positional tolerance on **Figure 10-48**.

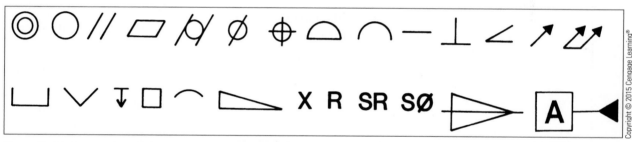

Figure 10-39 *Draw each symbol with correct proportions and label them.*

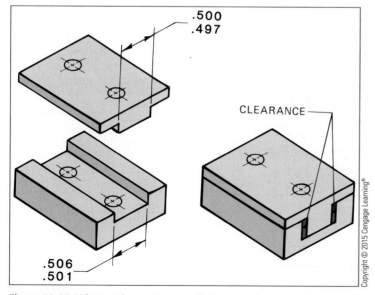

Figure 10-40 *What are the maximum and minimum tolerances and clearances for the mating parts?*

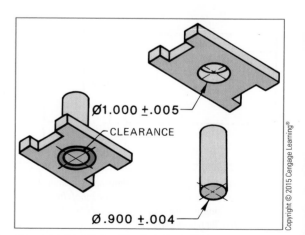

Figure 10-41 *What are the maximum and minimum tolerances and clearances for the hole and shaft.*

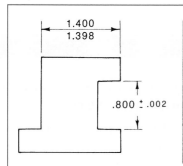
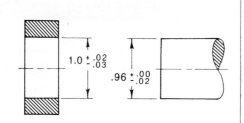

1. What is the tolerance at the top of the drawing?

2. What is the largest possible size of the slot?

3. What is the smallest possible size of the slot?

4. What is the basic size of the slot?

1. What is the tolerance of the hole?

2. What is the tolerance of the shaft?

3. What is the MMC of the hole?

4. What is the MMC of the shaft?

5. What is the LMC of the hole?

6. What is the LMC of the shaft?

7. What is the allowance of the tightest fit?

8. What is the allowance of the loosest fit?

1. What is the tolerance of the hole?

2. What is the tolerance of the shaft?

3. What is the MMC of the hole?

4. What is the MMC of the shaft?

5. What is the LMC of the hole?

6. What is the LMC of the shaft?

7. What is the allowance of the tightest fit?

8. What is the allowance of the loosest fit?

Figure 10-42 *Answer the following questions on tolerancing.*

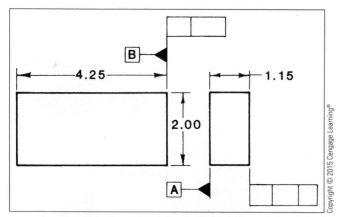

Figure 10-43 *Draw the multiview drawing and dimension with Grade 12 tolerances. Add geometric tolerancing symbols for the following:*
1. Datum A is parallel to its opposite side within .01″.
2. Datum B is flat within .005″.
3. Datum B is perpendicular to the bottom surface within .02″.

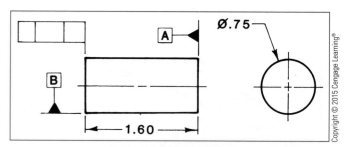

Figure 10-44 *Draw the multiview drawing and dimension with Grade 11 tolerances. Add geometric tolerancing symbols for the following:*
1. Datum A is parallel to its opposite side within .004″.
2. Roundness is within .008″.

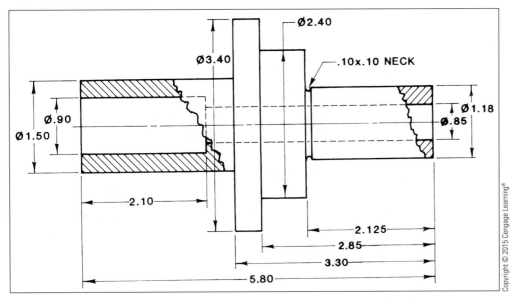

Figure 10-45 *As a designer, add tolerances of .002" and .008" to the appropriate dimensions. Add geometric characteristic symbols of flatness, straightness, perpendicularity, and circularity where they apply.*

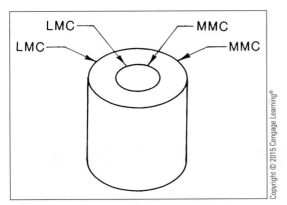

Figure 10-46 *Write a short description of how the geometric tolerance symbol affects the cylinder and the hole.*

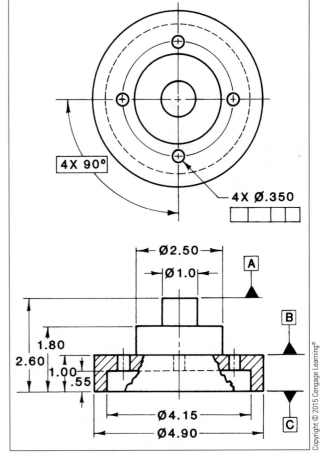

Figure 10-47 *Draw the working drawing and dimension with Grade 8 tolerances. Add geometric tolerancing symbols for true position within .005" for the holes, and flatness within .003" for datums A, B, C.*

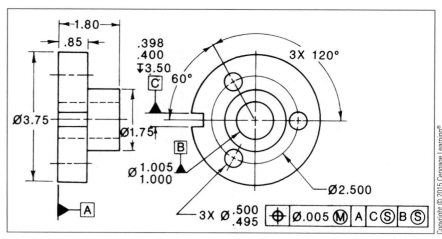

Figure 10-48 *Write a description for each geometric tolerance symbol.*

DESIGN EXERCISES

1. Use a tolerance dimension of Grade 13 for the axle and wheels for the toy car in Design Exercise 1 in Chapter 9.

2. Do a drawing of two steel pipes 6" long. The inside diameter of one pipe is .75". Design the second pipe to fit inside with a tolerance grade of 7. Add all required dimensions.

KEY TERMS

Basic (simple) dimensions

Basic hole size

Basic shaft size

Basic Size

Bilateral tolerance dimensions

Chain dimensions

Circularity (roundness)

Clearance fit

Cylindricity

Datum

Datum dimensioning

Datum identification box

Datum target

Direct dimension

Feature control frame

Flatness

Form tolerancing

General tolerance note

Geometric characteristic symbol

Geometric tolerancing

Interference fit

Least material condition (LMC)

Limit tolerance

Mating parts

Maximum material condition (MMC)

Minus tolerancing dimensions

Plus tolerancing dimensions

Positional tolerancing

Regardless of feature size (RFS)

Straightness

Tolerance

Tolerance zone

Transition fit

Unilateral tolerance dimension

Sectional Views

11

OBJECTIVES

The student will be able to:

- Draw a full section drawing
- Draw a half-section drawing
- Draw a broken-out section drawing
- Draw an offset section drawing
- Draw an assembly section drawing
- Draw a removed section drawing
- Draw a revolved section drawing
- Draw a thin material section drawing
- Draw a pictorial section drawing
- Rotate and align features for a section drawing

Introduction

Multiview drawings are adequate for describing the size and shape of an object when all-important features can be seen on the normal orthographic views. However, if the internal surfaces require the extensive use of hidden lines, the drawing may be difficult to understand. When this occurs, a sectional view is usually prepared to show the size and shape of the interior more clearly (**Figure 11-1**). A **sectional view**, or section, is a drawing of an object as it would appear if cut in half or quartered. Sectional views do not contain hidden lines because dashed lines can confuse the clarity of the sectional view (**Figure 11-2**).

Cutting Plane

To understand sectional views, imagine a saw cutting an object in half. Replace the position of the saw with an imaginary plane and remove the nearest part of the object. What remains is a sectional view.

To define the location of the imaginary sectional plane, a **cutting plane line** is drawn on the view where the cut was made. Cutting plane lines are very wide and dense, with arrows placed on the ends to indicate the line of sight direction for the sectional drawing (**Figure 11-3**). Several types of cutting plane lines are

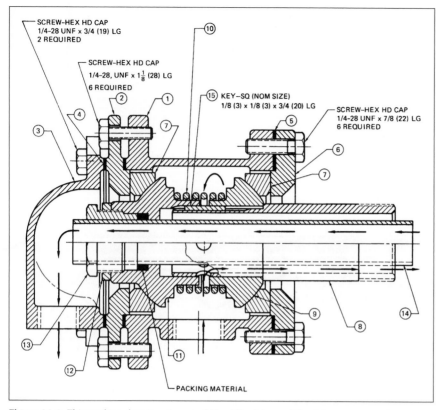

SCREW–HEX HD CAP
1/4–28 UNF x 3/4 (19) LG
2 REQUIRED

SCREW–HEX HD CAP
1/4–28, UNF x 1⅛ (28) LG
6 REQUIRED

KEY–SQ (NOM SIZE)
1/8 (3) x 1/8 (3) x 3/4 (20) LG

SCREW–HEX HD CAP
1/4–28 UNF x 7/8 (22) LG
6 REQUIRED

PACKING MATERIAL

Figure 11-1 *This working drawing view would be difficult to read if not drawn as a section.*

Reprinted from Technical Drawing and Engineering Communication by Goetsch, Chalk, Nelson, and Rickman, Delmar/Cengage Learning

used, depending on the type of drawing, size, and location of the section (**Figure 11-4**). **Figure 11-5** shows the relationship between the cutting plane, the cutting plane line, and the resulting sectional view.

Section Lining

Where the cutting plane passes through solid material, **section-lining** lines are drawn. **Crosshatching** is the universal symbol for section lining (**Figure 11-5**); however, section-lining symbols are standardized for most manufacturing and construction materials. When the universal crosshatching symbol is used rather than the standard material section-lining symbol, a note near the section is used to indicate the type of material. The spacing of the crosshatching will vary depending on the size of the section (**Figure 11-6**).

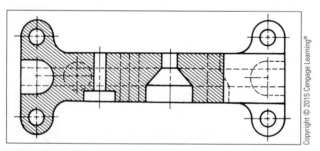

Figure 11-2 *Hidden lines will confuse the clarity of a sectional view.*

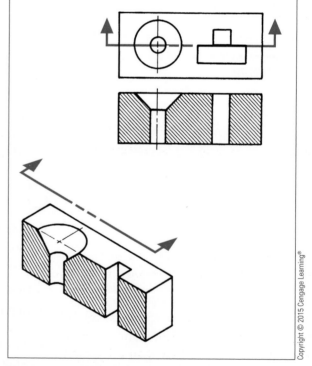

Figure 11-3 *The arrows on the cutting plane line indicate the direction of the line of sight for the sectional view.*

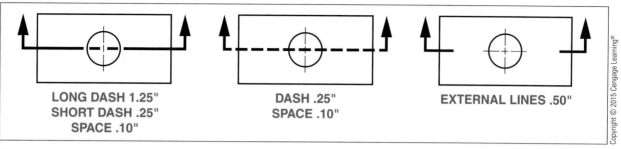

LONG DASH 1.25"
SHORT DASH .25"
SPACE .10"

DASH .25"
SPACE .10"

EXTERNAL LINES .50"

Figure 11-4 *Average size and spacing recommended for the cutting plane lines.*

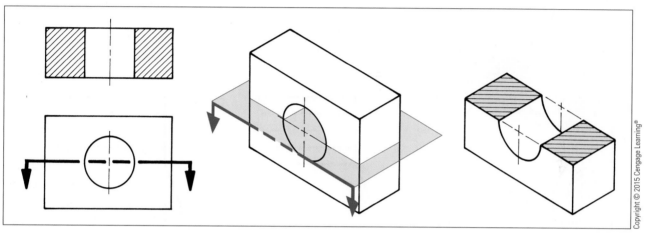

Figure 11-5 *The cutting plane defines the surface to be sectioned.*

Do not take the time to measure the spacing. Estimate with your eye when drawing. When adjacent parts are crosshatched, the direction of the lines is reversed or the angle of the crosshatching is changed to differentiate the separate pieces (**Figures 11-7** and **11-1**).

Figure 11-8 shows the method of interrupting the section lining to place dimensions and notes. This should only be done in large sectioned areas. It is preferable to place notes and dimensions on the outside of the view.

SECTION-LINING EXCEPTIONS

Although section lining is added to areas intersected by the cutting plane, there are some exceptions to this principle.

Ribs, webs, spokes, gear teeth, lugs, and ball bearings are not section lined, because they are either comparatively thinner than the remainder of the object or they appear intermittently (**Figure 11-9**).

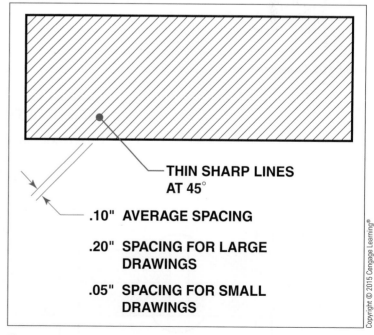

THIN SHARP LINES AT 45°

.10" AVERAGE SPACING

.20" SPACING FOR LARGE DRAWINGS

.05" SPACING FOR SMALL DRAWINGS

Figure 11-6 *Crosshatch spacing will vary according to the drawing size. Space the crosshatching by "eye" when drawing.*

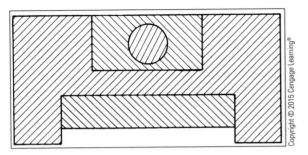

Figure 11-7 *Reverse the direction of the crosshatching for adjacent parts.*

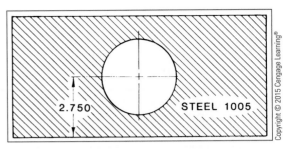

Figure 11-8 *If necessary, notes and dimensions may be placed on the sectional view as shown.*

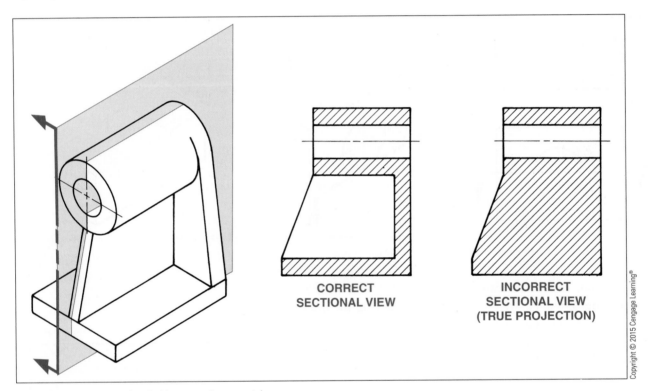

CORRECT
SECTIONAL VIEW

INCORRECT
SECTIONAL VIEW
(TRUE PROJECTION)

Figure 11-9 *Do not crosshatch thin supporting materials.*

Shafts and fasteners, such as nuts, bolts, rivets, pins, and screws, are also not section lined (**Figure 11-10**).

Objects with webs and spokes that do not align with the cutting plane line are rotated (**Figure 11-11**) before projecting the sectional view. The spoke is not section lined.

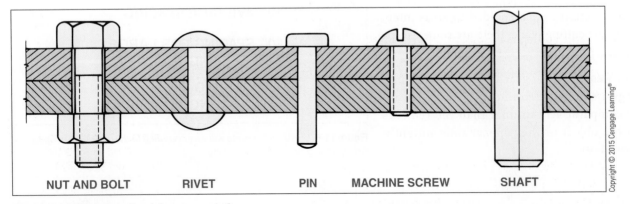

NUT AND BOLT   RIVET   PIN   MACHINE SCREW   SHAFT

Figure 11-10 *Do not crosshatch fasteners or shafts.*

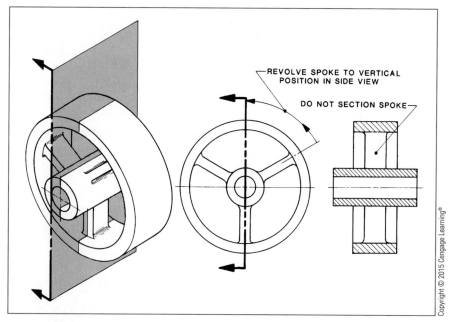

Figure 11-11 *Revolve a spoke to a vertical position in a sectional view.*

Holes that do not intersect the cutting plane are also rotated to reveal a hole through the section view (**Figure 11-12**). Objects with protrusions are rotated and sectioned in the same manner (**Figure 11-13**).

When objects containing curved surfaces are sectioned, the true projection may reveal curved lines outlining the section lined areas. When this occurs, the section is drawn using straight line separations (**Figure 11-14**).

Sectional View Types

The type of sectional view selected depends on the size, scale, and complexity of the area to be sectioned.

FULL SECTIONS

When the cutting plane extends entirely through an object, the resulting section is a **full section**. **Figure 11-15** shows a multiview drawing with a cutting plane on the top view, and the corresponding full sectional view in the front-view position. Full sections of a front view will place the cutting plane on either horizontal (top) views or vertical (profile) views. In architectural drawings, a full-section top view is called a plan view or floor plan. A full section for a profile view is an elevation section.

HALF-SECTIONS

A **half-section** is one-half of a full section. Half-sections are prepared for symmetrical objects where the details on both sides of a centerline are identical. Two perpendicular cutting plane lines,

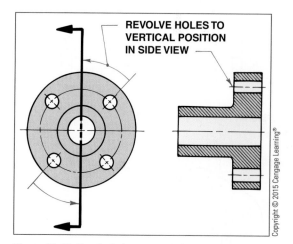

Figure 11-12 *Revolve holes to a vertical position in a sectional view.*

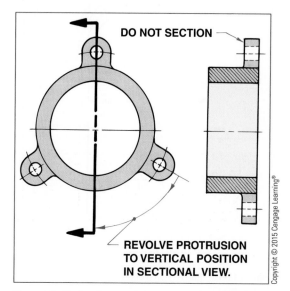

Figure 11-13 *Revolve protrusions to a vertical position in a sectional view.*

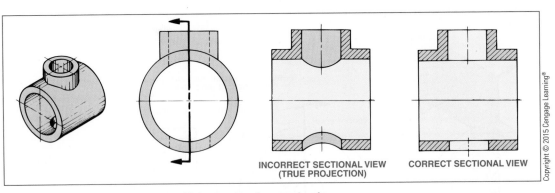

Figure 11-14 *Drafting conventions simplify intersections for curved surfaces.*

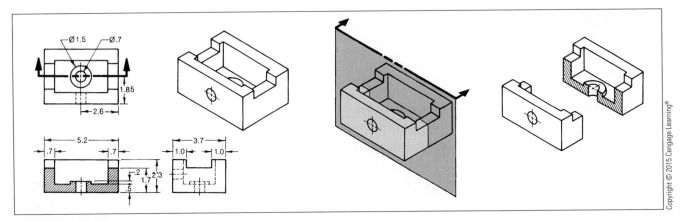

Figure 11-15 *Full-section working drawings for a jig base.*

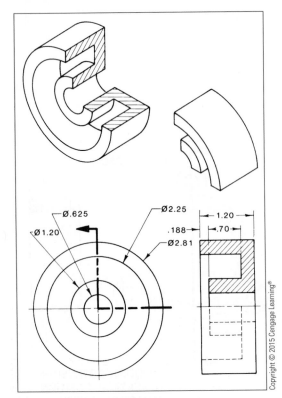

Figure 11-16 *Half-section working drawing of a belt pulley.*

one on the vertical plane and one on the horizontal plane, define the section area. These align with centerlines and remove an imaginary quarter of the object (**Figure 11-16**).

BROKEN-OUT SECTIONS

When only a small part of an object needs to be sectioned, a **broken-out section** of that part is prepared. An irregular short break line separates the sectioned area from the remainder of the drawing. **Figure 11-17a** shows a broken-out section in both pictorial and orthographic views. Note how a broken-out section facilitates the dimensions to visible lines in **Figure 11-17b**.

OFFSET SECTIONS

A cutting plane normally extends through an object in a straight line. However, a straight cutting plane may miss key features that should be shown for maximum clarity. When this occurs, the cutting plane is offset to align with special features. The cutting plane line in **Figure 11-18a** is **offset** to align with a countersunk hole, a counterbored hole, and a slot, none of which fall on a straight line. When offsetting cutting plane lines, always bend the line at right angles to the cutting plane and perpendicular to the plane of projection. Note that the bends in the cutting plane are never shown in the sectioned view (**Figure 11-18b**).

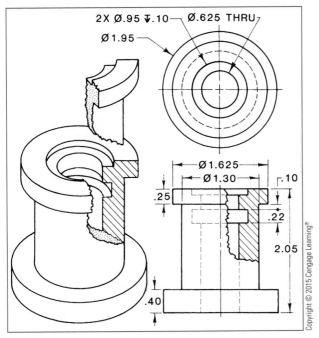

Figure 11-17a *Broken-out section working drawing of a sleeve bearing.*

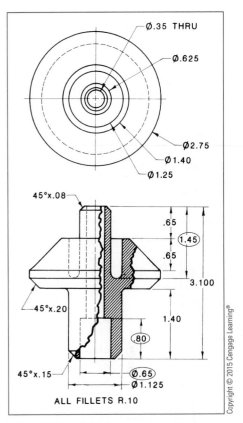

Figure 11-17b *A broken-out section will permit dimensioning to object lines instead of hidden lines.*

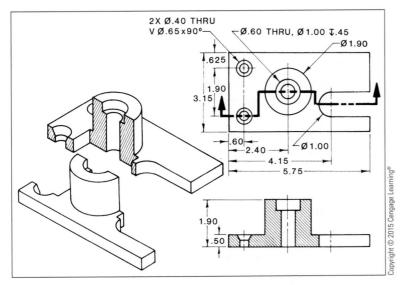

Figure 11-18a *Offset-section working drawing of a jig bearing locator.*

REMOVED SECTIONS

When sections are drawn in a location other than in one of the six multiview positions, they are **removed sections**. Sections are often removed to be drawn at a different scale or to dimension some portion in detail. When sections are removed, they must be indexed to a cutting plane line containing identifying letters (**Figure 11-19**).

REVOLVED SECTIONS

Sectional views are often revolved 90° to reveal a section of an object perpendicular to the plane of projection. In visualizing a revolved section, imagine a cutting plane

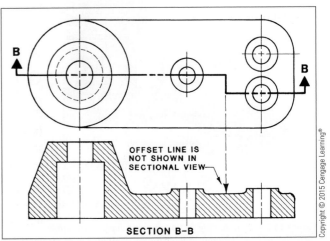

Figure 11-18b *The edge of the offset section is not shown in the section view.*

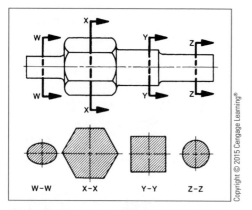

Figure 11-19 *Removed sections in a drawing of a spindle arbor.*

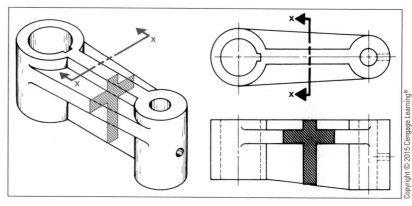

Figure 11-20a *Example of a revolved section in a working drawing and an isometric drawing.*

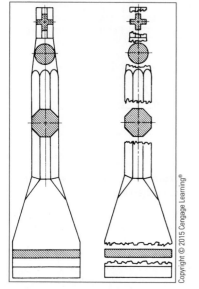

Figure 11-20b *Two methods for drawing revolved section views for a jackhammer chisel.*

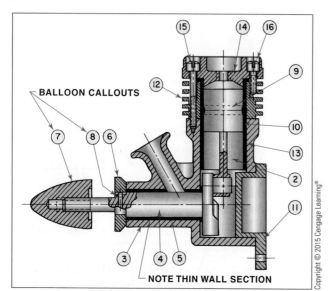

Figure 11-21 *An assembly section drawing depicting multiple sectioned parts of a model airplane motor. (Reprinted from Technical Drawing and Engineering Communication by Goetsch, Chalk, Nelson, and Rickman, Delmar/Cengage Learning.)*

passing through an object creating a slice of the object. Now imagine rotating the slice 90°. The rotated slice is a **revolved section**. Some revolved sections break the object to show the section, and others show the section directly on the view (**Figures 11-20a** and **11-20b**).

ASSEMBLY SECTIONS

One common use for sectional drawings is to show the assembly of parts. When multiple-part assembly drawings are sectioned, the adjoining parts are section lined at different angles. **Figure 11-21** shows an example of an **assembly section** with multiple parts sectioned.

THIN WALL SECTIONS

The outlines of ribs, spokes, and other thin walls are drawn on sectional views. However, section lining is not added to these parts. **Figure 11-22** shows an example of a typical **thin wall section** for a corner bracket.

Materials too thin to section with crosshatching, such as paper and sheet metal, are sectioned solid.

AUXILIARY SECTIONS

Most sectional views are aligned with the normal planes of projection. When a section must be drawn of an auxiliary view, follow the guidelines covered in Chapter 12. **Auxiliary sections** should be aligned with the projection lines of the auxiliary views (**Figure 11-23**).

ENLARGED SECTIONS

Sectional details are often too small to be interpreted and dimensioned. In such a case, **enlarged sections** are prepared to clarify details and for ease of dimensioning (**Figure 11-24**).

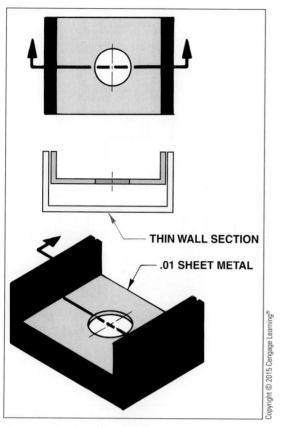

Figure 11-22 *Thin wall section for a corner bracket.*

THIN WALL SECTION

.01 SHEET METAL

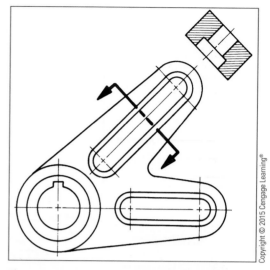

Figure 11-23 *Auxiliary section for a locking lever arm.*

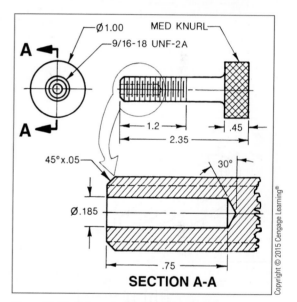

Ø1.00   MED KNURL

9/16-18 UNF-2A

A

A

1.2

.45

2.35

45°x.05

30°

Ø.185

.75

SECTION A-A

Figure 11-24 *Enlarged section of a screw adjuster.*

DRAFTING EXERCISES

1. Sketch the sections in **Figure 11-25**.

2. With instruments, draw the sections noted in **Figure 11-26**. Have your instructor add new cutting planes for additional sections to be drawn in the working drawings.

3. Draw the working drawings and the isometric sections for **Figures 11-27** through **11-29**.

4. Make the following ECOs (refer to "Engineering Change Orders," in Chapter 17) to **Figure 11-30**:

   a. Change the base dimension 1.60" to 1.50".

   b. Change the base dimension 2.10" to 2.00".

   c. Change the counterbore depth to .200".

   d. Change the vertical section to a horizontal section through the bearing.

5. With a CAD system, draw the sections in **Figure 11-30**.

6. Draw the sectional segments as removed segments in **Figure 11-31**.

7. Complete the multiview sectional drawings in **Figure 11-32**.

8. Draw a front and a side view of the rod hanger in **Figure 11-33**. Make the side view a full section.

9. Draw the views of the bearing hanger in **Figure 11-34**. Change the right-side view to a full section.

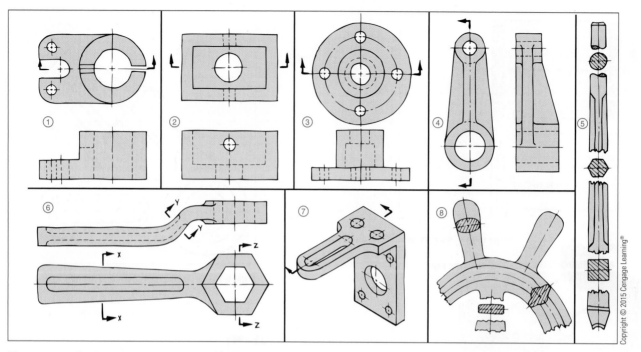

Figure 11-25 *Sketch the sectional views for each exercise.*

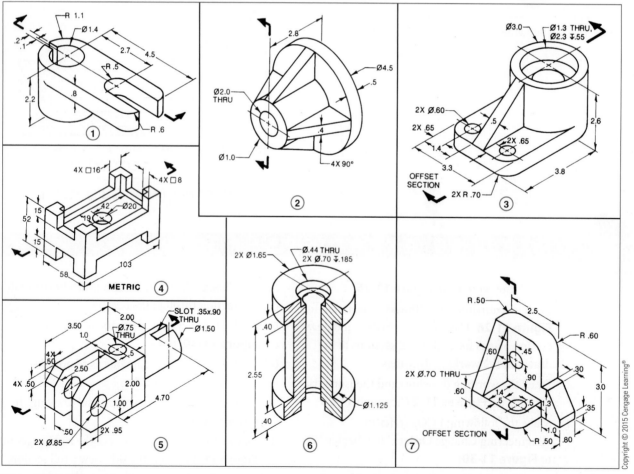

Figure 11-26 *With drawing instruments, draw the sectional views as required.*

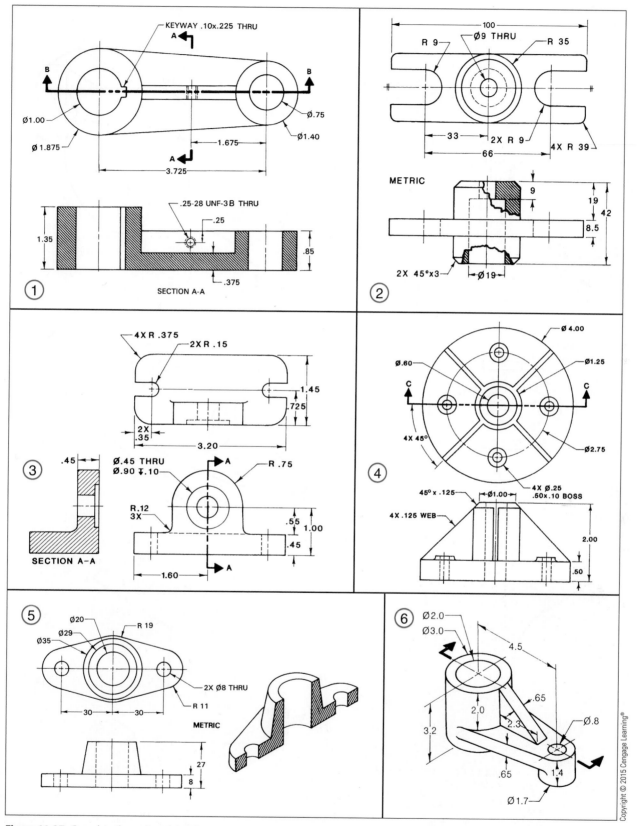

Figure 11-27 *Complete the sectional drawings.*

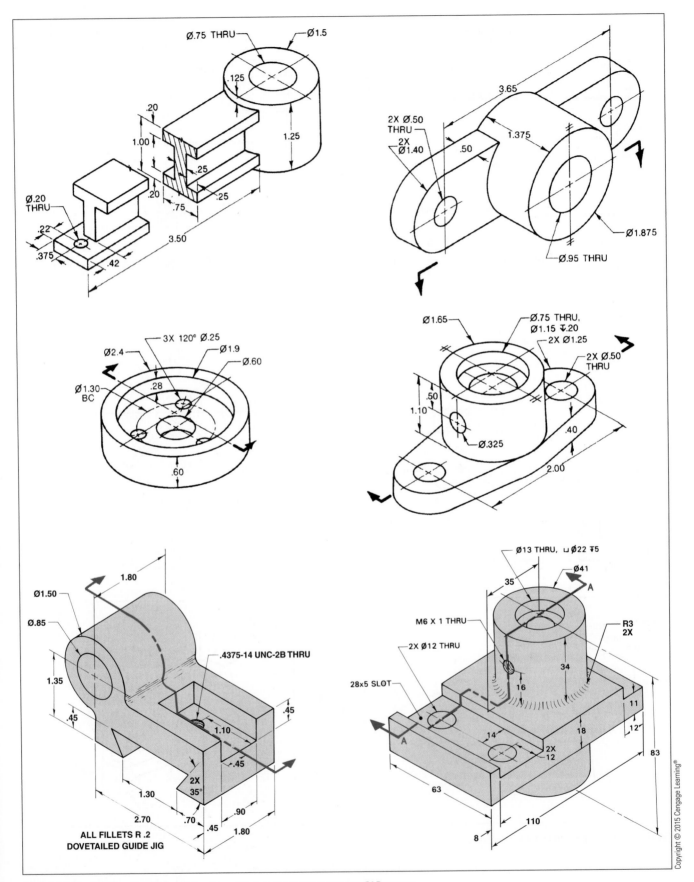

Figure 11-28 *Complete the sectional drawings with drawing instruments or a CAD system.*

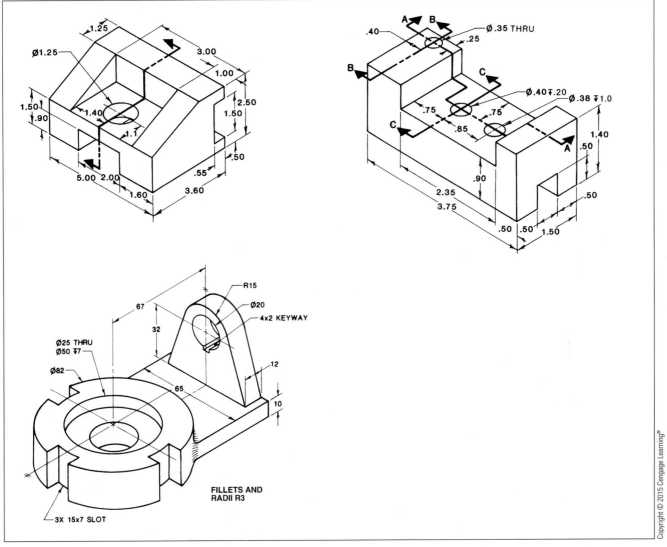

Figure 11-29 *Complete the sectional drawings with drawing instruments or a CAD system.*

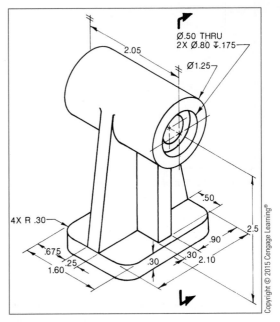

Figure 11-30 *Make the ECOs (refer to "Engineering Change Orders" in Chapter 17) listed in Exercise 4.*

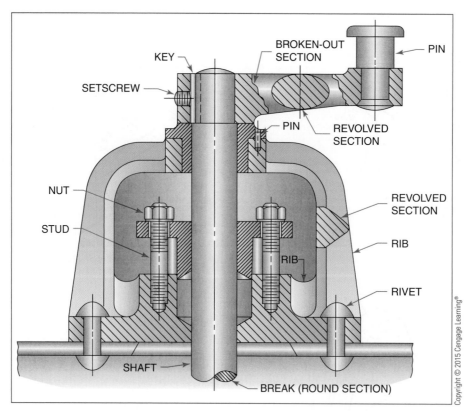

Figure 11-31 *Examples of various types of sections.*

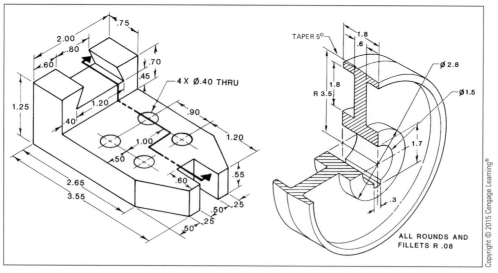

Figure 11-32 *Complete the multiview sectional drawings.*

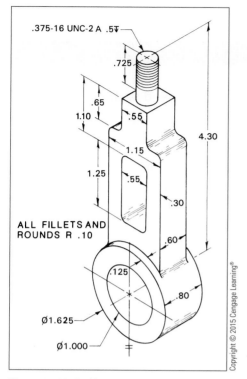

.375-16 UNC-2 A .5

.725

.65
1.10

.55

1.15

1.25

.55

.30

ALL FILLETS AND
ROUNDS R .10

.60

4.30

125

.80

Ø1.625

Ø1.000

Figure 11-33 *Rod hanger.*

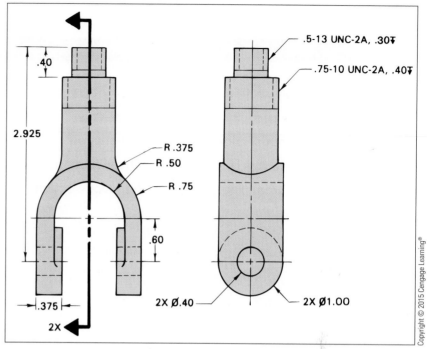

.40

2.925

.5-13 UNC-2A, .30

.75-10 UNC-2A, .40

R .375
R .50
R .75

.60

.375

2X

2X Ø.40

2X Ø1.00

Figure 11-34 *Bearing hanger.*

DESIGN EXERCISES

1. Design a tape dispenser for a standard-size roll of Scotch tape. Draw a set of working drawings, including a full sectional drawing.

2. Design a coffee cup. Draw the working drawings, including a sectional drawing.

3. Design a stand to support a video monitor over a microcomputer. Research the dimensions for any computer and monitor. Draw a set of working drawings, including sectional drawings.

4. Design a wheel cover (hubcap) for a sports car. Include a section drawing.

5. Design a steering wheel for a sports car.

6. Redesign the rod hanger in **Figure 11-33** so it will support a 1.5" pipe. Add strength with the design to the screw support. Draw an isometric section of your new design.

7. Redesign the bearing support in **Figure 11-34** so it will support a .5" steel rod. Add a new bracket support for the top. Draw an isometric section of your new design.

KEY TERMS

Assembly section

Auxiliary section

Broken-out section

Crosshatching

Cutting plane line

Enlarged section

Full section

Half-section

Offset section

Removed section

Revolved section

Section line

Sectional view

Thin wall section

Auxiliary Views and Revolutions

OBJECTIVES

The student will be able to:

- Project foreshortened lines
- Project an auxiliary view
- Project an auxiliary view from the front view
- Project an auxiliary view from the top view
- Project an auxiliary view from the side view
- Draw primary and secondary auxiliary views
- Draw a simple auxiliary view on a CAD system

Introduction

In Chapter 8, you learned that when each surface of an object is either parallel or perpendicular (90°) to other surfaces, one or more of the six normal orthographic views (front, top, right side, left side, back, or bottom) can be used to accurately describe the object. **Figure 12-1** shows an object of this type in which all angles are perpendicular. The three views are drawn as viewed through the normal orthographic planes of projection. Each projection plane is parallel to the corresponding face of the object. In this drawing, all lines and surfaces are shown as true size.

When an inclined surface is viewed from a point perpendicular to the surface plane, the true shape of the inclined surface will appear (**Figure 12-2**). These views are known as **auxiliary views**.

Foreshortening

Many objects contain surfaces that are not parallel or perpendicular to other surfaces. When objects of this type are drawn using the normal planes of projection, any inclined or sloped surface will be **foreshortened**. A foreshortened line or surface is one that appears shorter than actual size, because the viewing angle is not perpendicular to the line of sight. For example, hold a 12" ruler perpendicular

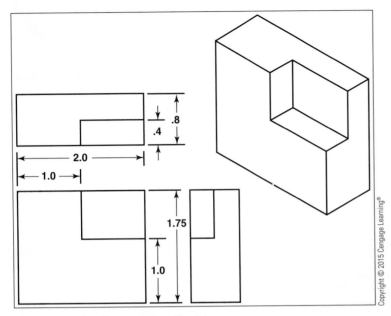

Figure 12-1 *All surfaces and edges (lines) that are perpendicular and parallel to the planes of projection will appear true size.*

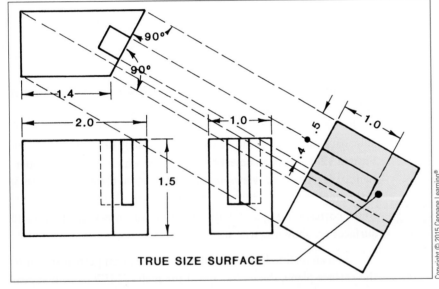

Figure 12-2 *An auxiliary view is projected perpendicular to its inclined surface.*

to your line of sight (**Figure 12-3**). Note the true vertical distance covered by the ruler. Now, tilt the ruler from the bottom to move the top of the ruler away from you. The ruler no longer appears 12" long, nor does it cover the same amount of vertical distance. It appears foreshortened.

Foreshortened lines are also shown in **Figure 12-4**. When the music box lid is closed (A), the front view shows all true dimensions (width and height). Likewise, when the lid is fully opened to 90° (B), the 2" × 4" lid measures exactly 2" × 4" on the front orthographic view. When the front view of the music box is drawn with the lid partially opened, the 4" width doesn't change, but the 2" depth is foreshortened (C). You can also partially open your book cover and observe the same effects.

To understand this concept better, look at **Figure 12-5**. Surface D is foreshortened in both the top and right-side views because this surface is not perpendicular to either the top or right-side planes of projection. Surfaces A, B, and C are shown in true size because these surfaces are perpendicular to the normal orthographic planes of projection. Only the true depth of surface D is shown on the top and right-side views. Only the true length of surface D is shown, as an inclined line representing the edge, on the front view. Even when all six orthographic planes are used (**Figure 12-6**), the true shape of surface D in **Figure 12-5** will not appear. This surface only appears, in foreshortened form, on the top and right-side views. It appears hidden, in foreshortened form, in the bottom and left-side views.

Auxiliary Planes

The plane through which the inclined surface is viewed is the **auxiliary plane**. The auxiliary plane must always be parallel to the inclined surface to be drawn. **Figure 12-7** shows the use of an auxiliary plane compared to the normal orthographic planes of projection. You can see that viewing a surface through a parallel auxiliary plane eliminates foreshortened lines and reveals the true geometric shape and size of the surface. Viewing an inclined surface through an auxiliary plane is similar to viewing the sides of a right-angled object through one of the six planes of an orthographic projection box.

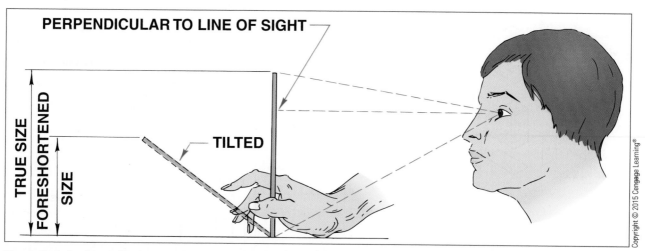

PERPENDICULAR TO LINE OF SIGHT

TRUE SIZE
FORESHORTENED
SIZE

TILTED

Figure 12-3 *An object will only appear true size when the line of sight is perpendicular to that object.*

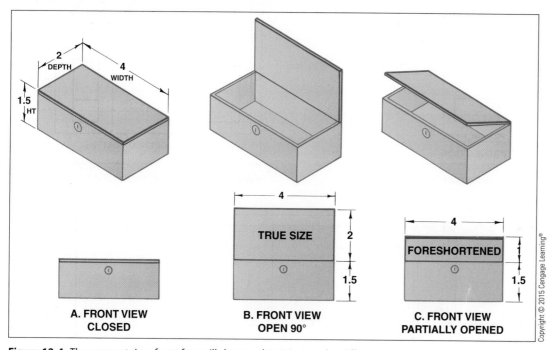

2 DEPTH

4 WIDTH

1.5 HT

A. FRONT VIEW CLOSED

4

TRUE SIZE

2

1.5

B. FRONT VIEW OPEN 90°

4

FORESHORTENED

1.5

C. FRONT VIEW PARTIALLY OPENED

Figure 12-4 *The apparent size of a surface will change when it is viewed at different angles.*

The auxiliary planes can also be envisioned as **hinged planes** that are folded from the normal orthographic planes. **Figure 12-7** shows the auxiliary plane on a projection box, and an auxiliary plane folded from the normal orthographic planes. Both methods reveal the true shape of the inclined surface, which cannot be seen through any of the normal orthographic planes.

Figure 12-8 shows an orthographic drawing containing two auxiliary views of a manufactured jig guide. In this drawing, the auxiliary views are needed because the true shape of the angled surfaces does not show in any of the orthographic views. The front view does show the exact shape and size of this part. The top view shows round holes as ellipses rather than true circles. It also shows

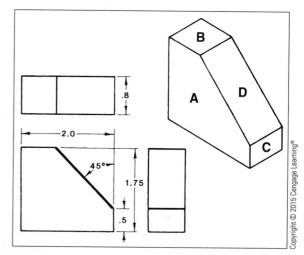

.8

2.0

45°

1.75

.5

A B C D

Figure 12-5 *Inclined surfaces will appear foreshortened (not true size).*

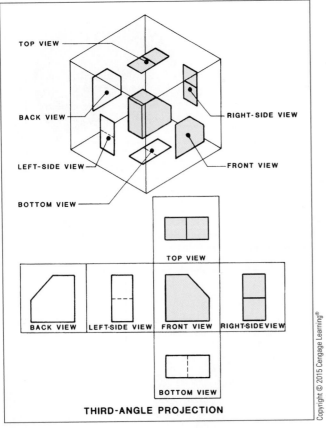

Figure 12-6 *The true size of the inclined surface is not shown in any of the six orthographic projections.*

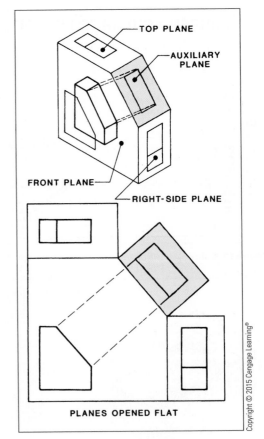

Figure 12-7 *Auxiliary projection plane.*

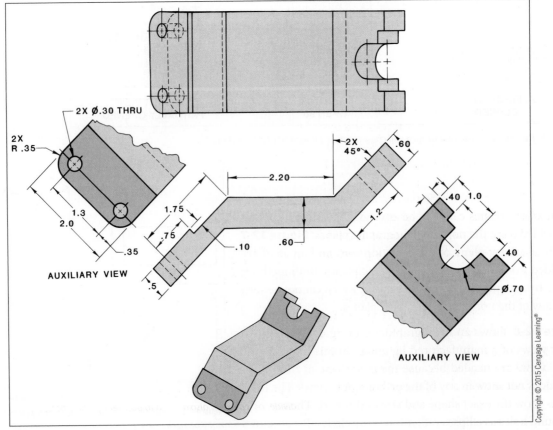

Figure 12-8 *An auxiliary view shows the offset surface in true size and form.*

foreshortened overall lengths. An auxiliary view can show the true shape of this part with the proper relationship between holes and radii.

Projection

When drawing auxiliary views, always project perpendicularly from a **true-size edge** on a normal view. In **Figure 12-9**, the true length of the inclined surface is shown as an inclined line representing the surface edge on the front view. The length of this line is true (not fore-shortened); therefore, the auxiliary view length can be projected from this line. The true depth of the inclined surface is shown on the top and side views. This depth can be transferred at 90° angles from the length projection line to form the outline of the auxiliary view. Two procedures for transfer-ring width dimensions for an auxiliary view are shown in **Figure 12-10**.

Primary Auxiliary Views

Primary auxiliary views are drawings projected directly from any of the six normal orthographic views. The three most common auxiliary views are projected from the front, side, or top normal views. **Primary auxiliary views** are used to define either depth, height, or width details not found in true form in one of the principal views. The true height and width of the object, plus the true length of the inclined plane edge, are shown on the orthographic front view in **Figure 12-11**. The true-size surface of the sloped surface is not shown; therefore, a primary front auxiliary view is projected from the front view to show the sloped surface in true size. This view is projected per-pendicularly from the inclined plane edge on the front view.

The normal side view shows the true height and depth of the object, plus the true length of the inclined plane edge (**Figure 12-12**). A true-size surface of the sloped surface is not shown. A pri-mary side auxiliary view is projected from the side orthographic view to show the sloped surface in true size.

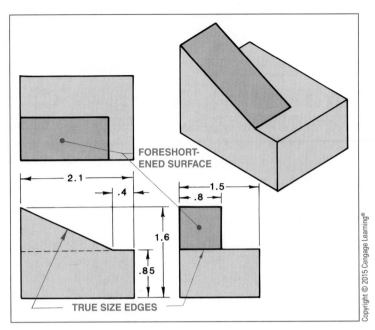

Figure 12-9 *Only the edge view of an inclined surface will appear true length.*

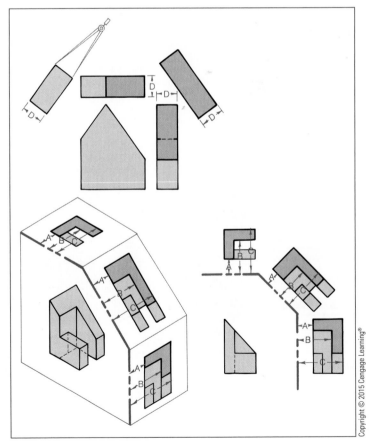

Figure 12-10 *Procedure for depth measurements of an auxiliary view. A divider or scale may also be used to transfer measurements.*

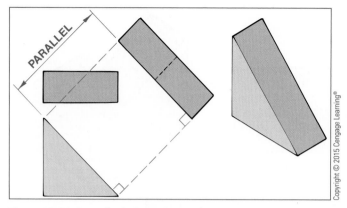

Figure 12-11 *Example of a front-view auxiliary.*

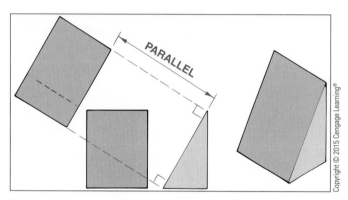

Figure 12-12 *An example of a side-view auxiliary.*

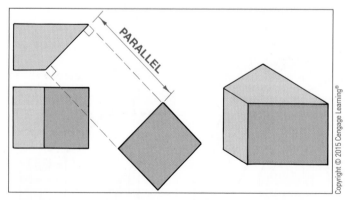

Figure 12-13 *An example of a top-view auxiliary.*

True width and depth dimensions are on the orthographic top view (**Figure 12-13**). A primary top auxiliary view is needed to show the sloped surface in true size.

More than one primary auxiliary may be drawn off a multiview view as shown in **Figure 12-14**.

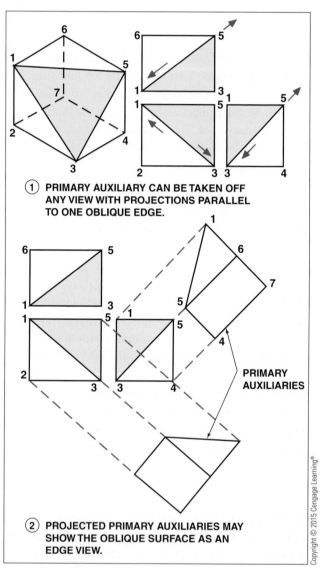

① **PRIMARY AUXILIARY CAN BE TAKEN OFF ANY VIEW WITH PROJECTIONS PARALLEL TO ONE OBLIQUE EDGE.**

② **PROJECTED PRIMARY AUXILIARIES MAY SHOW THE OBLIQUE SURFACE AS AN EDGE VIEW.**

Figure 12-14 *The first projected auxiliary is called a primary auxiliary view.*

Secondary Auxiliary Views

When surfaces are sloped in several primary auxiliary views, a secondary auxiliary view may be necessary to show the true shape of the inclined surfaces. A **secondary auxiliary view** is a view projected from a primary auxiliary view. Before you can draw a secondary auxiliary view, a primary auxiliary view must be drawn. The secondary auxiliary view is then projected perpendicularly from a selected true-size edge on the primary auxiliary view (**Figures 12-15** and **12-16**).

Additional auxiliaries may be projected from the secondary auxiliary. They are called successive auxiliaries and are shown in **Figure 12-17**.

Completeness of Auxiliary Views

In orthographic views, all parallel or perpendicular surfaces are shown in true scale. Inclined surfaces are distorted or foreshortened. In an auxiliary view, just the opposite is true. Because the auxiliary plane is parallel to the inclined surface, it will not be parallel to the normal right-angle surfaces of the object. Therefore, all surfaces, other than the inclined plane, will appear foreshortened when viewed from the auxiliary plane (**Figures 12-18** and **12-19**). For this reason, only the actual inclined surface is usually shown as a partial auxiliary view (**Figure 12-20**). **Figure 12-21** shows a partial auxiliary view of a digital clock.

OTHER ANGLES

In addition to being projected from normal orthographic views and other auxiliary views, auxiliary views can be projected from any surface. In all cases, the view must be projected perpendicular to the surface to be shown to ensure a true shape description. **Figure 12-22** shows a nut with auxiliary views projected from surfaces that are not normal orthographic views.

Notice that the auxiliary planes of projection are all parallel to the surfaces to be drawn. The auxiliary projections are perpendicular to the inclined surface.

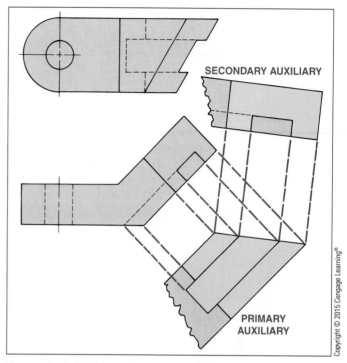

Figure 12-15 *Multiple auxiliary views may be drawn.*

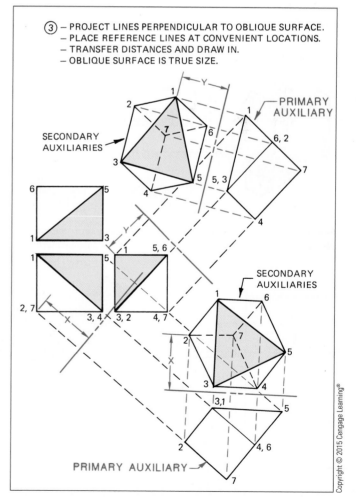

③ – PROJECT LINES PERPENDICULAR TO OBLIQUE SURFACE.
– PLACE REFERENCE LINES AT CONVENIENT LOCATIONS.
– TRANSFER DISTANCES AND DRAW IN.
– OBLIQUE SURFACE IS TRUE SIZE.

Figure 12-16 *The auxiliary view projected from a primary auxiliary is called a secondary auxiliary.*

- USE REFERENCE LINE FOR PROJECTION.
- AUXILIARIES CAN BE PROJECTED AT ANY POSITION.
- PROJECTION LINES ARE PERPENDICULAR TO REFERENCE LINE.
- SUCCESSIVE AUXILIARIES CAN BE CONTINUED INDEFINITELY.

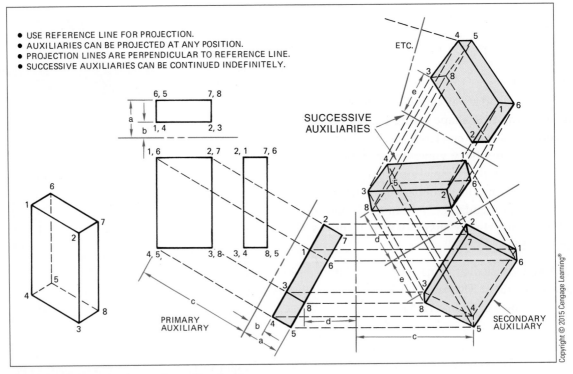

Figure 12-17 *Examples of successive auxiliary drawings.*

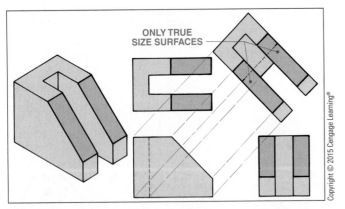

Figure 12-18 *Only the inclined surface will appear true size in the auxiliary view, because the auxiliary is viewed perpendicular to its inclined surface.*

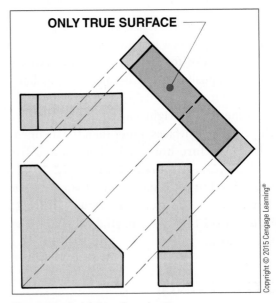

Figure 12-19 *A full auxiliary drawing.*

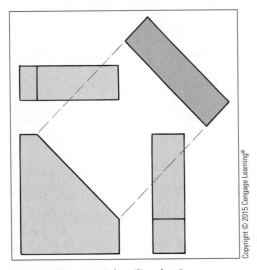

Figure 12-20 *A partial auxiliary drawing.*

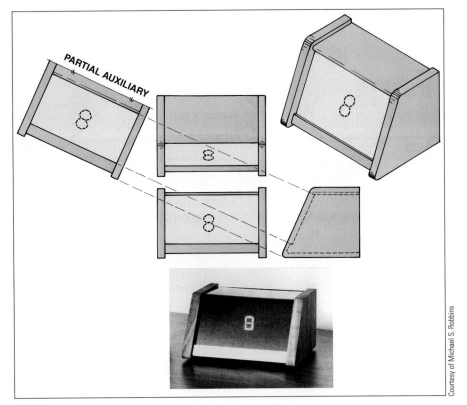

Figure 12-21 *An auxiliary drawing of a single-number digital clock.*

HIDDEN LINES

Normally, hidden (dashed) lines are omitted from auxiliary views as their use may clutter the drawing and make interpretation difficult. However, if needed for clarity, hidden lines can be added to auxiliary views.

HOLES AND ELLIPSES

When a hole is projected from an inclined plane on an orthographic drawing, it will appear as an ellipse (**Figure 12-23**). To show the hole in its true form, an auxiliary view must be drawn.

DIMENSIONING

A basic rule for dimensioning is to dimension only true sizes. Never dimension a foreshortened line or surface. Draw the auxiliary view, and place the required dimensions on the true-size auxiliary view (**Figure 12-24**).

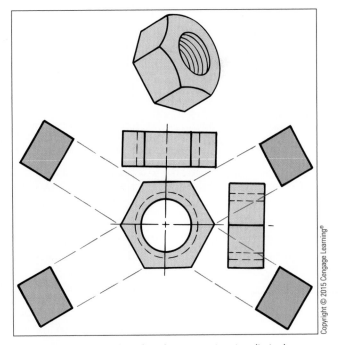

Figure 12-22 *The number of auxiliary projections is unlimited.*

Procedures

The following illustrations will show the procedure to develop several types of auxiliary views. **Figure 12-25** shows the steps for an auxiliary view of a wedge stop.

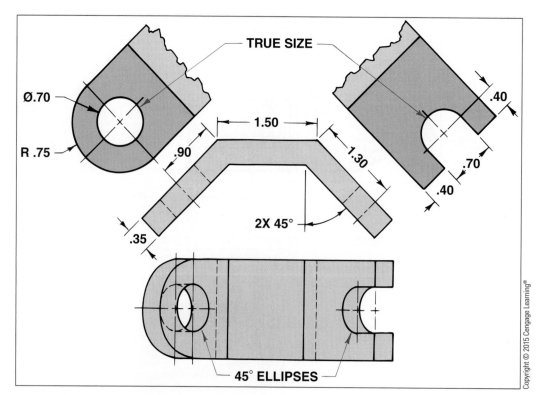

Figure 12-23 *Projecting circles and ellipses.*

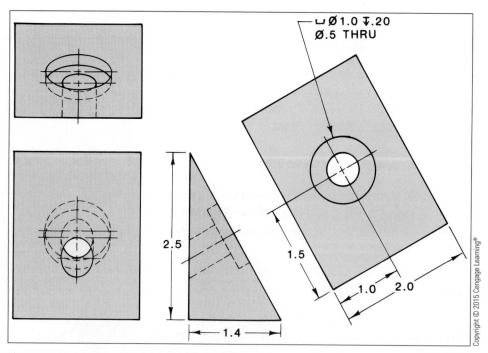

Figure 12-24 *Dimension only true sizes.*

Figure 12-26 develops the auxiliary view of one surface of a rectangular pyramid.

Figure 12-27 develops the auxiliary view of a truncated hexagonal prism.

Figure 12-28, the last example, shows the steps to draw the auxiliary view of a truncated cylinder.

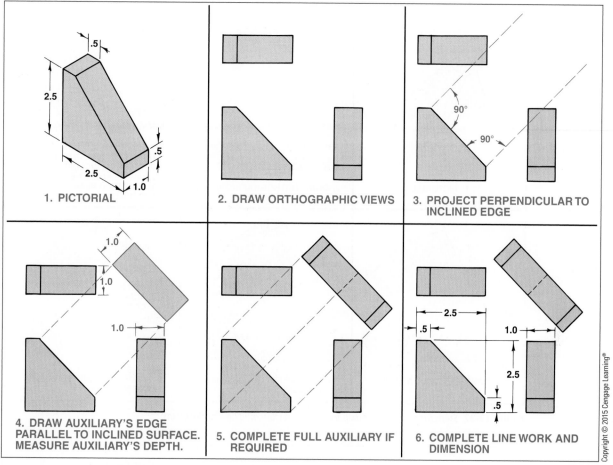

Figure 12-25 *Steps for drawing an auxiliary view for a wedge stop.*

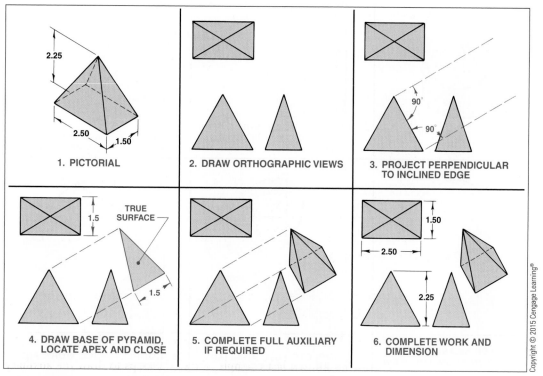

Figure 12-26 *Steps for drawing a full auxiliary view for a rectangular pyramid.*

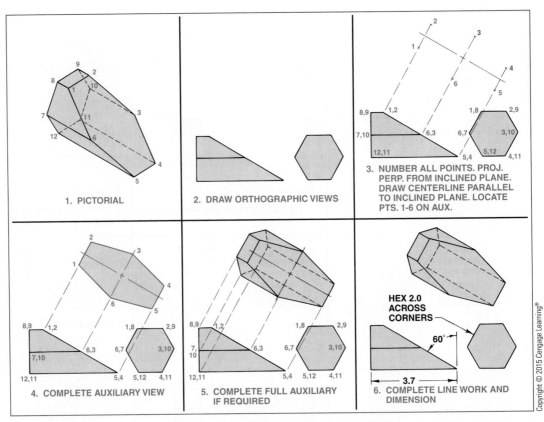

Figure 12-27 *Steps for drawing an auxiliary view for a truncated hexagonal prism.*

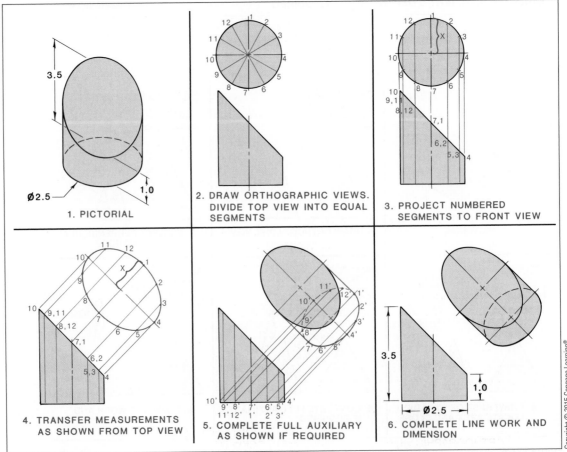

Figure 12-28 *Steps for drawing an auxiliary view for a truncated cylinder.*

Revolutions

The positioning of an object's multiview drawing for clarity may be done with three procedures. Orthographic projection is the basis for all multi-view drawings (see Chapter 8). All projections are perpendicular between adjacent views. Descriptive geometry is the theory of doing graphical solutions of points, lines, and planes in space using the principles of orthographic projection (see Chapter 13). Auxiliary projections can be drawn at any angle off a primary orthographic view or auxiliary view, as explained earlier in this chapter.

A **revolution drawing** is a fourth procedure used to reposition a view from the primary orthographic drawing and project its repositioned view (**Figure 12-29**). The drafter must decide the most advantageous viewing positions.

Any view can be rotated about a theoretical axis placed on the view (**Figure 12-30**) or on a corner of the view (**Figure 12-29**). The views may be rotated any number of degrees clockwise or counterclockwise.

The first view projected from a rotated primary orthographic view is called a **primary revolution** (**Figure 12-31**). Rotating a primary revolution and projecting a new view is called a **successive revolution**. Successive revolutions may be used to rotate an object to any desired position (**Figures 12-32a** and **12-32b**). There is no limit to the number of successive revolutions that can be drawn.

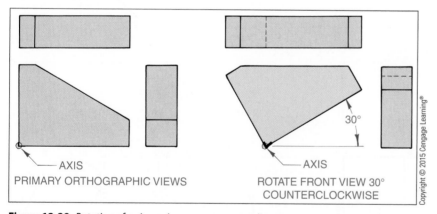

Figure 12-29 *Rotation of a view using a corner as an axis.*

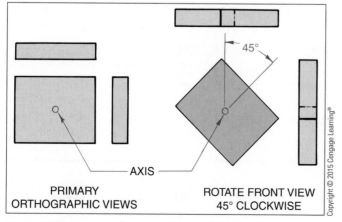

Figure 12-30 *Rotation of a view with the axis on the view.*

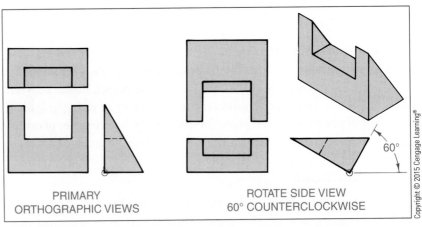

Figure 12-31 *The first rotation of a primary orthographic view is called a primary revolution.*

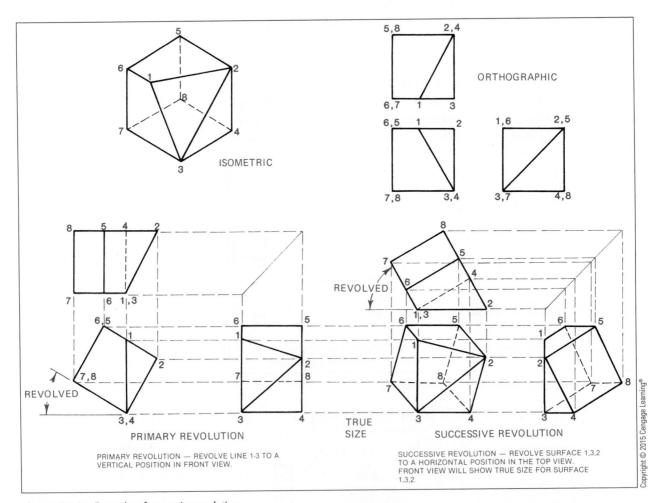

Figure 12-32a *Examples of successive revolutions.*

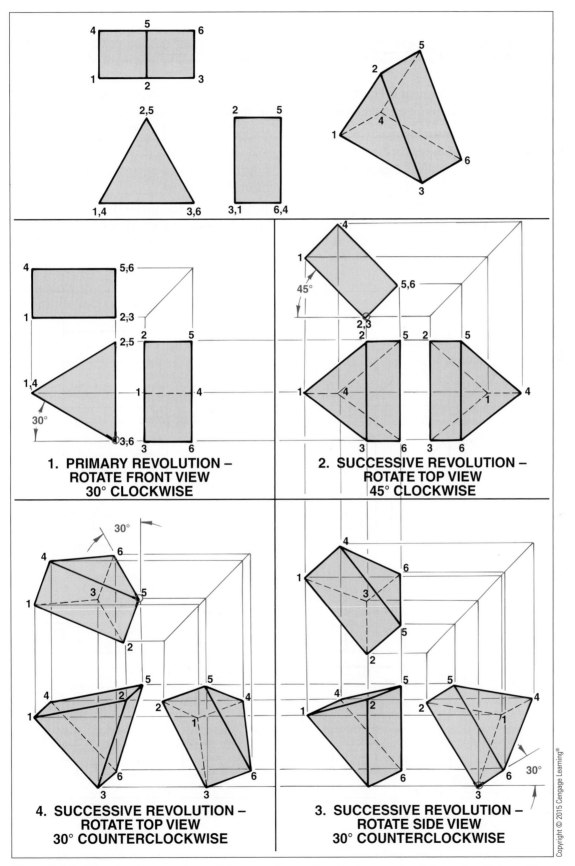

Figure 12-32b *When drawing successive revolutions, any view may be rotated any number of degrees to achieve specific placement of revolved views.*

DRAFTING EXERCISES

1. Sketch the multiview drawings in **Figure 12-33**. Sketch the auxiliary views for each problem.

2. Using your drawing instruments or CAD, draw the multiview drawings of the octagon, hexagon, and cylinder in **Figure 12-34**. Draw their auxiliary views. Numbering all the corners will help with the development of the auxiliary views.

3. With drafting instruments or CAD, draw the multiview drawings and their auxiliary views as needed for **Figures 12-35** through **12-51**.

4. With a CAD system, duplicate as closely as possible the procedures in **Figures 12-52** through **12-56**.

5. Follow the instructions for the revolution exercises in **Figures 12-57** and **12-58**.

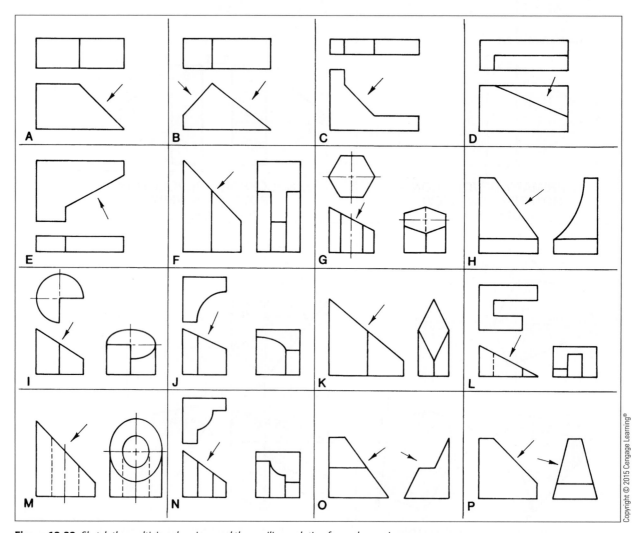

Figure 12-33 *Sketch the multiview drawings and the auxiliary solution for each exercise.*

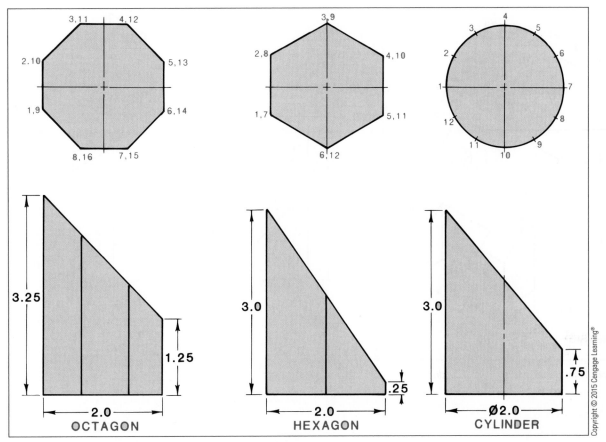

Figure 12-34 *Complete the multiview drawing and the auxiliary view for each problem.*

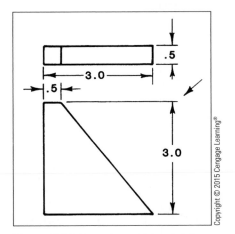

Figure 12-35 *Draw the multiview drawings and auxiliary view.*

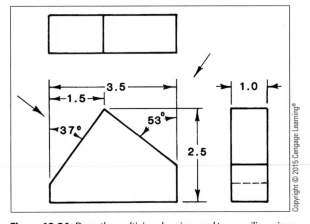

Figure 12-36 *Draw the multiview drawings and two auxiliary views.*

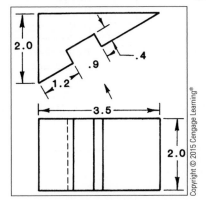

Figure 12-37 *Draw three-view multiview drawings, isometric, and auxiliary views.*

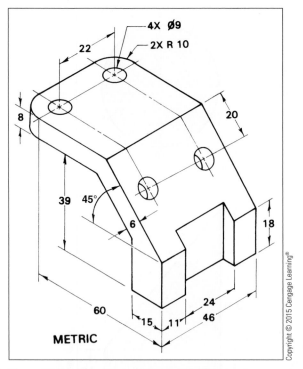

Figure 12-38 *Draw the multiview and auxiliary drawings.*

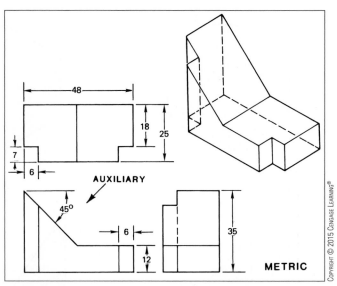

Figure 12-39 *Draw the multiview, isometric, and auxiliary drawings.*

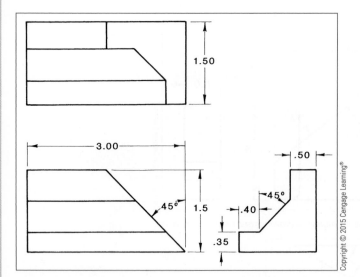

Figure 12-40 *Draw the multiview drawing and auxiliary views of the front and side views.*

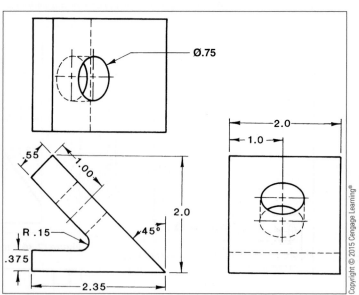

Figure 12-41 *Draw the multiview drawing and an auxiliary view of the inclined surface.*

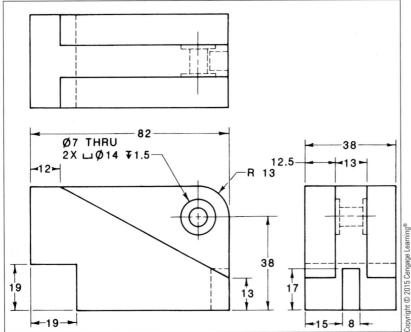

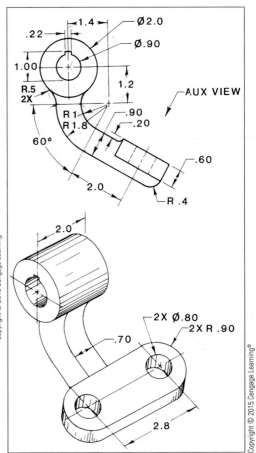

Figure 12-42 *Draw the multiview drawing and an auxiliary view to show the true size of the inclined surfaces.*

Figure 12-43 *Draw the multiview drawing with an auxiliary view of the base pad.*

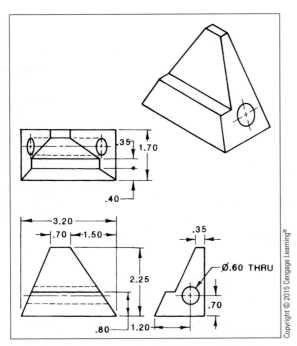

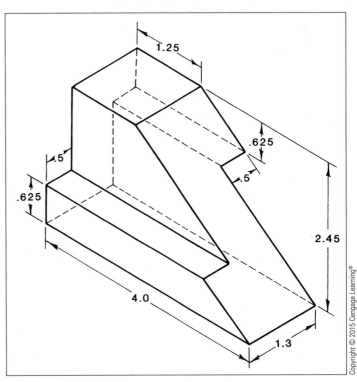

Figure 12-44 *Draw the multiview drawing and right-side auxiliary view of the angle insert guide.*

Figure 12-45 *Draw the multiview drawing and auxiliary view of the inclined surface.*

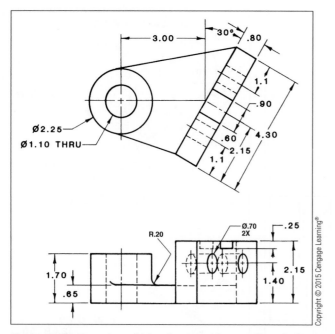

Figure 12-46 *Draw the multiview drawing with an auxiliary view of the angled base. Select the best direction for auxiliary view.*

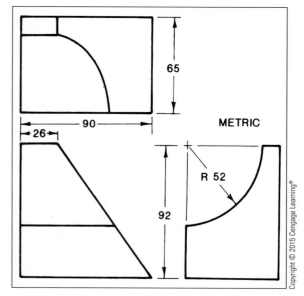

Figure 12-47 *Draw the multiview, isometric, and auxiliary view.*

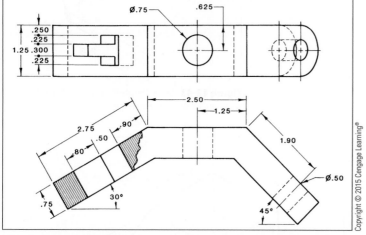

Figure 12-48 *Draw the multiview drawing and two partial auxiliary views for the angled clamping jig.*

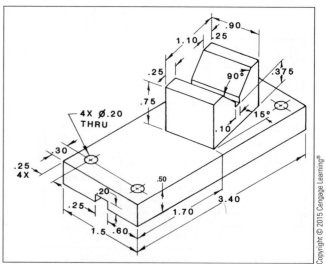

Figure 12-49 *Draw the multiview drawing and a partial auxiliary view for the angle slot's face.*

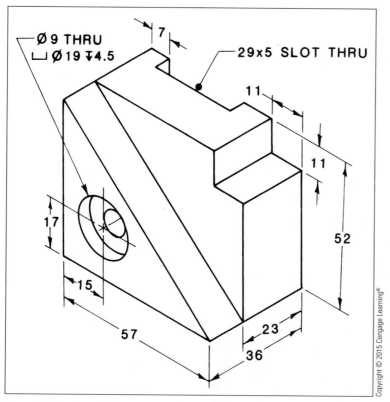

Figure 12-50 *Draw the multiview, isometric, and auxiliary view.*

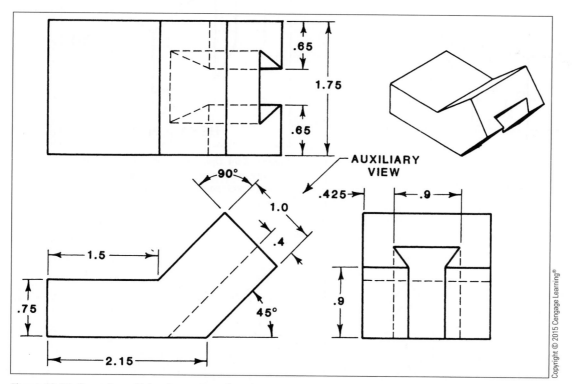

Figure 12-51 *Draw the multiview, isometric, and auxiliary view.*

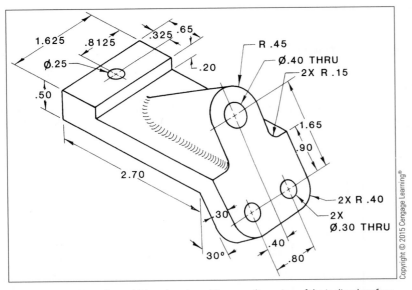

Figure 12-52 *Draw the multiview drawing with an auxiliary view of the inclined surface.*

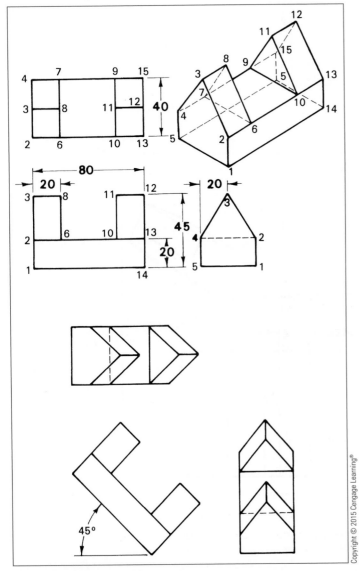

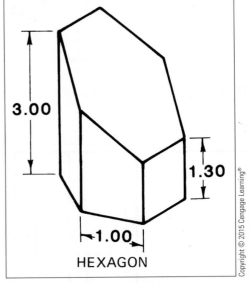

Figure 12-54 *Draw the isometric, multiview, and auxiliary view drawings. Rotate the front view 45° clockwise and complete the other view(s).*

Figure 12-53 *Draw the orthographic, isometric, and front-view revolution as shown. Revolve the side view 15° counter-clockwise and complete the front and top views. (See Chapter 15 for isometric procedure.)*

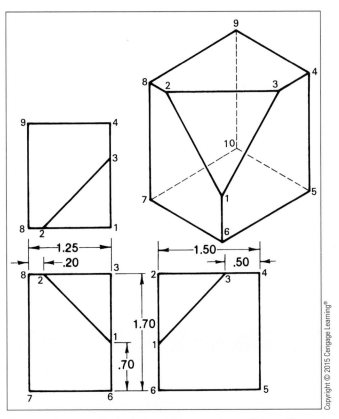

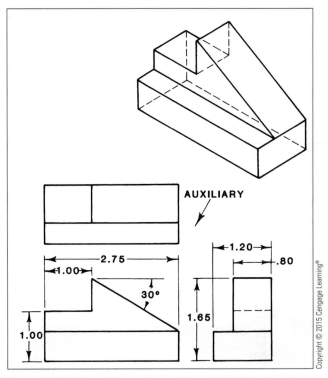

Figure 12-55 *Draw the multiview, primary, and secondary auxiliary views of the oblique surface to draw its true size.*

Figure 12-56 *Draw the isometric, multiview, and auxiliary drawings. Rotate the top view 15° and complete the other two views.*

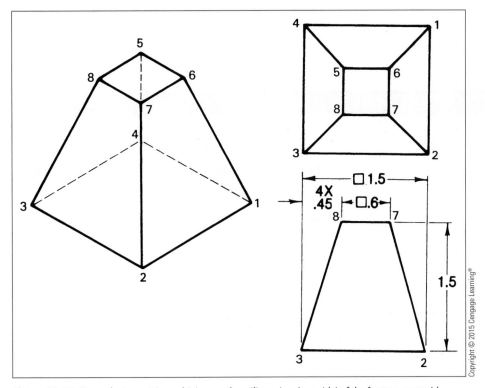

Figure 12-57 *Draw the isometric, multiview, and auxiliary view (one side) of the frustum pyramid. Revolve the front view 45° counter-clockwise and complete the other two views.*

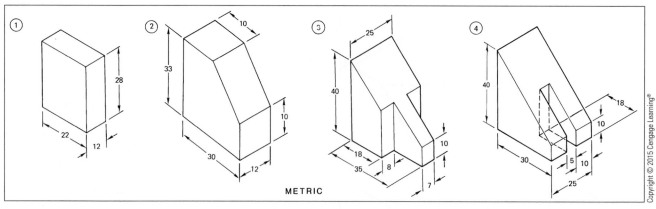

Figure 12-58 *1. Draw the isometric and number all corners. 2. Draw orthographic and number all corners. 3. Select and draw a primary auxiliary (use reference plane). 4. Select and draw a secondary auxiliary (use reference plane).*

DESIGN EXERCISES

1. Design a recipe book holder for use when cooking in the kitchen. Draw a multiview working drawing and an auxiliary view of the slanted position of the slanted support for the cookbook.

2. Design a wall-mounted mirror and frame holder that will reflect a back view of your head from the mirror in your bathroom medicine cabinet.

KEY TERMS

Auxiliary plane

Auxiliary view

Foreshortened

Hinged planes

Primary auxiliary view

Primary revolution

Revolution drawing

Secondary auxiliary view

Successive revolutions

True-size edge

Descriptive Geometry

13

OBJECTIVES

The student will be able to:

- Draw a point in space onto various planes of projection
- Draw a line in space onto various planes of projection
- Draw a point view of a line
- Draw the true length of a line
- Draw an edge view of a surface
- Draw a true size surface
- Lay out and solve a descriptive geometry exercise on a CAD system

Introduction

Descriptive geometry is a study of the three-dimensional relationship of points, lines, angles, and surfaces as drawn on two-dimensional surfaces. The principles of descriptive geometry are used to graphically solve problems and gain information about geometric forms and their position in space. Since descriptive geometry is directly related to orthographic and auxiliary projection, a basic understanding of these concepts is a vital prerequisite to understanding the material presented in this chapter. **Figure 13-1** graphically defines the geometric terms commonly used in the development of graphic solutions using descriptive geometry. Most graphic problems can be solved by finding the true length and point projection of lines, and the true size and edge view of planes.

Points in Space

Geometric forms are all constructed from a series of points. Understanding the precise location of a single point is the first step in understanding the principles of descriptive geometry.

When a **point in space** is theoretically surrounded by the **frontal plane**, **horizontal plane**, and **profile plane**, it may be projected perpendicularly to the face of the planes (**Figure 13-2**).

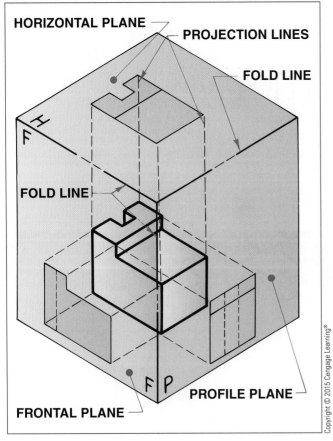

Figure 13-1 *Graphic terms for descriptive geometry.*

It is extremely important to understand why X = X and Y = Y. Study **Figure 13-2** until it is clear. When the planes are opened flat onto a two-dimensional surface, the distances of X = X and Y = Y remain the same. It is possible to solve the location of a point to a third plane when the point's location is known on two planes. This principle in **Figure 13-2** is the basis of descriptive geometry. When you can solve for one point, you can solve for many points, and points make up drawings.

Geometric Planes

Planes used in descriptive geometry include the normal orthographic planes, plus auxiliary planes. There are only six normal orthographic planes; however, the number of auxiliary planes is unlimited. An auxiliary plane (**Figure 13-3**) is a plane that lies on any angle other than the normal orthographic planes. **Figure 13-4** shows several examples of auxiliary planes projected from orthographic planes. Any angle may be used with a **fold line** of an auxiliary plane.

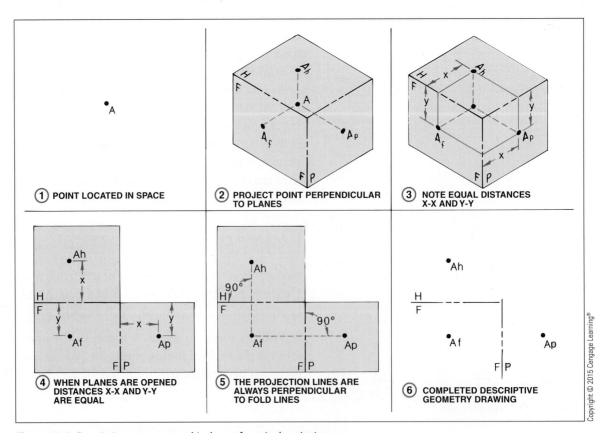

① POINT LOCATED IN SPACE

② PROJECT POINT PERPENDICULAR TO PLANES

③ NOTE EQUAL DISTANCES X-X AND Y-Y

④ WHEN PLANES ARE OPENED DISTANCES X-X AND Y-Y ARE EQUAL

⑤ THE PROJECTION LINES ARE ALWAYS PERPENDICULAR TO FOLD LINES

⑥ COMPLETED DESCRIPTIVE GEOMETRY DRAWING

Figure 13-2 *Descriptive geometry graphics layout for a single point in space.*

Geometric Lines

Think of a line as a series of connected points. A straight line is the shortest distance between any two points. Only two points are needed to create a straight line, since all points located on a straight line are aligned. **Figure 13-5** shows the projection of two points in space to create a straight line.

The connection of three points in space creates a triangle (**Figure 13-6**). It logically follows that any two-dimensional plane surface or three-dimensional solid object can be created by the projection and connection of multiple points in space (**Figure 13-7**).

Curved lines are composed of a series of unaligned points. A curved line is actually a series of very short straight lines connecting closely spaced points. Closely spaced points create smooth curves, and widely spaced points create ragged curves.

TRUE LINE LENGTH

Lines perpendicular and parallel to normal orthographic planes are all true length. Lines on auxiliary planes that are oblique or inclined are not true length when viewed from a normal plane of projection. However, lines on auxiliary planes are true length when viewed perpendicular to the adjacent plane.

When a fold line is parallel to a line in space, the line on that plane will be true size. Therefore, to find the true length of a line on any plane, locate the fold line parallel to the line. Project the line to the adjacent plane where the lines become true length (**Figure 13-8**).

A second procedure that may be used to find the true length of a line is to revolve the line on the horizontal plane until it is parallel to the reference line. Then, project the end of the line to the frontal plane as shown in **Figure 13-9**.

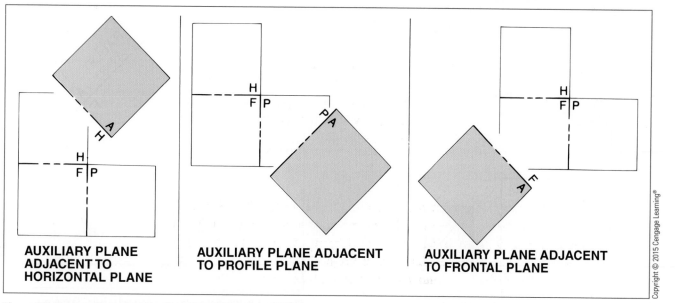

AUXILIARY PLANE ADJACENT TO HORIZONTAL PLANE

AUXILIARY PLANE ADJACENT TO PROFILE PLANE

AUXILIARY PLANE ADJACENT TO FRONTAL PLANE

Figure 13-3 *An auxiliary plane.*

Figure 13-4 *An auxiliary plane may be taken off any plane at any position.*

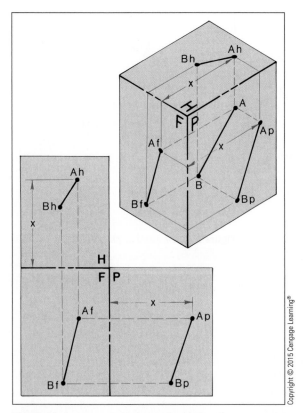

Figure 13-5 *Descriptive geometry layout for a line in space (two points).*

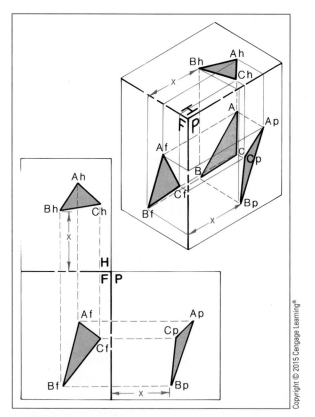

Figure 13-6 *Descriptive geometry layout for a triangular surface in space (three points).*

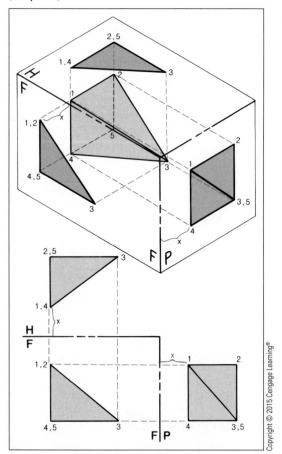

Figure 13-7 *Descriptive geometry layout for a five-point solid in space planes.*

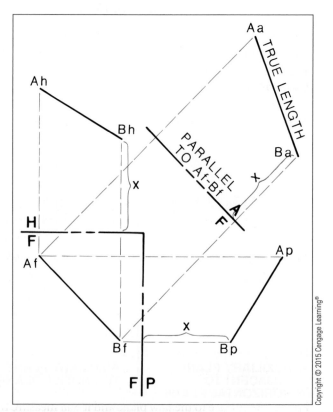

Figure 13-8 *When the fold line is parallel to a line in space, the line on that projected plane will be true size.*

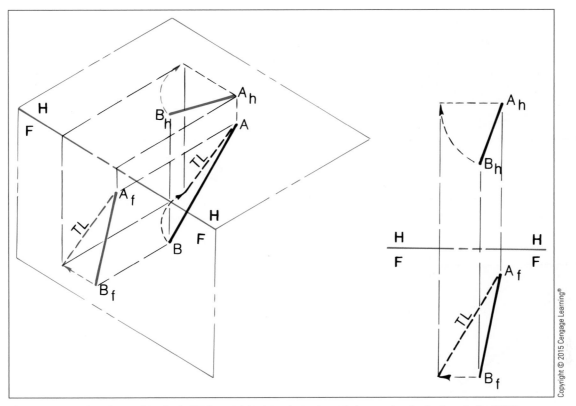

Figure 13-9 *Finding the true length of a line by revolving it on the horizontal plane and projecting it to the frontal plane.*

POINT VIEW OF A LINE

Lines on a plane are represented by points on all other perpendicular planes. Think of a point on a line as a cross section of the line. The point view of a line can be drawn by placing the fold line perpendicular to the true length line. Project the two points of the line that will appear as a single point (**Figure 13-10**). In **Figure 13-10**, the fold line A/B is perpendicular to the true length line. All X and Y distances are equal.

EDGE VIEW OF A PLANE

To find the edge view of a plane, follow the steps in **Figure 13-11**. Establish a true length line within the plane's surface. Locate a point view of the true length line using the procedures in **Figure 13-10**. Locate the third point and connect the points with a line. This creates an edge view of the triangular plane.

TRUE SIZE OF A SURFACE

Drawing the true size of a plane is an important graphic solution procedure used in descriptive geometry. To draw the true size of a surface, find an edge view of the surface. Draw a fold line parallel to the edge view. Project the surface to the new plane and it will measure true size and shape (**Figure 13-12**).

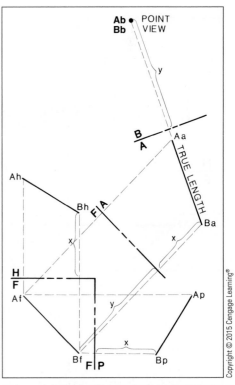

Figure 13-10 *When the fold line is perpendicular to the true length line, the projection will be a point view. Note that all X distances are equal and all Y distances are equal.*

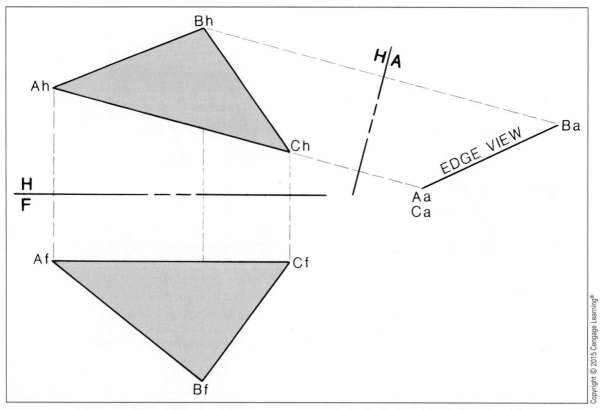

Figure 13-11 *The fold line F-H is parallel to edge Af-Cf; therefore, edge ah-Ch is true length on the horizontal plane. Fold line H-A is perpendicular to the true length line Ah-Ch; therefore, true length edge Aa-Ca will appear as a point view. The surface Aa-Ba-Ca will appear as an edge view on auxiliary plane A.*

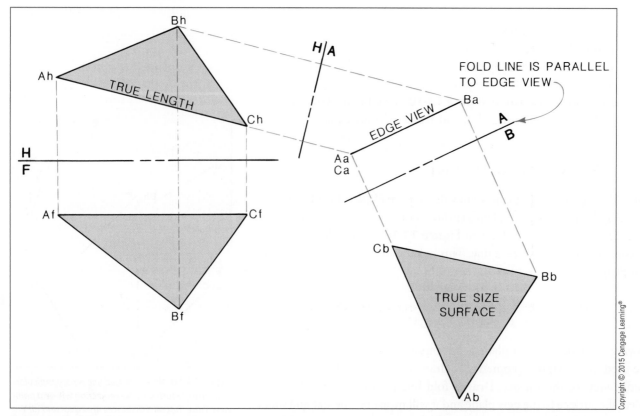

Figure 13-12 *Once the edge view is drawn, the placement of a fold line parallel to it will show the true size of the surface on auxiliary plane B.*

DRAFTING EXERCISES

1. Sketch the graphic solutions for the exercises shown in **Figure 13-13**. Sketch the exercises two or three times larger than shown.

2. With instruments, find the graphic solutions for the problems in **Figure 13-14**. Triple the size of the exercises with a scale or dividers.

3. Solve the problems in **Figure 13-14** using a CAD system.

4. Find the point in space on the profile plane if the point on the frontal plane is in the exact center of the plane, and the point on the horizontal plane is .75" above the fold line. Draw all planes 2" square.

5. Complete the horizontal plane for **Figure 13-15**. Find the edge view of the plane and then the true size.

6. Complete the profile plane for **Figure 13-16**. Find the edge view of the plane and then the true size.

7. Note in **Figure 13-17** on the frontal plane the surface A, B, C, and point P. Find the shortest distance from surface A, B, C, to point P.

8. Complete the auxiliary planes A, B, C, and E for the rectangular prism in **Figure 13-18**.

9. Complete the true shape drawing for **Figure 13-19**.

10. Follow the instructions on page 218 for **Figure 13-20**.

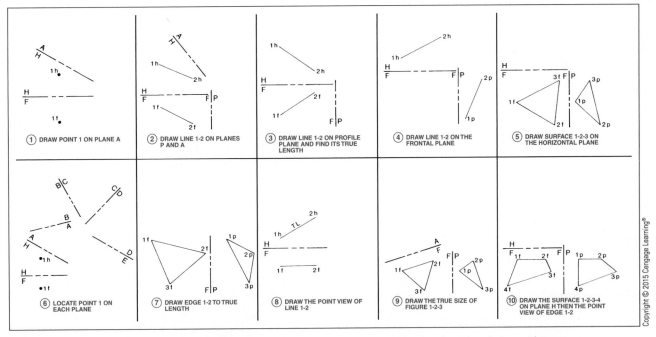

Figure 13-13 *Triple each exercise with a scale or dividers on a separate drawing format. Sketch or draw the solutions with instruments.*

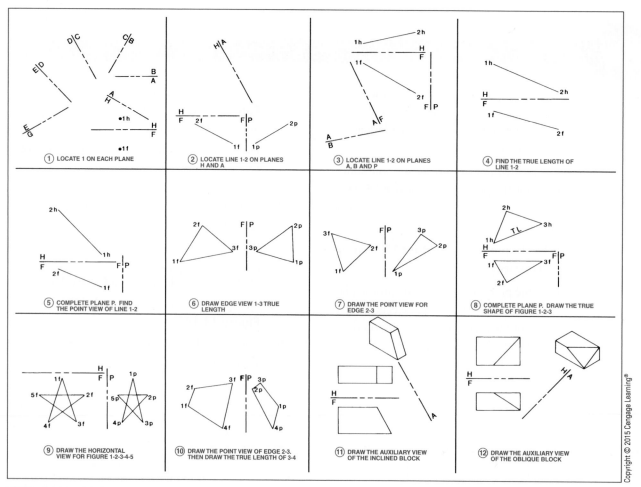

Figure 13-14 *With drawing instruments, triple the size of each exercise and solve.*

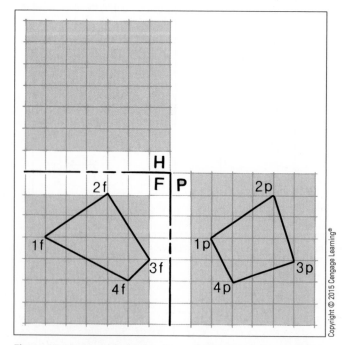

Figure 13-15 *Complete the drawing on the horizontal plane, and then find the edge view and true size of the plane.*

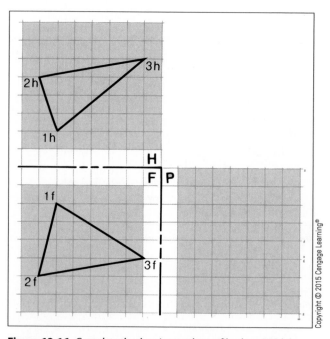

Figure 13-16 *Complete the drawing on the profile plane. Find the edge view of the plane and then the true size.*

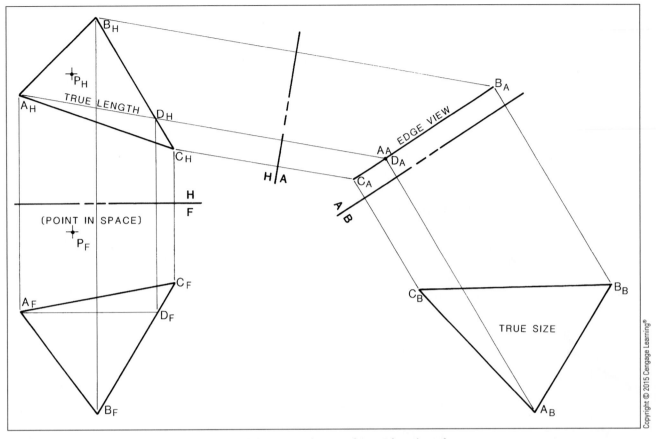

Figure 13-17 *1. Locate point P in views A and B. 2. Find the shortest distance of point P from the surface.*

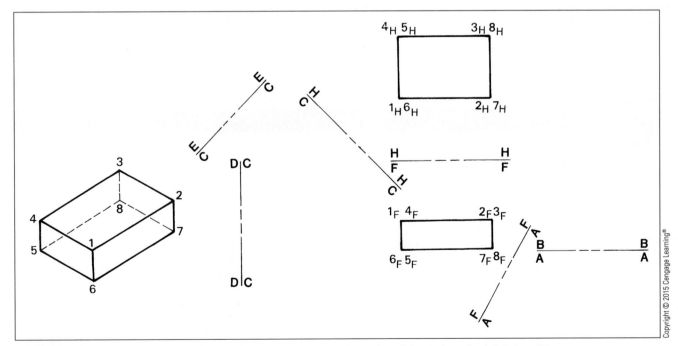

Figure 13-18 *Double the size of this exercise (dividers or photocopying) and complete all the auxiliary planes A, B, C, and E.*

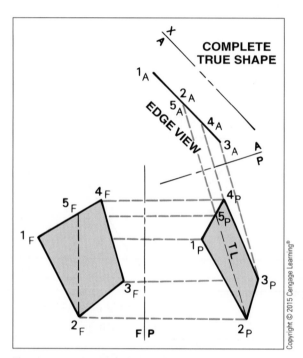

Figure 13-19 *Complete the true shape drawing of the polygon.*

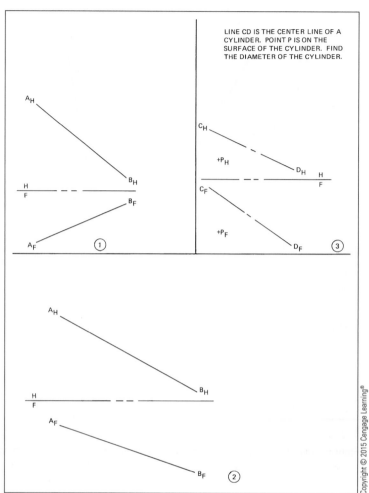

LINE CD IS THE CENTER LINE OF A CYLINDER. POINT P IS ON THE SURFACE OF THE CYLINDER. FIND THE DIAMETER OF THE CYLINDER.

Figure 13-20 *Find the point views for exercises 1 and 2. Find the diameter of the cylinder. (Redraw or photocopy this page.)*

KEY TERMS

Descriptive geometry

Fold line

Frontal plane

Horizontal plane

Point in space

Profile plane

Development Drawings

14

OBJECTIVES

The student will be able to:

- Recognize the basic geometric shapes
- Find true-length lines
- Draw patterns using the parallel line method
- Draw patterns using the radial line method
- Draw patterns using the triangulation method
- Draw patterns using a CAD system

Introduction

Products with thin, continuous surfaces, such as sheet metal, paper, and plastic sheets, are manufactured with the aid of flat **patterns**. The drawings used to develop these patterns are called **development drawings**, **flat surface patterns**, surface developments, stretchout drawings, and sheet metal patterns. This field of drafting is called pattern drafting. Pattern drafting is used extensively in the automotive, aircraft, packaging, and HVAC (heating, ventilation, and air conditioning) industries (**Figure 14-1**).

Although finished products are three-dimensional, development drawings (patterns) are drawn on a two-dimensional surface. All dimensions on a pattern drawing must be true size, since the pattern is used as a template to cut material to an exact size and shape. The material, once cut in the shape of the pattern, is formed into a three-dimensional shape and fastened into its final form. **Figure 14-2** shows a container box pattern outlined as a flat surface. **Figure 14-3** shows the container material cut to the outline of the pattern. **Figure 14-4** shows how the material is folded into a three-dimensional form and fastened into position (**Figure 14-5**).

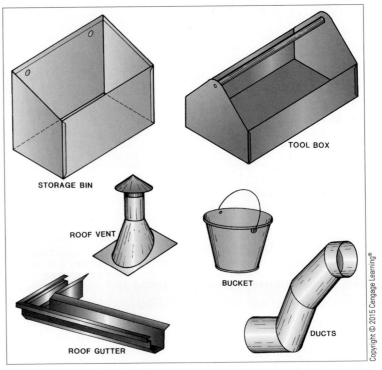

Figure 14-1 *Many products are manufactured from pattern developments.*

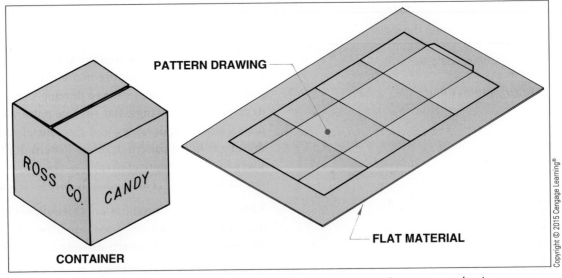

Figure 14-2 *A container box and its pattern on a flat material. All dimensions are true size on a pattern drawing.*

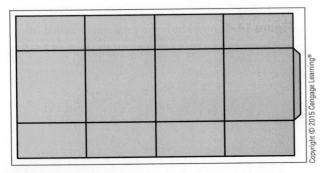

Figure 14-3 *Cut-out pattern.*

Surface Forms

Surface planes, for which development drawings are prepared, include flat, curved, warped, regular solid, and basic geometric forms. Plane (flat) surfaces are simple to lay out because there are no inclined or oblique surfaces to project (**Figure 14-6**).

Surfaces with a single curve include cylinders and cones (**Figure 14-7**). The development of pattern drawings for single-curve surfaces is also simple, since all lines are either parallel or radiate from a common apex. **Warped surfaces** are curved in two directions, which makes pattern layout more complex. Warped surfaces include spheres, paraboloid, and hyperboloid forms (**Figure 14-8**). The surfaces of regular solids, such as the tetrahedron, hexahedron, octahedron, dodecahedron, and icosahedron, form patterns as shown in **Figure 14-9**. The surfaces of basic forms, such as prisms, pyramids, cylinders, and cones, are created through the development of pattern shapes as shown in **Figure 14-10**.

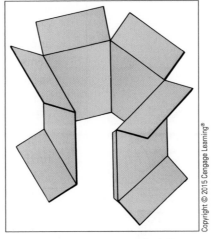

Figure 14-4 *Pattern is formed by folding.*

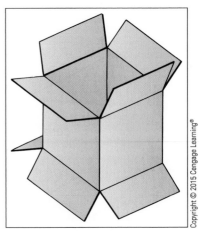

Figure 14-5 *Fasten together by taping or gluing.*

Pattern Drawing Terminology

A pattern drawing is used as the guide for cutting the surface outline of an object from a sheet. Pattern drawings are also used to show the location of lines for bending, folding, or rolling flat forms into three-dimensional shapes. In addition, they are used for outlining the position

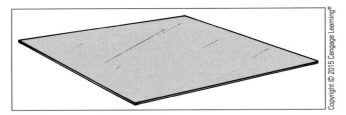

Figure 14-6 *The flat or plane surface.*

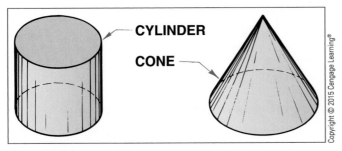

Figure 14-7 *Examples of single-curve surfaces.*

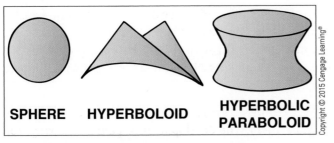

Figure 14-8 *Examples of warped (double-curve) surfaces.*

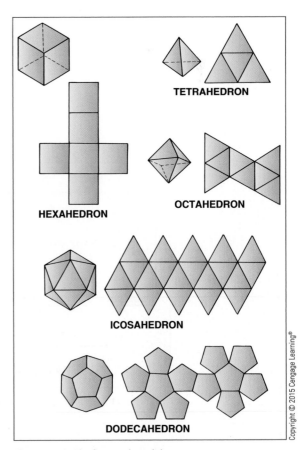

Figure 14-9 *The five regular solids.*

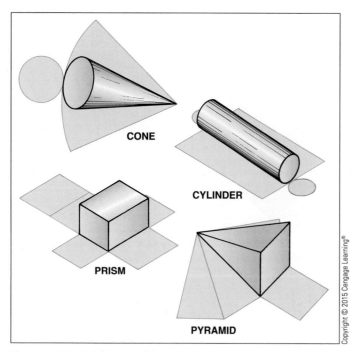

Figure 14-10 *Basic forms and their patterns.*

of seams and hems for fastening. This is all accomplished through **stretchout lines**, **fold (bend) lines**, and **tab lines** (**Figure 14-11**). Tabs are used to form hems and seams. **Hems** are formed along the edges of sheet metal pieces to seal in the sharp edges. Seams are used to connect two or more pieces of material (**Figure 14-12**).

In addition to the basic outline of the pattern, further material must be added to provide a base for permanent connections. The type of connection depends on the size and type of material, plus the fastening method. Common methods for fastening flat material include soldering, welding, riveting, gluing, stapling, and sewing.

All lines on a pattern drawing must be true size; otherwise, the size and shape of the object cut from the pattern will be incorrect. All lines that fall on normal orthographic planes, perpendicular to the plane of projection, are **true length**. All lines on inclined or oblique surfaces are not true length (**foreshortened**) on normal orthographic views. The true length of these foreshortened lines must be determined before they can be used in a pattern drawing. The true length of foreshortened lines can be obtained by revolvement, using one of two methods (**Figure 14-13**).

There are three basic methods of developing pattern drawings, depending on the geometric shape of the finished form. These include parallel line development, radial line development, and triangulation.

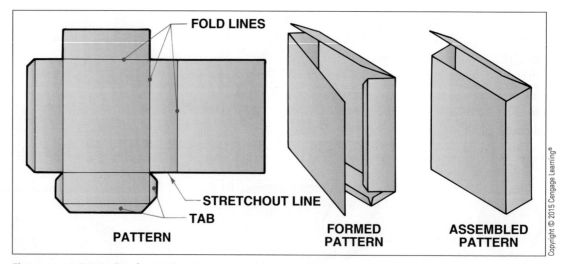

Figure 14-11 *Terminology for a pattern drawing.*

Parallel Line Development

Parallel line development is a method used to develop patterns for objects that contain parallel lines, such as prisms and cylinders.

PRISMS

All lines in a multiview drawing of a prism are true length. Therefore, parallel lines are extended from the top and bottom of the front view, which represents the true height of the pattern (**Figure 14-14**). These lines are called "stretchout lines" because they represent the pattern's stretched out form. The true length of the pattern is determined by measuring, projecting, or transferring the width of each side to the pattern. A fold line marks the location of each corner. The fold lines are extended upward using construction lines, and the height of each corner is projected from the front view to the stretchout lines. Lastly, the tabs necessary for making hems and seams are added.

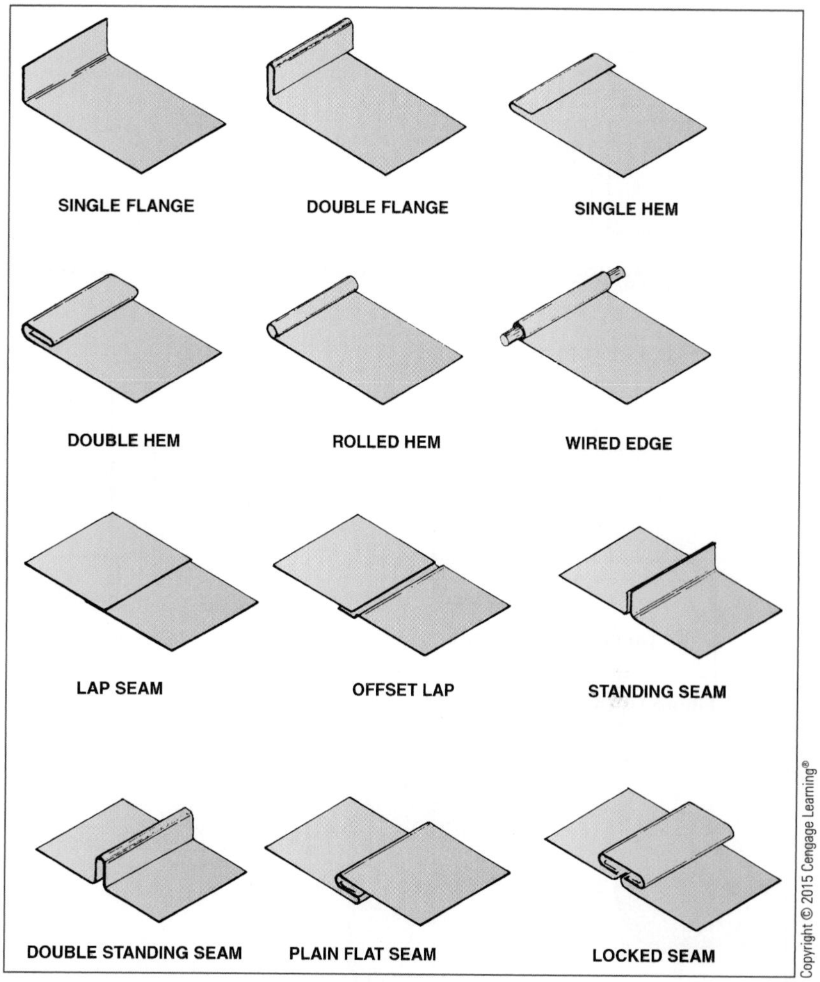

SINGLE FLANGE DOUBLE FLANGE SINGLE HEM

DOUBLE HEM ROLLED HEM WIRED EDGE

LAP SEAM OFFSET LAP STANDING SEAM

DOUBLE STANDING SEAM PLAIN FLAT SEAM LOCKED SEAM

Figure 14-12 *Various types of seams and hems.*

CYLINDERS

In developing a pattern for a cylinder, parallel stretchout lines are projected from the top and bottom of the front view (**Figure 14-15**). The length of the stretchout line is equal to the circumference of the cylinder. The circumference is determined by multiplying the diameter by 3.1416 (π), or by stepping off the **chordal distances** around the circumference of the top view.

CURVED SURFACES

When the object lines of a geometric form are parallel, the pattern outline of irregular cuts in its surface can also be determined, using parallel line development methods. Curved surfaces

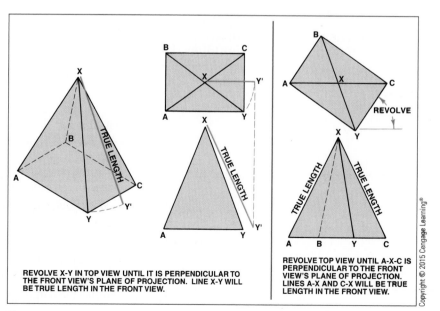

REVOLVE X-Y IN TOP VIEW UNTIL IT IS PERPENDICULAR TO THE FRONT VIEW'S PLANE OF PROJECTION. LINE X-Y WILL BE TRUE LENGTH IN THE FRONT VIEW.

REVOLVE TOP VIEW UNTIL A-X-C IS PERPENDICULAR TO THE FRONT VIEW'S PLANE OF PROJECTION. LINES A-X AND C-X WILL BE TRUE LENGTH IN THE FRONT VIEW.

Figure 14-13 *Two procedures used to find the true length of a line.*

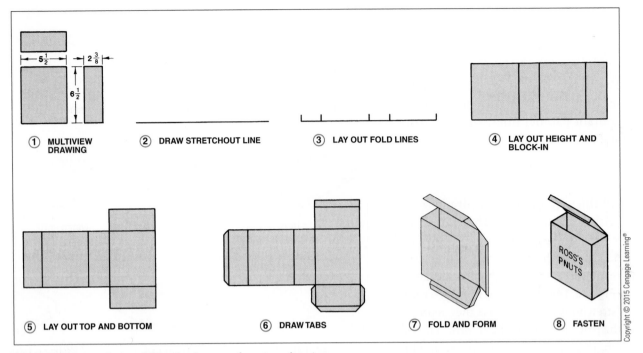

Figure 14-14 *Steps for parallel line development of a rectangular prism.*

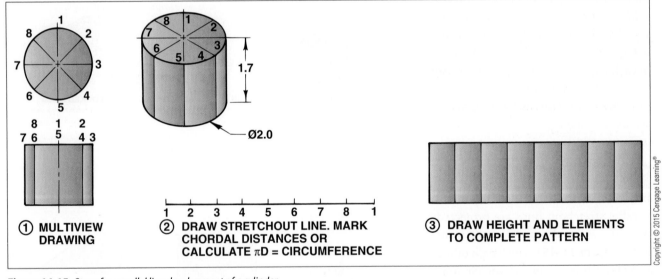

Figure 14-15 *Steps for parallel line development of a cylinder.*

may consist of a series of small, flat surfaces called **elements**. Closely spaced elements create smooth curves, and widely spaced elements create flat-sided curves.

When projecting an irregular outline on a parallel surface, such as a cylinder, first draw a stretchout line and calculate the circumference (**Figure 14-16**). Next, divide the circumference into an equal number of **chords**, and transfer these chordal distances to the stretchout line. Finally, measure or project the height of each element on a multiview front or side view, and transfer the heights to similar elements on the stretchout line. Connecting the intersecting points creates the outline of the completed pattern.

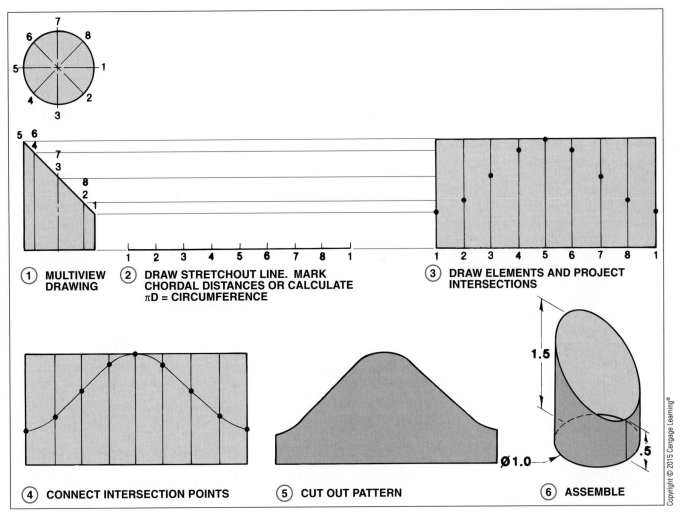

Figure 14-16 *Steps for parallel line development of a truncated cylinder.*

Radial Line Development

Geometric forms, such as pyramids and cones, do not contain parallel lines. These forms contain lines that radiate to a common apex and are developed through the use of **radial lines**. Since the inclined vertical edges of cones and pyramids are not parallel, stretchout lines are not parallel. In **radial line developments**, stretchout lines are arcs, and element lines become triangles that radiate from an apex to the **stretchout arc** (**Figure 14-17**).

PYRAMIDS

The orthographic top view of a pyramid's base reveals true width and length measurements. The orthographic front or side view contains receding object lines that are not true length, depending on the orientation of the view. To find the true length of a pyramid's side, the pyramid's

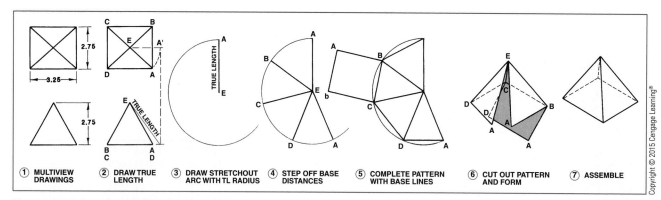

1. **MULTIVIEW DRAWINGS** 2. **DRAW TRUE LENGTH** 3. **DRAW STRETCHOUT ARC WITH TL RADIUS** 4. **STEP OFF BASE DISTANCES** 5. **COMPLETE PATTERN WITH BASE LINES** 6. **CUT OUT PATTERN AND FORM** 7. **ASSEMBLE**

Figure 14-17 *Steps for radial line development of a rectangular pyramid.*

corners must be revolved to convert inclined lines into normal orthographic true height lines. This is shown in **Figure 14-17** by rotating corner A to position A', aligning corner A–E on a normal orthographic plane. This true-length line (A'–E) is the true length of the radius from the apex to the stretchout arc. Once the stretchout arc is drawn, the base distances (which are true length in the top view) are located on the arc and connected to the apex. The pattern is then completed with the addition of the base.

TRUNCATED AND TRIANGULAR PYRAMIDS

To develop a pattern for a truncated (cut) pyramid, follow the first five steps demonstrated in **Figure 14-17**. Then, determine the true length of the lines connecting the base corners with the corners of the cutoff by revolving a corner (**Figure 14-18**). This involves finding the true length of lines H–C, G–D, F–E, and I–B. These lengths are projected to the true length line A–H', then transferred to finish the outline of the truncated portion of the pyramid.

The development of patterns for triangular pyramids is similar to that for square pyramids. The exception is that only three base sides are projected on the stretchout line (**Figure 14-19**) after the true length for the stretchout arc is found.

CONES

Unlike pyramids, the elements of a cone are all equal in length from base to apex. The side and front view of a cone are identical and contain true length object lines (**Figure 14-20a**). The radial stretchout arc for the development of a cone pattern is equal to the distance from the base to the apex, as found on the side view. The top view of a cone reveals the true circumference, but the element lines are foreshortened.

To complete a pattern drawing of a cone, draw an arc with a radius equal to the distance from the base to the true length apex (**Figure 14-20a**). Divide the circumference of the base into evenly spaced distances, and

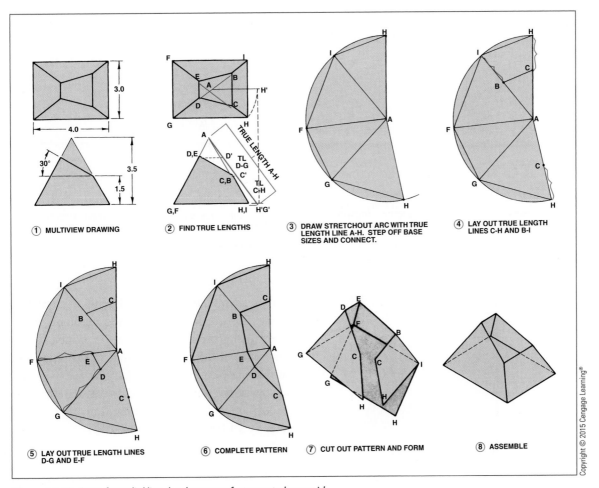

Figure 14-18 *Steps for radial line development of a truncated pyramid.*

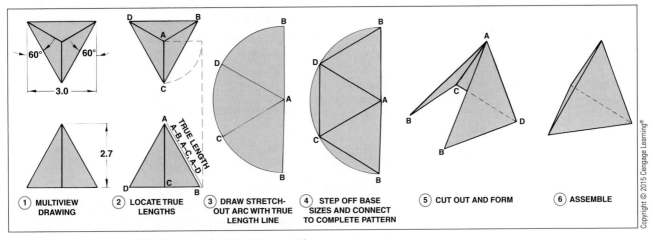

Figure 14-19 *Steps for radial development of a triangular pyramid.*

transfer these distances to the radial stretchout line. These points are then connected to the apex, forming triangular elements. There are eight elements in **Figure 14-20a**. More elements will create a smoother and more accurate cone.

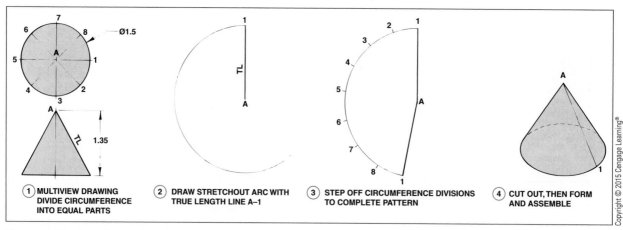

① MULTIVIEW DRAWING
DIVIDE CIRCUMFERENCE
INTO EQUAL PARTS

② DRAW STRETCHOUT ARC WITH
TRUE LENGTH LINE A–1

③ STEP OFF CIRCUMFERENCE DIVISIONS
TO COMPLETE PATTERN

④ CUT OUT, THEN FORM
AND ASSEMBLE

Figure 14-20a *Steps for radial line development of a cone.*

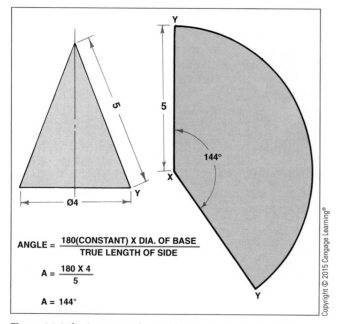

ANGLE = $\dfrac{180(\text{CONSTANT}) \times \text{DIA. OF BASE}}{\text{TRUE LENGTH OF SIDE}}$

A = $\dfrac{180 \times 4}{5}$

A = 144°

Figure 14-20b *Computing the stretchout angle for the development of a cone's pattern.*

A simple, symmetrically shaped cone, without cuts or intersections, can also be plotted mathematically by determining the angle of the stretchout and the circumference of the base. The angle of the stretchout is determined by multiplying the base diameter by 180, divided by the true length of the side. The length of the stretchout line is determined by multiplying the base diameter by 3.1416 (π) (**Figure 14-20b**).

TRUNCATED CONES

Truncated cones are developed like truncated pyramids. The top opening is divided into equal parts, since the truncated portion intersects a curved surface (**Figure 14-21**). These points are then projected to a side element to show the true lengths. After a true-length radial stretchout is drawn, each true distance from base to the intersection point is transferred to equivalent points on the stretchout arc. Connecting these points completes the pattern.

Triangulation Development

Parallel line and radial line developments are used to develop patterns for objects that contain flat or single-curved surfaces. These methods cannot be used to develop patterns for warped surfaces, such as spheres, paraboloids, or hyperboloids, including oblique pyramids, oblique cones, and square-to-round intersections. Since warped surfaces are curved in two directions, the even distribution of elements used in parallel and radial line development is not possible. The process of triangulation is used to develop patterns for warped surfaces.

Triangulation development is a method of dividing a warped surface into a series of triangles and transferring the true size of each triangle to a flat pattern. When flat triangulated patterns are bent or folded

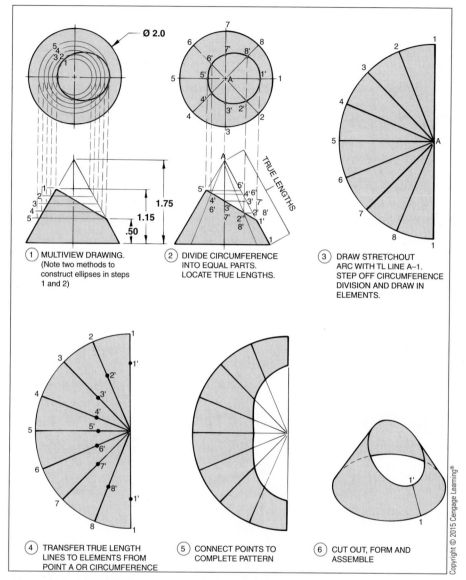

Figure 14-21 *Steps for radial line development of a truncated cone.*

into a three-dimensional form, the desired warped form is created. Triangulation is an approximate method of pattern development. The use of many small triangles creates a closer approximation of the desired form than a few large triangles.

The process of triangulation involves dividing the top and front views of an object into a series of triangles. This is done by dividing the top and bottom outlines into a series of parts and connecting them with element lines on both views. The intersections of element lines are evenly spaced on the top view. They converge at common points on the bottom view, creating a succession of triangles (**Figure 14-22**). Next, true lengths are determined for each element line. These distances are transferred progressively to the stretchout. In the development of triangulated patterns, one triangle at a time is added to the stretchout. Each element line is connected with the chordal divisions laid out on the circumference of the cone's base in the top view.

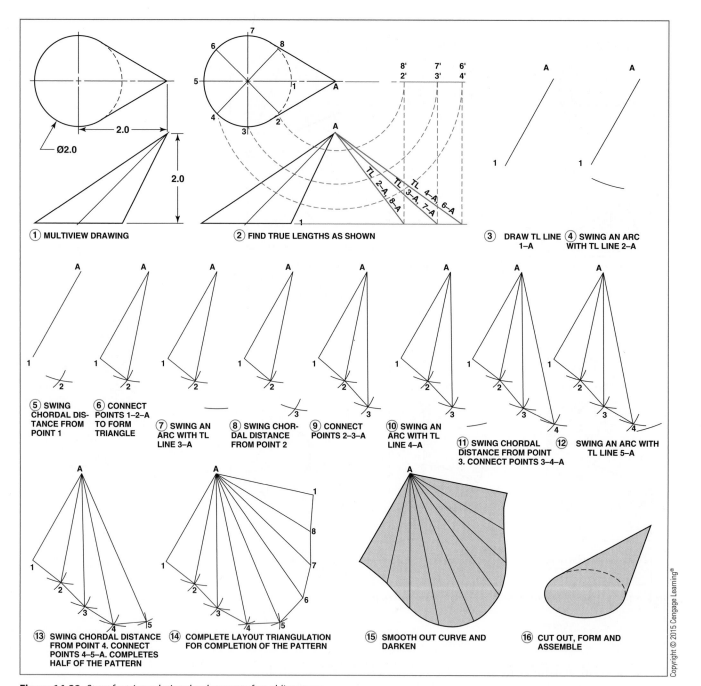

Figure 14-22 *Steps for triangulation development of an oblique cone.*

Triangulation can also be used to develop flat or single-curved surfaces. Since the process is very time-consuming, triangulation is normally used only when necessary. **Figure 14-22** shows the development of an oblique cone pattern through triangulation methods. **Figure 14-23** shows the triangulation of an oblique pyramid pattern.

When flat surfaces meet and the intersection lines are straight, the related pattern can be developed using parallel line development. When single-curved surfaces meet, the intersection lines can be developed

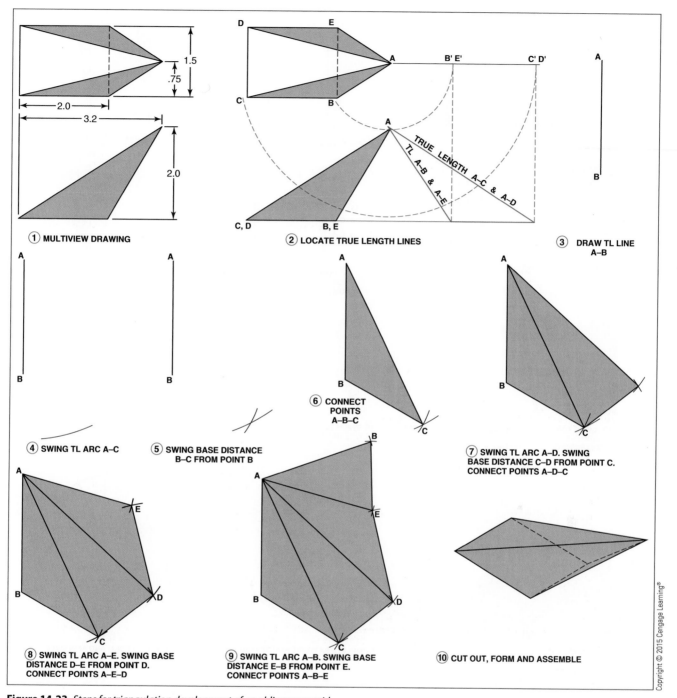

Figure 14-23 *Steps for triangulation development of an oblique pyramid.*

using radial line development methods. However, when warped surfaces meet, or when different geometric forms intersect at varying angles, triangulation must be used to develop the pattern for both parts. **Figure 14-24** shows the development of a pattern drawing for a square form intersecting a round form. This same procedure can be used to develop patterns for any intersecting geometric forms.

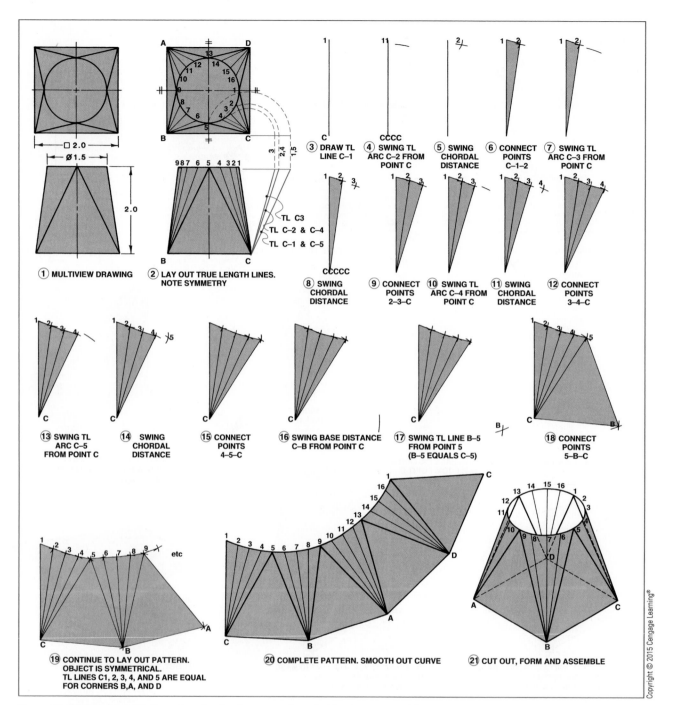

Figure 14-24 *Development of a pattern drawing for a square form that intersects a round form.*

DRAFTING EXERCISES

1. Sketch the approximate pattern of each item in **Figure 14-25** using parallel line development.

2. Sketch the approximate pattern of each item in **Figure 14-26** using radial line development.

3. Sketch the approximate pattern of each item in **Figure 14-27** using triangulation development.

4. With instruments or a CAD system, draw the patterns for **Figures 14-28** through **14-41**.

5. Make the following ECO (refer to "Engineering Change Order" in Chapter 17) changes for **Figure 14-42**.
   a. Change the vertex offset from 1.1" to 1.25".
   b. Increase the base diameter to 4.00".
   c. Change the overall height to 2.5".

6. Design a two-piece sheet metal pattern for a transitional fit in **Figure 14-43**.

7. Develop patterns 1–6 in **Figure 14-44** by parallel line development.

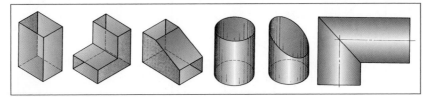

Figure 14-25 *Sketch an approximate parallel line pattern for each item.*
Copyright © 2015 Cengage Learning®

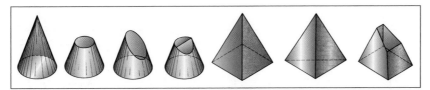

Figure 14-26 *Sketch an approximate radial line pattern for each item.* Copyright © 2015 Cengage Learning®

Figure 14-27 *Sketch an approximate triangulation pattern for each item.*
Copyright © 2015 Cengage Learning®

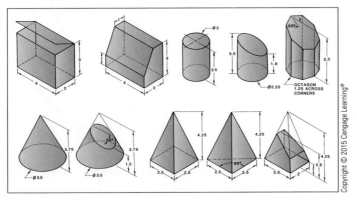

Figure 14-28 *Draw the pattern for each item.*

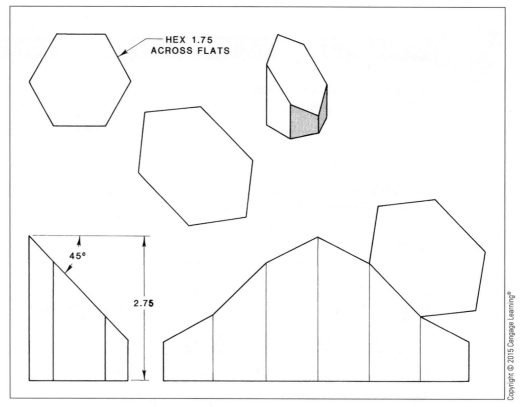

HEX 1.75
ACROSS FLATS

45°

2.75

Figure 14-29 *Draw the multiview, auxiliary view, and pattern drawings.*

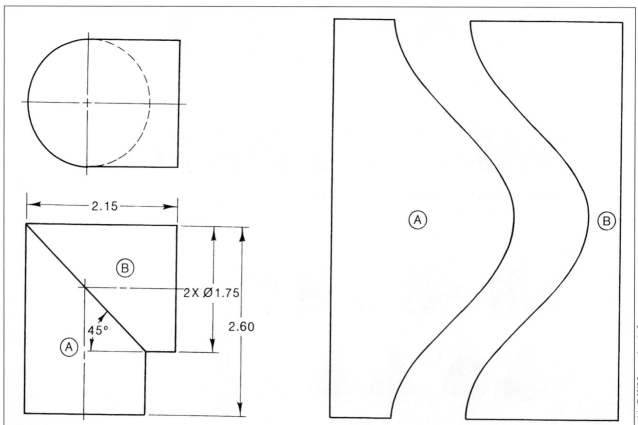

2.15

Ⓑ

2X Ø1.75

45°

2.60

Ⓐ

Ⓐ

Ⓑ

Figure 14-30 *Redraw the multiview drawing and patterns for the 90° sheet metal elbow.*

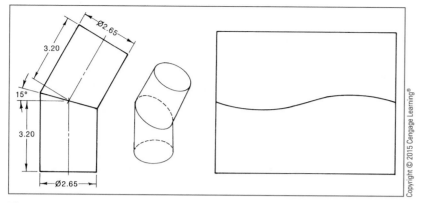

Figure 14-31 *Draw the multiview drawing and patterns for the elbow.*

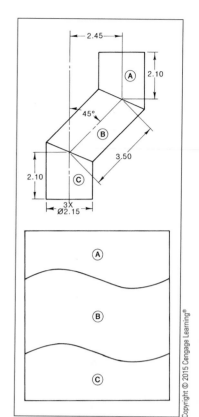

Figure 14-32 *Draw the working drawing and patterns for the three-piece elbow.*

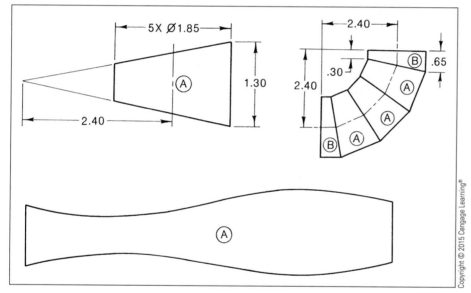

Figure 14-33 *Draw the working drawing and patterns A and B to make the five-piece elbow. Pattern A is already drawn.*

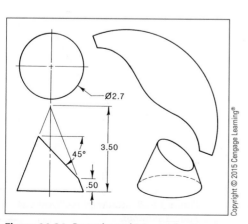

Figure 14-34 *Draw the multiview and pattern drawings.*

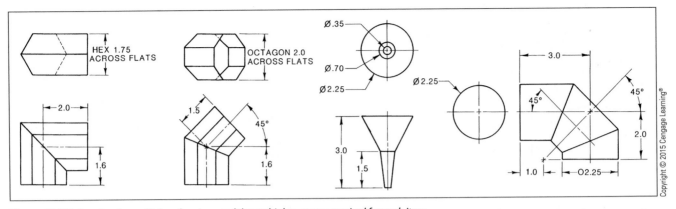

Figure 14-35 *Develop the multiview drawings and the multiple patterns required for each item.*

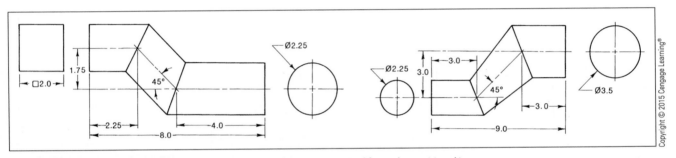

Figure 14-36 *Develop the multiview drawings and the multiple patterns required for each transitional item.*

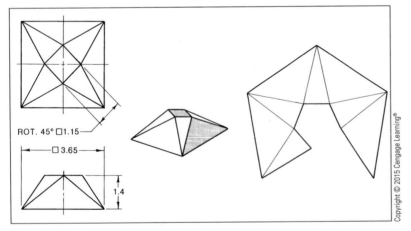

Figure 14-37 *Draw the multiview and pattern drawings.*

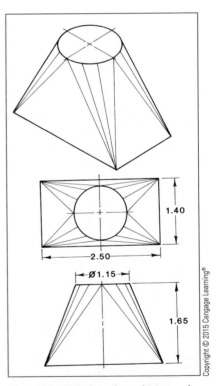

Figure 14-38 *Redraw the multiview and pattern drawings.*

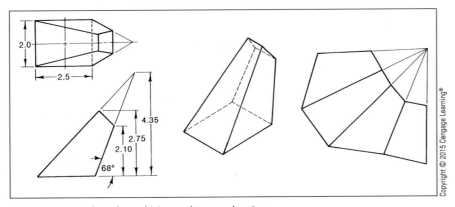

Figure 14-39 *Redraw the multiview and pattern drawings.*

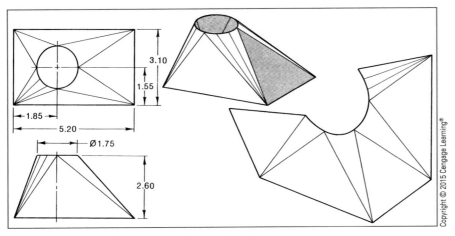

Figure 14-40 *Redraw the multiview and pattern drawings.*

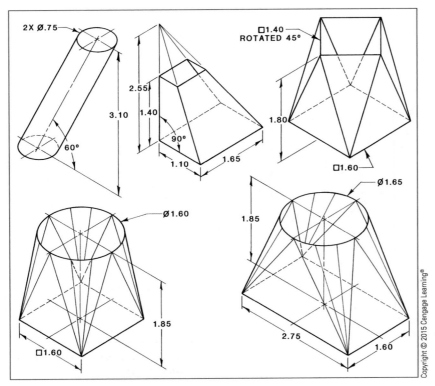

Figure 14-41 *Develop the multiview and pattern for each item.*

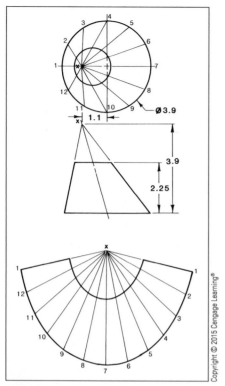

Figure 14-42 *Make the ECOs (refer to "Engineering Change Orders") as specified in Drafting Exercise 5, page 233, and complete the multiview and pattern.*

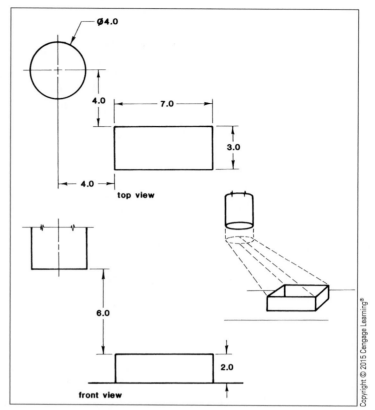

Figure 14-43 *Design a two-piece sheet metal pattern for a transitional fit.*

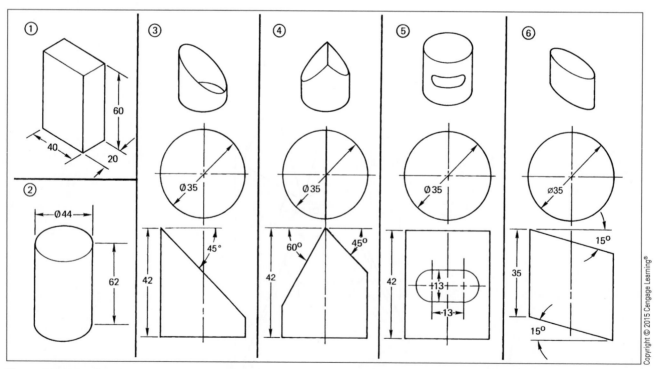

Figure 14-44 *Develop patterns 1-6 by parallel line development.*

Copyright © 2015 Cengage Learning®

DESIGN EXERCISES

1. Design a one-piece cardboard candy box with the dimensions of 4" × 2" × 6". Add the following design information to the outside surfaces:
   Company name
   Logo design

2. Design a heavy cardboard or wood storage case to hold 50 CD-ROMs in their individual containers. Draw and cut out a paper test pattern. The discs should be loosely held in position for easy access.

3. Design a cardboard carrying case for eight standard-size cans of soda.

4. Design a cardboard carrying case or envelope to carry an 18" × 24" drafting board, a 24" T square, and assorted drafting supplies, such as pencils, erasers, triangles, irregular curves, and compasses.

5. Design a sheet metal mailbox for the front curb. Draw a multiview working drawing and the required pattern drawings.

6. Design an automatic signaling device for a mailbox that will alert the residents that the mail has been delivered.

KEY TERMS

Chordal distance

Chord

Development drawing

Element

Flat surface pattern

Fold (bend) lines

Foreshortened

Hem

Pattern

Radial line development

Radial line

Stretchout arc

Stretchout lines

Tab lines

Triangulation development

True length

Warped surfaces

15

Pictorial Drawings

The student will be able to:

- Block in an isometric square
- Draw nonisometric lines
- Construct isometric circles
- Draw isometric circles with a template
- Draw isometric drawings with axes in various positions
- Center an isometric drawing
- Plot an isometric curve
- Draw isometric radii and rounds
- Draw various types of isometric sections
- Draw an isometric assembly drawing
- Draw an isometric exploded-view drawing
- Dimension an isometric drawing with the aligned system
- Dimension an isometric drawing with the unidirectional system
- Sketch an isometric drawing on isometric grid paper
- Draw a dimetric drawing
- Draw a trimetric drawing
- Draw an oblique cabinet drawing
- Draw an oblique cavalier drawing
- Draw circles and curves on an oblique drawing
- Draw simple pictorial drawings on a CAD system
- Draw a one-, two-, and three-point perspective drawing

Introduction to Pictorial Drawings

Pictorial drawings have been used throughout history to communicate ideas that cannot be described easily with words. Early attempts to show depth and detail were very crude and often distorted the appearance of objects. **Figure 15-1** shows primitive pictorial drawings that lack correct proportion and realism. The methods and techniques used to prepare realistic pictorial drawings have been continually refined through the years. In addition to their function as an art form, pictorial drawings are used today by engineers, architects, designers, and drafters to illustrate the size and shape of an object. The pictorials in **Figure 15-2** clearly defines the shapes of each manufactured item.

Pictorial drawings are multidimensional; that is, they show more than one side of an object (**Figure 15-3**). They can show width, height, and length in one drawing. For this reason, they are easier to understand and interpret than orthographic multiview drawings, which shows only one two-dimensional face of an object in each view. Since pictorial are not always dimensionally accurate, they are rarely used as detailed working drawings for manufacturing or construction purposes.

Figure 15-1 *Primitive drawings usually appeared two-dimensional and distorted.*

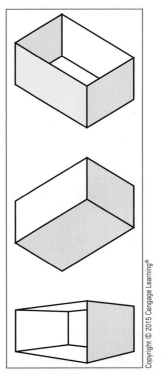

Figure 15-3 *Pictorial drawings will show two or more sides on each drawing.*

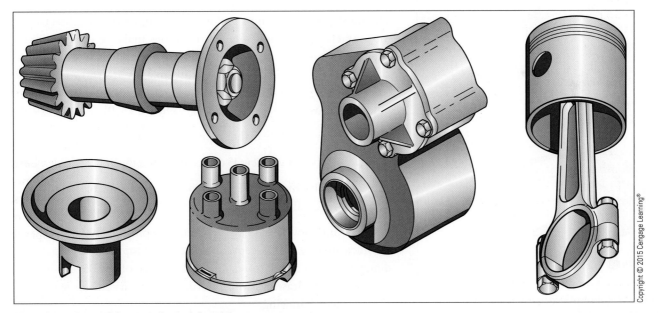

Figure 15-2 *Pictorial drawings clearly define objects.*

In engineering and architecture, pictorial drawings are used to improve the visualization skills of nontechnical personnel who may have difficulty reading orthographic multiview drawings. Pictorial drawings are used extensively:

- As ideas recorded in the design process
- To clarify assembly and repair instructions
- For new product sales training
- For sales promotion

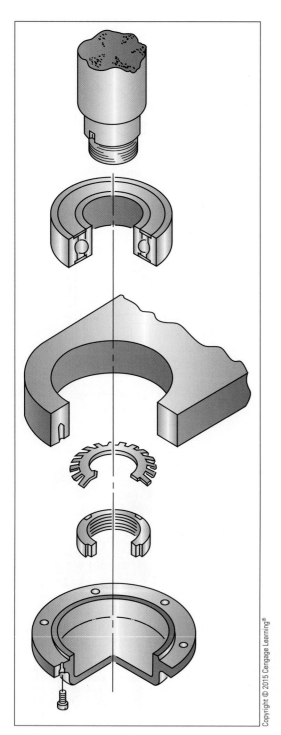

Figure 15-4 *An exploded pictorial will show all of the item's parts and assembly process.*

Copyright © 2015 Cengage Learning®

Once a product is manufactured, a photograph may be used for these purposes. Only pictorial drawings or models can be used to show the appearance of a product before manufacture. Unlike photographs, pictorial drawings can be used to create **sectional** or **exploded views** that show details or the interior of an object (**Figure 15-4**). Presenting a pictorial drawing with a two-dimensional working drawing will clearly define the external parts of the air-pressure valve in **Figure 15-5**.

In the fields of engineering and architectural, there are three types of pictorial drawings: **axonometric**, **oblique**, and **perspective** (**Figures 15-6** and **15-7**).

Axonometric drawings are the most frequently used for engineering drawings. They may be drawn as an **isometric**, **dimetric**, or **trimetric**. The receding lines in these drawings are always parallel.

Oblique drawings are used mostly for simple thin objects or for progressive design sketches. They may be drawn as a **cavalier** or **cabinet** drawing.

Perspective drawings are used extensively to create realistic renderings of objects usually for sales purposes. They may be drawn as a one-point, two-point, or three-point perspective.

Isometric Drawings

The most popular form of pictorial drawing used in engineering is the **isometric drawing**. Isometric drawings makes it easier to draw, add details, and add dimensions than with the other types of pictorial drawings. This is because the three planes are shown true size (**Figure 15-8**). To visualize an object in a true isometric form, the following drawing steps are necessary (**Figure 15-9**).

Step 1. Draw the three orthographic multiviews.

Step 2. Revolve the top view 45 degrees and draw the front and side views.

Step 3. Revolve the side view in Step 2, 35 degrees. Project and draw the remaining two views.

The front view will be a true isometric projection. Note that all the lines are true size, except for the revolved top view in Step 3. The dimensions will appear foreshortened because of the 35° revolution. Most isometric drawings will use true size dimensions for speed and convenience of drawing.

ISOMETRIC AXIS

The **isometric axis** is the basis for all isometric drawings. This axis consists of three lines, 120° apart. These three lines represent the front corner and the two receding edges of an object (**Figure 15-10**). Although the three axis lines are always 120° apart, the axis can be positioned to show either the top, bottom, or side views as the dominant view (**Figure 15-11**).

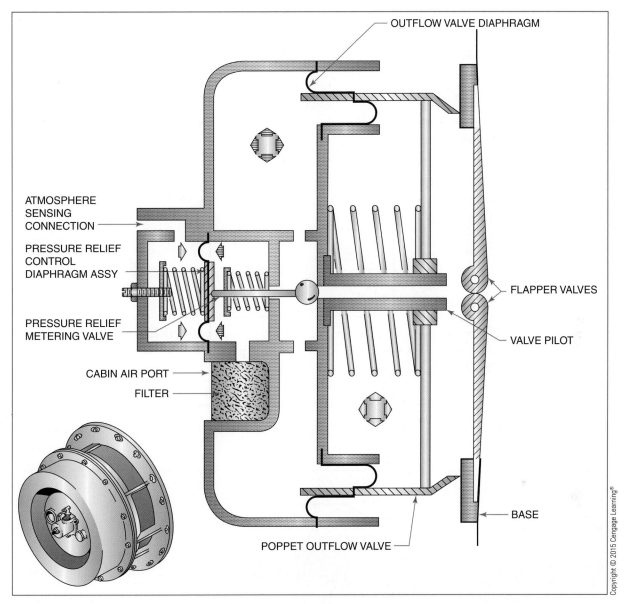

Figure 15-5 *A 2D and 3D drawing will clearly define this air-pressure valve.*

Labels in figure:
OUTFLOW VALVE DIAPHRAGM
ATMOSPHERE SENSING CONNECTION
PRESSURE RELIEF CONTROL DIAPHRAGM ASSY
PRESSURE RELIEF METERING VALVE
CABIN AIR PORT
FILTER
FLAPPER VALVES
VALVE PILOT
BASE
POPPET OUTFLOW VALVE

ISOMETRIC LINES

All lines drawn parallel to any of the isometric axis lines are **isometric lines**. These lines are always true size. To prepare an isometric drawing, all isometric lines are drawn 30° or vertical from the horizontal (**Figure 15-12**).

NONISOMETRIC LINES

As you learned in developing auxiliary views, objects may contain lines that are not parallel to the major axes of the object. These lines, representing sloping surfaces, are called **nonisometric lines**. Nonisometric lines are

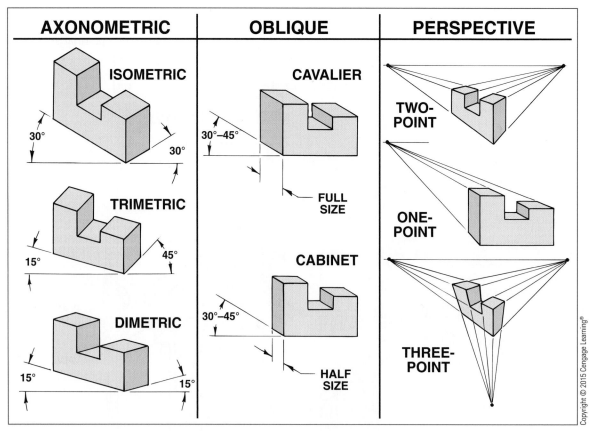

Figure 15-6 *Various types of pictorial drawings.*

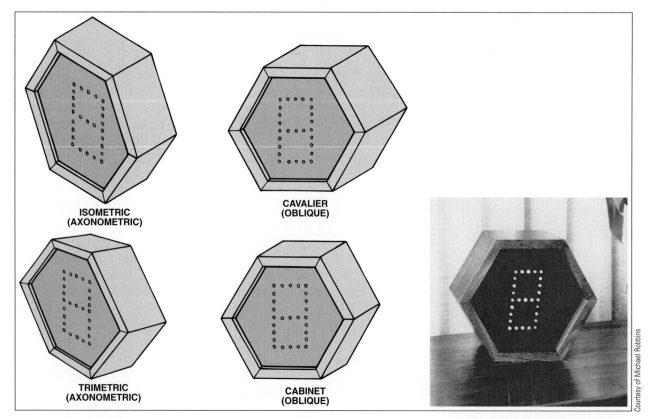

Figure 15-7 *Various types of pictorial representation of a digital clock.*

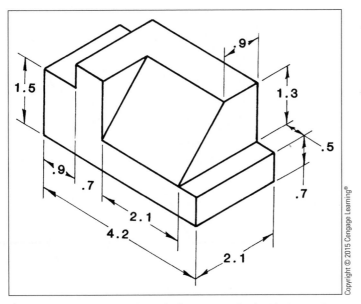

Figure 15-8 *This isometric shows the front, top, and right sides in true size dimensions.*

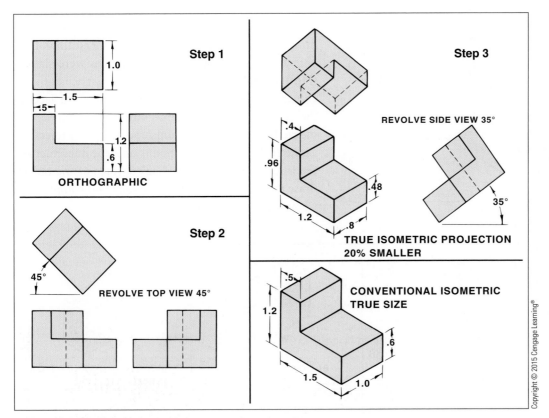

Figure 15-9 *The true isometric and the conventional isometric drawings.*

lines that are not parallel to isometric lines. The angle and length of nonisometric lines can only be established by locating points of intersection with isometric lines. This is because actual measurements can only be made on true-size isometric lines.

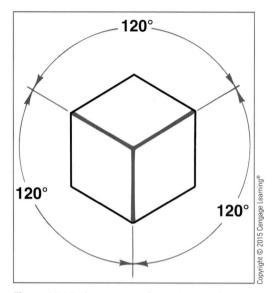

Figure 15-10 *The major axes for an isometric drawing are spaced at 120°.*

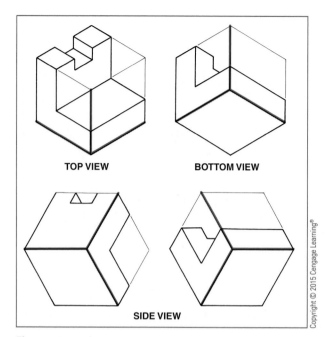

Figure 15-11 *Alternate positions for isometric axes.*

Before drawing nonisometric lines, first lightly draw all isometric lines, completing the outline of the object as it would appear if no surfaces were slanted (**Figure 15-13**). Next, transfer the distances of each corner from an orthographic view to one of the faces of the isometric drawing. Connect these points on the isometric lines to form the nonisometric lines. Project 30° isometric lines from these points until the lines intersect the lines representing the back surface of the object. Erase the isometric construction lines and complete the isometric drawing as shown.

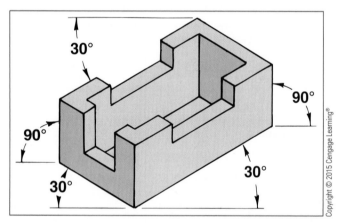

Figure 15-12 *Isometric lines are 30° from the horizontal.*

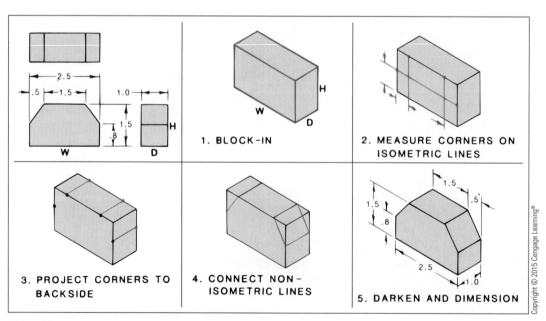

Figure 15-13 *Procedural steps to draw non-isometric lines.*

CENTERING ISOMETRIC DRAWINGS

To center an isometric drawing, follow the steps in **Figure 15-14**:

Step 1. Check the overall dimensions of the object to be drawn.

Step 2. Outline the area needed for the drawing.

Step 3. Locate the center of the area.

Step 4. Draw a vertical line upward from the center of the area equal to one-half the height of the object (.75").

Step 5. From the top of this line, draw a line 30° downward to the right, equal to one-half the length (1.25") of the object.

Step 6. From the right end of this line, draw a line 30° downward to the left, equal to one-half the thickness of the object (.5"). The left end of this line represents the front, top corner (X) of the object.

Step 7. Extend a light vertical line down from corner X to represent the front edge of the object. The length of this line is the height of the object at its highest point (1.5"). Next, extend 30° lines upward to the left and right to represent the top-right and top-left edges of the object. This creates the isometric axis. These lines should be true size overall dimensions of the object. Now, block in all remaining isometric lines. The blocked-in isometric will be centered horizontally and vertically.

Step 8. Complete the isometric drawing. **Figure 15-15** reviews the steps to draw an isometric drawing.

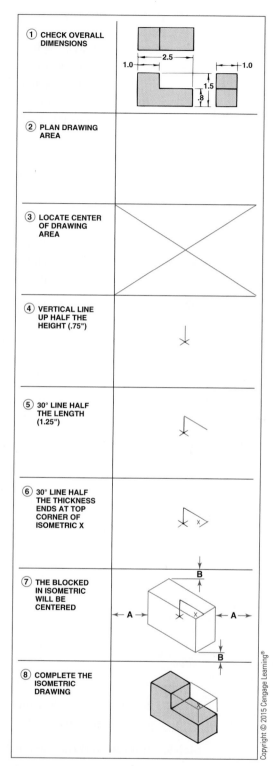

Figure 15-14 *Steps to center an isometric drawing.*

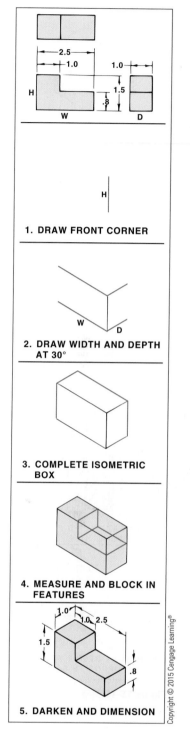

Figure 15-15 *Steps in drawing an isometric drawing.*

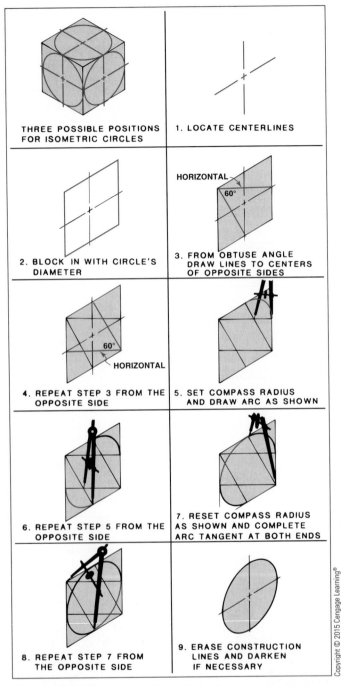

Figure 15-16 *Steps in constructing an isometric circle.*

ISOMETRIC CIRCLES

Circles will appear as **ellipses** on isometric drawings because they are located on receding planes that are not perpendicular to the line of sight. Ellipses are oriented differently on each of the three isometric planes (**Figure 15-16**). To draw **isometric circles** (which are actually 35° ellipses), follow the steps outlined in **Figure 15-16**:

Step 1. Locate the center of the circle, then draw centerlines that are parallel to the isometric lines of the object.

Step 2. Establish the size of the circle by drawing an isometric square with sides equal to the diameter of the circle.

Step 3. Draw two lines from the vertex of the obtuse angle of the square to the point where the centerlines intersect the opposite edges of the square. One line will be horizontal and the other will be 60°.

Step 4. Repeat Step 3 in reverse from the obtuse angle on the opposite side of the square.

Step 5. Place the point of a compass on the intersection of these two lines. Set the compass opening to align with the intersection of the centerline and the outer edge of the square. Draw an arc from this point to the corresponding point on the adjacent side of the square.

Step 6. Repeat Step 5 on the opposite end of the square.

Step 7. Place the compass point on the obtuse angle corner of the square. Draw a large arc on the opposite side to connect with the edges of the small axis drawn in Steps 5 and 6. Be certain to draw smooth tangent points.

Step 8. Repeat this procedure on the opposite side of the square to complete the isometric circle.

Step 9. Erase all construction lines.

FILLETS AND ROUNDS

Fillets are continuous internal (concave) arcs. **Rounds** are continuous external (convex) arcs on the surface of an object. Since arcs are partial circles, to draw an isometric arc, draw part of an isometric circle. To visualize this, observe the circles drawn in the corners of the object in **Figure 15-17**. To draw isometric arcs, use the same procedure as drawing isometric circles on the three different planes. Draw only the part of the circle that describes the fillets or rounds.

ELLIPSE TEMPLATES

Ellipse templates eliminate much of the tedious work of drawing isometric circles and arcs. An **ellipse template** (**Figure 15-18**) is the most efficient method of drawing isometric circles. The isometric ellipse template uses 35° ellipses. To use an isometric template, first locate and draw the intersecting centerlines of the circle. Choose the correct isometric circle size. Align the marks on the edge of the isometric template opening with the centerlines on the drawing. Trace around the inside of the hole with a pencil.

ELLIPSES ON NONISOMETRIC PLANES

The template and construction method of drawing ellipses can only be used on isometric planes that are at 30°. These methods cannot be used on nonisometric planes that do not align with the isometric axis. Ellipses, or any projection or depression on a nonisometric plane, may be plotted directly from a normal orthographic view.

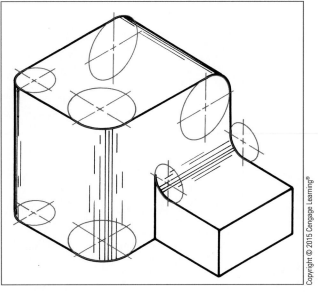

Figure 15-17 *Positioning of isometric circles for fillets and rounds.*

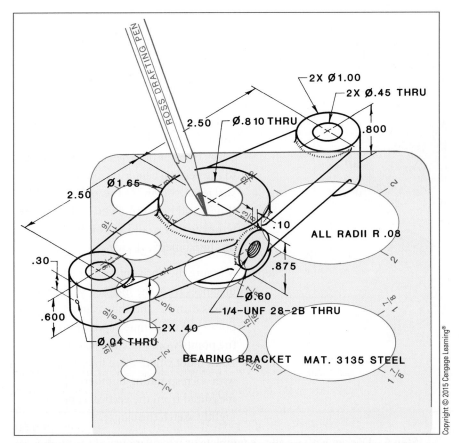

Figure 15-18 *Line up the diagonals on the isometric template's circle on the drawing's isometric centerlines, and trace around to complete the isometric circle.*

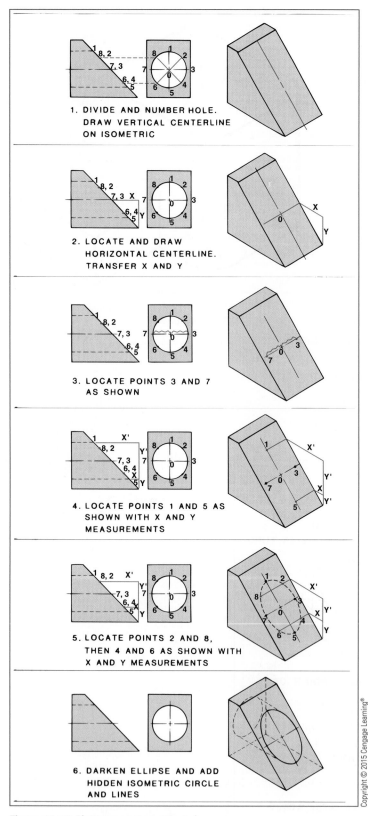

Figure 15-19 *Plotting a nonisometric circle.*

To plot ellipses on nonisometric surfaces, first prepare an orthographic two-view drawing. Show the sloped surface profiled on one view, and the true shape of the circle on the other. Follow the steps outlined in **Figure 15-19**:

Step 1. Divide the circle into eight or more equal parts as shown on the side view. Number each equally spaced mark on the circumference of the true circle. The accuracy and smoothness of the final ellipse will increase as the number of circle divisions is increased. Project these points horizontally to intersect the line representing the edge of the inclined plane on the front view. Label the points of intersection on the profile plane. Draw the isometric form. Draw a centerline representing the vertical center of the ellipse on the inclined surface.

Step 2. Measure the X and Y distances to find the position of the horizontal centerline on the inclined surface. Label the intersection of the vertical and horizontal centerline with a 0.

Step 3. Measure the distance from 0 to 3, and from 0 to 7 on the front view. Transfer these distances to the horizontal centerline on the inclined surface as shown.

Step 4. Determine the distance from 0 to 1, and from 0 to 5 on the side view sloping edge. Transfer or project these distances to the vertical centerline on the inclined surface as shown. At this point, much time could be saved if you had an ellipse template that would fit points 1-5 to 3-7 on the isometric drawing.

Step 5. Measure the horizontal and vertical X-Y distances from the inclined surface for points 8 and 2, and then 6 and 4. If more points are used, lines representing all of the points would be drawn at this time. Draw a dot where these lines intersect with corresponding numbered lines measured from the vertical centerline. Lightly sketch the ellipse.

Step 6. Darken with an ellipse template or irregular curve to form the ellipse as shown.

PLOTTING CURVES

Curved nonisometric lines are drawn by transferring points from an orthographic view to an isometric plane, as in the following steps (**Figure 15-20**):

Step 1. Draw two orthographic views of the object containing the curved surface.

Step 2. Draw random horizontal and vertical lines covering the curved areas to form a square grid pattern. The closer the grid line spacing, the greater the accuracy of the final drawing. Mark each point of intersection between the grid lines and the curved line with a dot.

Step 3. Draw the isometric plane containing the curved areas. Draw the same grid spacing on the isometric plane. Transfer the position of the dots from the orthographic view to similar points on the isometric grid.

Step 4. Connect the points with a smooth, freehand-sketched line.

Step 5. From each point on the curve, project a 30° line upward to the right.

Step 6. Measure and transfer the thickness of the object at each point, to points representing the back side of the curved surfaces. If the thickness of the object is the same throughout, these distances will all be equal.

Step 7. Connect the points on the back surface of the curved areas with a freehand-sketched line to complete the layout.

Step 8. Complete the drawing with instruments and an irregular curve.

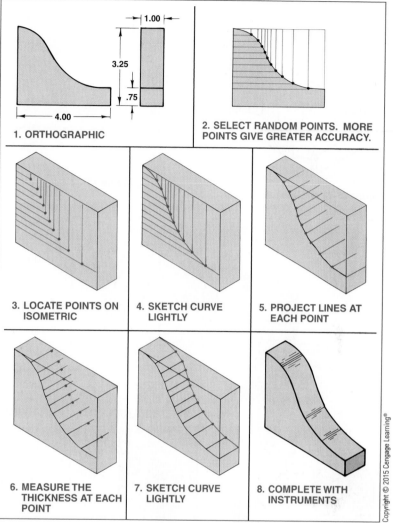

Figure 15-20 *Plotting isometric irregular curves.*

ISOMETRIC SECTIONS

Isometric drawings are normally used to show only the exterior appearance of an object. When the inside of an object must be shown in pictorial form, an isometric section is also prepared. The two basic types of isometric sections are the full isometric section and the half isometric section.

Isometric sections, like orthographic sections, follow a cutting plane line that defines the edge of the area to be sectioned. In a **full isometric section**, the cutting plane extends across the entire width and height of an object (**Figure 15-21**). A full isometric section cuts the object in half to show its inside half. In a **half isometric section**, the cutting plane extends only halfway through the object. Half-sections show only a one-fourth **cutaway view** of the object (**Figure 15-22**).

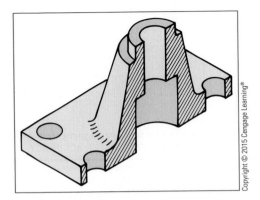

Figure 15-21 *Full isometric section.*

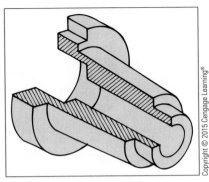

Figure 15-22 *Half isometric section.*

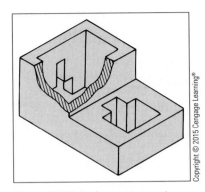

Figure 15-23 *Broken-out isometric section.*

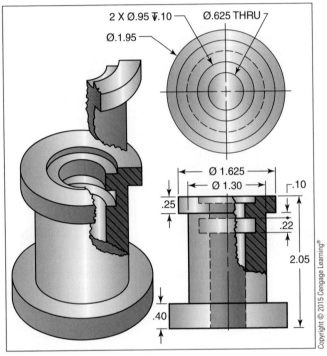

Figure 15-24 *A broken-out section will define interior details.*

Symmetrical objects are usually half-sectioned to their centerline. When objects are not symmetrical, or do not contain a centerline, a separate broken-out isometric section may be prepared. In **Figure 15-23**, part of the wall is removed to reveal details on the back surface. **Broken-out sections** are also used to show the shape of an object that may be difficult to interpret from the isometric drawing. The drawing of a shaft bearing includes a two-view multiview drawing and a pictorial broken-out section showing the interior details (**Figure 15-24**).

Sometimes a sectional view is removed to show the shape of an object without changing the form of the isometric drawing (**Figure 15-25**). There are two methods of drawing isometric sections. The first involves drawing the complete object lightly (**Figure 15-26**). The cutting plane line or outline of the cutaway area is drawn. Lines representing the cutaway are erased to show the section. In the second method, only the sectioned area part is drawn (**Figure 15-27**).

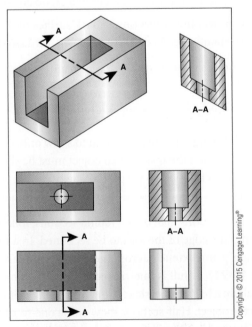

Figure 15-25 *Isometric removed section.*

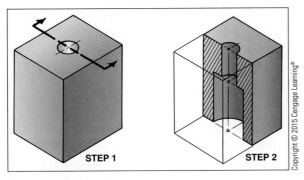

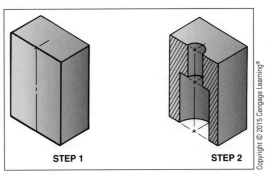

Figure 15-26 *Isometric section.*

Figure 15-27 *Lay out only the isometric section.*

ISOMETRIC EXPLODED DRAWINGS

Isometric drawings are ideal for showing various parts of a multipart component in related position (**Figure 15-28**). Drawings of this type are exploded-view drawings because the parts are separated (disassembled). The parts are aligned and related to their assembled position. Exploded views allow each part to be drawn showing its full shape. This eliminates hidden areas that do not show when the parts are assembled.

Exploded-view assembly drawings also allow for the labeling and identification of parts (**Figure 15-29**). In very complex assembly drawings, connecting centerlines show the relationship of parts when assembled (**Figures 15-30** and **15-31**). This technique provides a guide for the assembly of parts to produce the finished component. Exploded views may be accompanied by an assembly drawing of the component with the parts all assembled (**Figure 15-32**).

ISOMETRIC DIMENSIONING

Since isometric drawings are not normally used as working drawings, detailed dimensions are not found on most of them. However, when isometric drawings of very basic objects are used for manufacturing or construction purposes, dimensions, notes, and specifications are included.

Dimension lines, extension lines, arrowheads, letters, and numerals on isometric drawings are similar to those on orthographic drawings. Isometric dimensions and extension lines all follow the isometric axis. Arrowheads are also drawn in the same plane as dimension and extension lines (**Figure 15-33**).

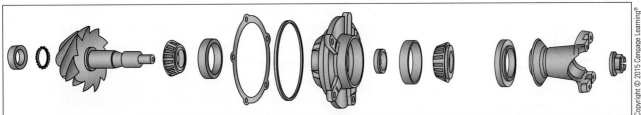

Figure 15-28 *Example of an exploded-view isometric drawing used for assembly instruction.*

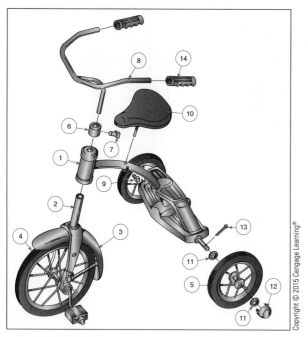

Figure 15-29 *Exploded views are used for labeling and assembly.*

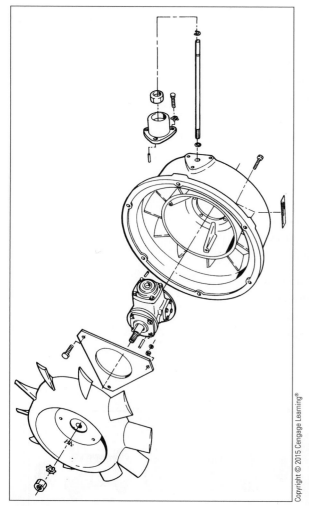

Figure 15-30 *Centerlines are used to align mating parts of an assembly.*

There are two methods of dimensioning isometric drawings: the aligned method and the unidirectional method. In **aligned dimensioning**, all isometric letters, numerals, and arrowheads are aligned with isometric planes (**Figure 15-33**).

Unidirectional dimensioning (**Figure 15-34**) is generally preferred for engineering working drawings. In the unidirectional method, all isometric numerals and letters are positioned vertically. A variation of unidirectional dimensioning is vertical plane dimensioning (**Figure 15-35**). In this method, all dimension and extension lines are drawn in the isometric plane, with the numerals placed horizontally. Straight horizontal lettering is simpler and faster to use on isometric drawings.

ISOMETRIC GRIDS

To save time in preparing isometric drawings, isometric grid paper is often used (**Figure 15-36**). The grid lines are printed on 30° angles representing the isometric axis. Grid lines are often printed in unreproducible blue. This enables an isometric drawing to be prepared directly on the grid paper without the grid lines appearing on the final print. The grid paper may be used as an underlay guide in preparing isometric drawings. The use of grids is a great timesaver for sketching and isometric layout work.

DIMETRIC DRAWINGS

In dimetric drawings, the receding angles may vary, but two of the three faces and axes are always equal (**Figure 15-37**). The method of drawing dimetric pictorials is similar to preparing isometric drawings. Once the axis angles are established, all lines extending in the same direction must be kept parallel.

TRIMETRIC DRAWINGS

In isometric drawings, all three angles around the isometric axis are the same. In dimetric drawings, two of the angles are the same. In trimetric drawings, all three of the angles around the axis are different (**Figure 15-38**). Trimetric drawings are prepared like isometric and dimetric drawings, except that each pictorial plane is projected on a different angle. All lines in each of the three planes must be parallel. This three-angle difference makes the preparation of trimetric drawings very time-consuming. Trimetric drawings are also difficult to dimension since each dimension and extension line must be drawn parallel to each of the three different planes of projection.

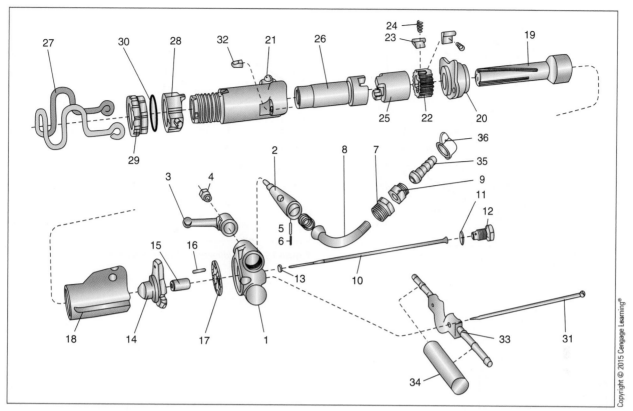

Figure 15-31 *Exploded views are necessary for complex assemblies.*

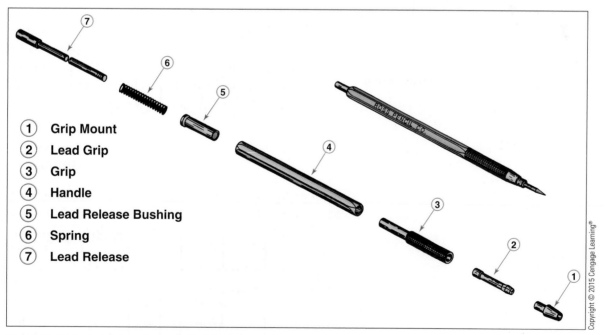

(1) **Grip Mount**

(2) **Lead Grip**

(3) **Grip**

(4) **Handle**

(5) **Lead Release Bushing**

(6) **Spring**

(7) **Lead Release**

Figure 15-32 *Exploded view of a mechanical drafting pencil.*

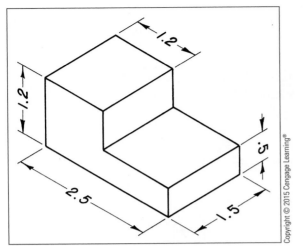

Figure 15-33 *Aligned isometric dimensions.*

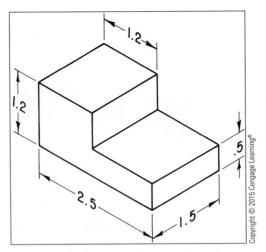

Figure 15-34 *Unidirectional isometric dimensions.*

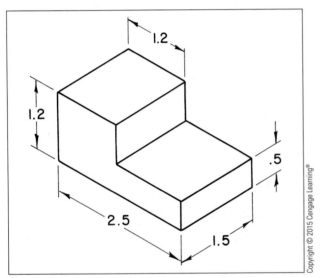

Figure 15-35 *Vertical plane dimensioning on an isometric drawing.*

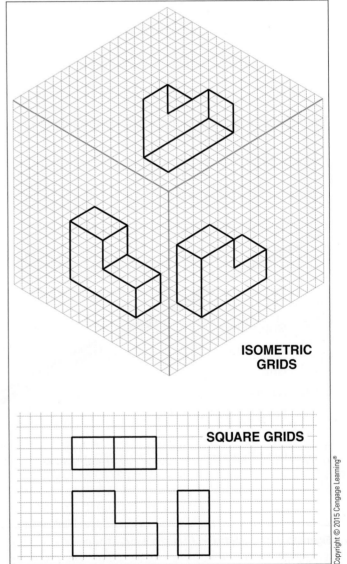

Figure 15-36 *Grid papers.*

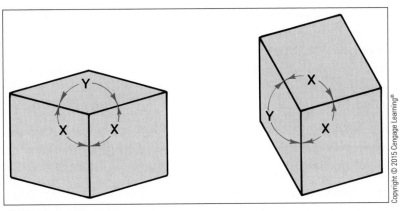

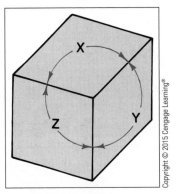

Figure 15-37 *In a dimension drawing the angles may vary, but two of the angles must be equal.*

Figure 15-38 *In a trimetric drawing angles may vary, but no angles will be equal.*

The angles of trimetric drawings, and to a lesser extent dimetric drawings, can be adjusted to any convenient angle to provide more realism and emphasize one side while making another side recede. This is closer to the way objects are normally viewed in isometric drawings.

Oblique Drawings

In axonometric drawings (isometric, dimetric, and trimetric), the object is drawn as viewed from its nearest edge. In oblique drawings, the object is drawn as viewed from its nearest surface. In oblique drawings, the object is drawn as viewed from one of its orthographic faces. The front face of an oblique drawing is parallel to the front plane of projection. This means that the front face is drawn to exact size and shape. The other two surfaces are projected from this front face at any convenient angle (usually 30° or 45°) and in any direction (**Figure 15-39**). The angle and direction depend on the nature of the object and the planes to be emphasized. There are two types of oblique drawings: cavalier and cabinet.

CAVALIER DRAWINGS

When the lengths of the receding angles of an oblique drawing are drawn to full scale, the drawing is an oblique cavalier drawing (**Figure 15-40**). If the length of the receding lines is excessive, a cavalier drawing will appear distorted.

Because the front face is both geometrically and dimensionally accurate, the front face can be dimensioned the same as in an orthographic drawing. The receding planes can be dimensioned as in isometric drawings with the dimension and extension lines drawn parallel to the angle of the plane.

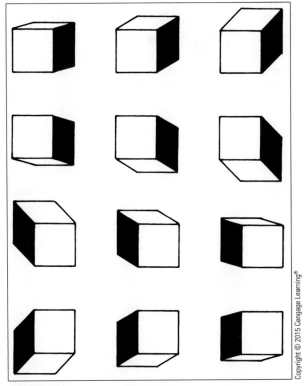

Figure 15-39 *Alternate positions for oblique drawings. Receding lines are usually at 15°, 30°, or 45°.*

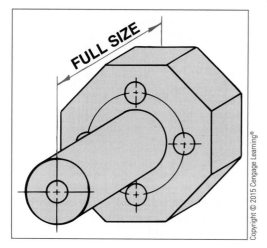

Figure 15-40 *Example of an oblique cavalier drawing. All dimensions are full size.*

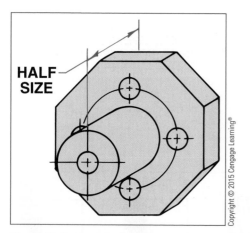

Figure 15-41 *Example of an oblique cabinet drawing. Only receding lines are drawn at half size.*

Cavalier drawings are especially appropriate for objects with a shallow depth dimension, since depth distortion is minimized. The largest area is usually selected as the front plane to avoid as much distortion as possible on the receding planes. Since the front plane of a cavalier drawing is geometrically accurate, the front face should contain the greatest amount of detail or irregularity.

CABINET DRAWINGS

To reduce the amount of visual distortion, long receding lines are sometimes reduced by one-half, but dimensioned true size. When this is done, the drawing is known as an oblique cabinet drawing (**Figure 15-41**). This creates a more realistic drawing but eliminates the use of true scale dimensions on the receding planes.

Figure 15-42 shows a comparison of the types of pictorial drawings. The drafter must be familiar with all pictorial drawing methods in order to select the best possible type for a set of working drawings.

OBLIQUE CIRCLES

Because the front face of an oblique drawing is not distorted, angles, circles, and arcs drawn on this surface are true and accurate. For this reason, the front face is chosen to show the circular end view of a cylinder hole or other curved surface (**Figure 15-43**). Circles and arcs shown on the receding planes of an oblique drawing will appear as ellipses. This is similar to the way they appear on axonometric receding planes (**Figure 15-44**).

Both isometric circles and oblique circles on receding planes are drawn as ellipses. When plotting circles on oblique receding planes, follow the steps in **Figure 15-45**.

The accuracy of the ellipse on an oblique drawing is not critical. A faster method of drawing an ellipse by estimation is shown in **Figure 15-46**.

OBLIQUE SECTIONS

Oblique full and half sections are drawn in a similar fashion to isometric sections. When oblique sections are drawn, they are usually cut through the receding plane (**Figure 15-47**). To draw an oblique section, draw the front orthographic view. Project the receding lines, beginning at the points

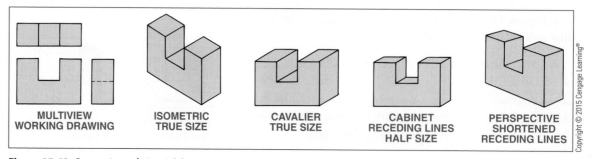

| MULTIVIEW WORKING DRAWING | ISOMETRIC TRUE SIZE | CAVALIER TRUE SIZE | CABINET RECEDING LINES HALF SIZE | PERSPECTIVE SHORTENED RECEDING LINES |

Figure 15-42 *Comparison of pictorial drawings.*

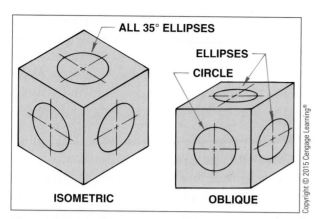

Figure 15-43 *Isometric and oblique circles.*

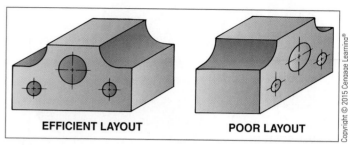

Figure 15-44 *Layout of isometric and oblique circles.*

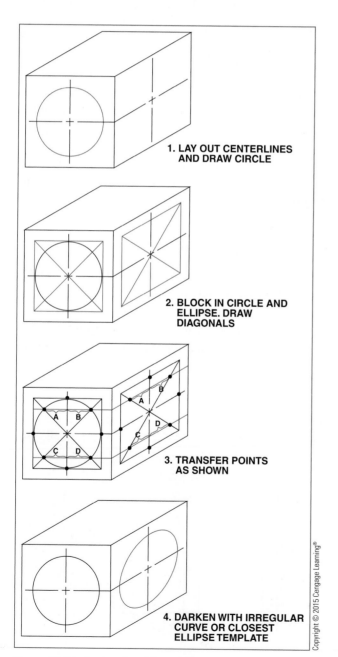

Figure 15-45 *Constructing oblique circles.*

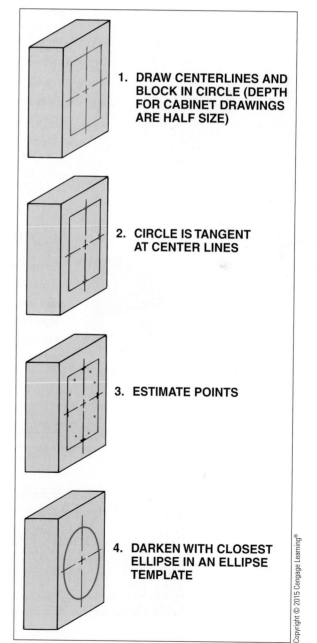

Figure 15-46 *Estimation procedure for drawing an ellipse.*

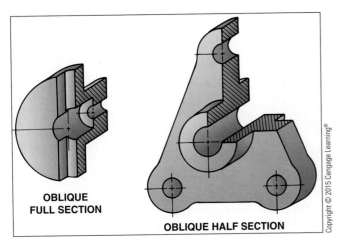

OBLIQUE FULL SECTION

OBLIQUE HALF SECTION

Figure 15-47 *Oblique sections.*

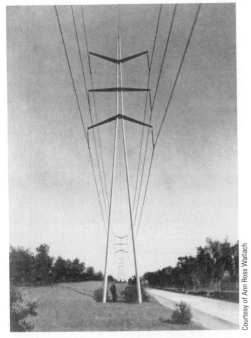

Figure 15-48 *Objects, in a two-dimensional illustration, will appear smaller as they approach the vanishing point.*

of intersection between the cutting plane line and front plane lines. Extend these lines on the established angle to the depth of each part of the object.

Perspective Drawings

The receding sides of axonometric and oblique drawings are parallel. This creates visual distortion because the human eye does not perceive objects in this manner. It sees close-up objects in true size, detail, and color. Distant objects are seen as smaller, less detailed, and paler in color. The combination of these factors gives the eye the ability to perceive the dimension of depth. Look at the perspective photo in **Figure 15-48**. Notice how the lines of the road are not parallel in the picture. Yet, in reality the two sides of the road are parallel. The camera, like the eye, sees objects in perspective. That is, parts of the object further from the eye appear to get progressively smaller as the viewing distance is increased. The reason objects appear to grow smaller and closer is the curvature of the earth. Therefore, on a perspective drawing the receding lines will converge to a vanishing point on the horizon. Perspective drawings, unlike axonometric drawings, are prepared with the more distant areas proportionally smaller to create the illusion of depth (**Figures 15-49** and **15-50**).

PERSPECTIVE USES

As previously mentioned, axonometric and oblique drawings of large objects appear greatly distorted because the large receding sides are parallel. For this reason, perspective drawings are often prepared for large objects, such as buildings, cars, boats, and machinery. **Figure 15-51** is a perspective drawing of a house. Notice the increased realism found in the perspective drawing.

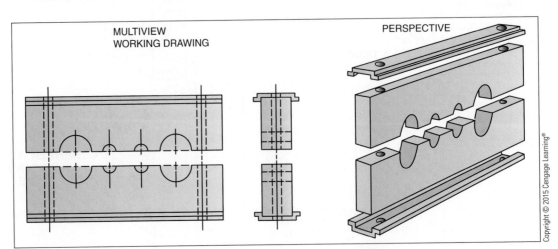

MULTIVIEW WORKING DRAWING

PERSPECTIVE

Figure 15-49 *Machined parts are drawn in perspective will appear realistic to the eye.*

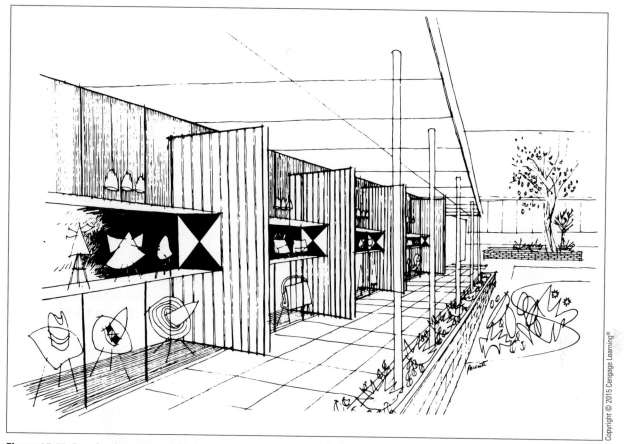

Figure 15-50 *Receding lines to a vanishing point, in an illustration creates the illusion of depth, as in a photograph.*

Figure 15-51 *Pictorials of large structures drawn in perspective will increase its realism to the eye.*

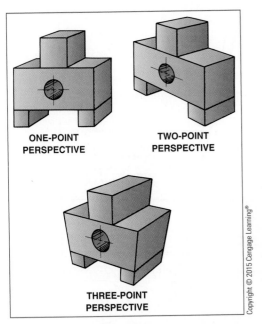

Figure 15-52 *Examples of one-, two-, and three-point perspective drawings.*

PERSPECTIVE TYPES

There are three basic types of perspective drawings: one-, two-, and three-point perspective (**Figure 15-52**). The type of drawing used depends on the size and shape of the object and the preferred viewing angle.

ONE-POINT PERSPECTIVE

A **one-point perspective** (parallel) drawing is similar to an oblique drawing, except the receding lines are not parallel. The receding lines of a one-point perspective drawing meet at a distant location called a **vanishing point** (**Figure 15-53**). As with oblique drawings, the front face is drawn at true scale. The receding planes can be projected in any direction from this true front plane.

The vanishing point can be located any distance on the horizon. When the distance from the horizon to the vanishing point is short, the angles of the receding sides will be more acute (**Figure 15-54**).

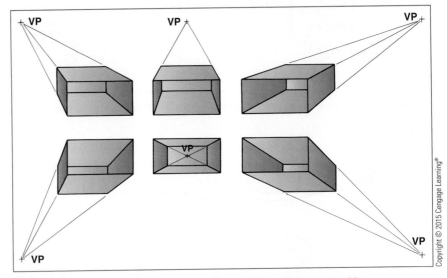

Figure 15-53 *Receding lines can be projected to vanishing points at any position.*

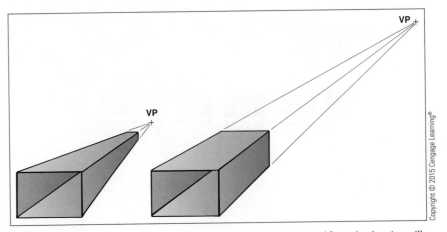

Figure 15-54 *The distance or position that the vanishing point is placed from the drawing will affect its final form.*

Before you can prepare a one-point perspective drawing, the concepts of station point, picture plane, and lines of vision must be understood (**Figure 15-55**). The **station point** represents the location of the observer. The final drawing will therefore appear as viewed from the station point. The **picture plane** represents the vertical plane upon which the object is viewed. The distance from the station point to the picture plane determines the size of the drawing. An object on a picture plane closer to the station point will appear smaller than the same object on a picture plane further from the station point. **Lines of vision** are imaginary lines connecting the station point with the picture plane and object.

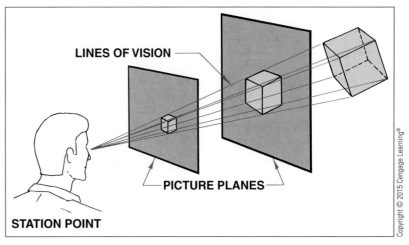

Figure 15-55 *The location of the picture plane controls the size of the perspective drawing.*

To draw a one-point perspective of the exterior of an object, follow the steps outlined in **Figure 15-56**:

Step 1. Draw the front and top orthographic views of the object. Draw horizontal lines representing the picture plane horizon and ground line. The **horizon** is a horizontal line upon which the vanishing points are placed. The ground line is the base line upon which the front plane is viewed.

Step 2. Locate the station point, which is the exact position from where the object (top view) will be viewed. The final drawing will appear as viewed from this point. Connect each back corner of the top view with the station point.

Step 3. Project a vertical line from the station point up to intersect the horizon line. This point of intersection is the vanishing point.

Step 4. Connect each corner of the front view with lines to the vanishing point.

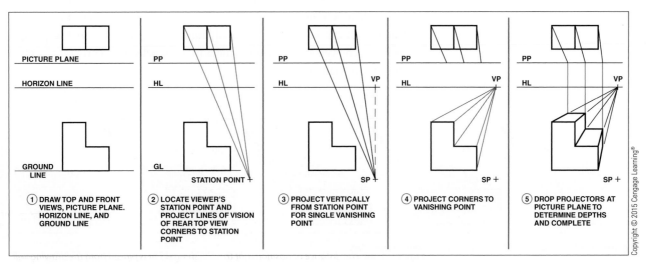

Figure 15-56 *Steps for drawing with one-point perspective.*

Step 5. Establish points of intersection between each of the station point connecting lines and the lines to the vanishing point. Draw vertical lines down from these points to intersect lines that connect similar points on the front plane with the vanishing point. Draw horizontal and vertical lines parallel with the lines on the front plane to represent the back of the object. To finish the drawing, connect all visible lines with an object line and erase all construction lines. Shade if required.

If part of the front plane's detail is missing, lightly block in the entire front plane, including the empty areas (**Figure 15-57**). This provides a base area from which to project light lines from the corners of the front face to the vanishing point. Draw the outline of the object on the receding sides using heavy lines. Erase the construction lines to complete the drawing.

One of the most common uses for a one-point perspective drawing is to show the interior view of a building (**Figure 15-58**).

TWO-POINT PERSPECTIVE

Two-point perspective (angular) drawings (**Figure 15-59**) are most popular because they provide the greatest amount of realism with a minimum amount of complex projection. Two-point perspective drawings are prepared when viewed from a corner, not from a front orthographic face. Each of the two sides is projected to separate vanishing points on the horizon. **Figure 15-60** shows how the two vanishing points relate to the object, picture plane, horizon, and station point.

VERTICAL POSITION ALTERNATIVES

The vertical position of the horizon line in relation to the object affects the amount of each surface on the final drawing (**Figure 15-61**). When the horizon is placed below the object, the bottom and two sides show (producing a worm's-eye view). When the horizon is placed above the object the top and two sides show (producing a bird's-eye view). When the horizon extends through the object, only the two sides show. This view more closely relates to an eye-level view, especially for large objects that are higher than eye level. The vertical position of the horizon is selected to provide the most appropriate viewing angle.

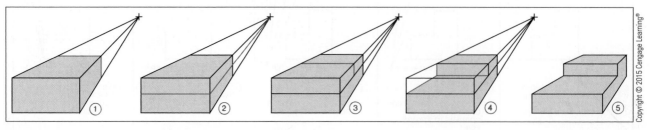

Figure 15-57 *Laying out and blocking in a one-point perspective.*

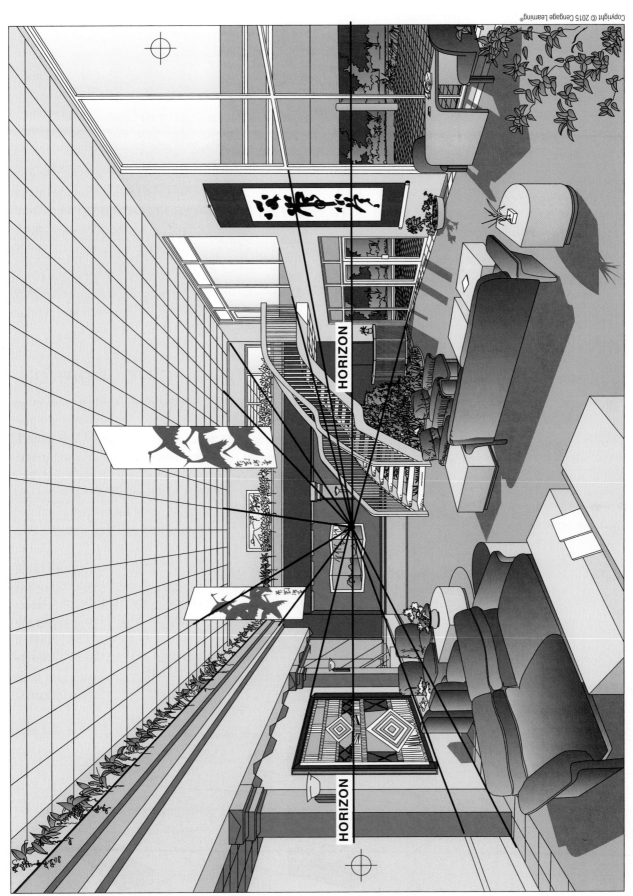

Figure 15-58 *Example of a one-point interior perspective.*

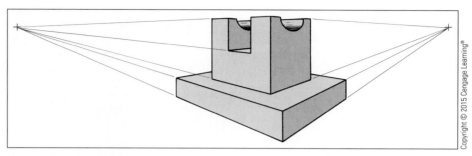

Figure 15-59 *Example of a two-point perspective.*

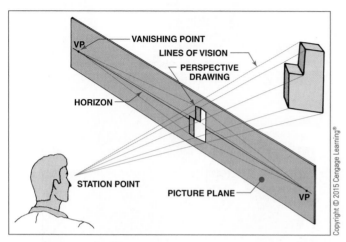

Figure 15-60 *Deciding factors that make up a two-point perspective drawing.*

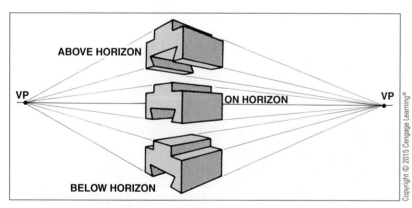

Figure 15-61 *The vertical position of the viewer affects the perspective drawing.*

STATION POINT POSITION

Just as the placement of the horizon greatly affects the vertical view of the object, the position of the station point changes the horizontal view of the object. Moving the station point closer to one vanishing point and farther away from the other results in the differences shown in **Figure 15-62**.

VANISHING POINT LOCATION

In addition to the vertical location of the horizon and the horizontal location of the station point, the final appearance of the drawing is greatly affected by the placement of the vanishing points on the horizon.

Figure 15-63 shows a perspective drawing with widely spaced vanishing points compared with a drawing with closely spaced vanishing points. To ensure that a base angle is more than 90°, spread the vanishing points apart on the horizon line and/or draw the perspective layout closer to the horizon line (**Figure 15-63**). The station point should be located below the picture plane at a distance of approximately two times the width of the object (**Figure 15-64**).

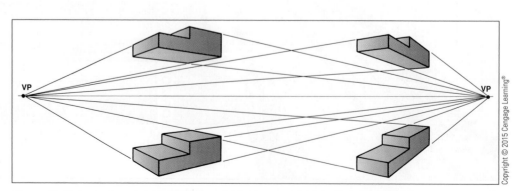

Figure 15-62 *As the viewer's position changes, the perspective drawing will change.*

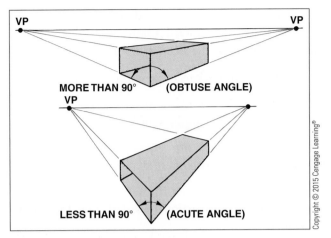

Figure 15-63 *Placing the vanishing points too close together will cause the drawing to appear distorted.*

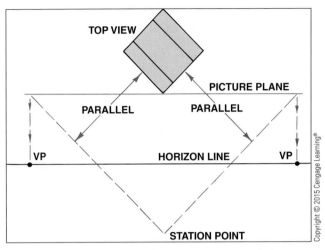

Figure 15-64 *Locating the vanishing point's position on the horizon.*

ESTIMATING PROCEDURE

Since perspective drawings are not used as working drawings, drawing to precise dimensional limits is usually not required. When only rough proportions are sufficient, a two-point perspective drawing can be prepared by using the following estimating procedures (**Figure 15-65**).

Step 1. Draw a horizontal line and locate the position of two vanishing points. Draw a vertical line representing the leading edge of the object above or below the horizon, depending on the view required.

Step 2. Draw lines connecting the front corner ends with each vanishing point.

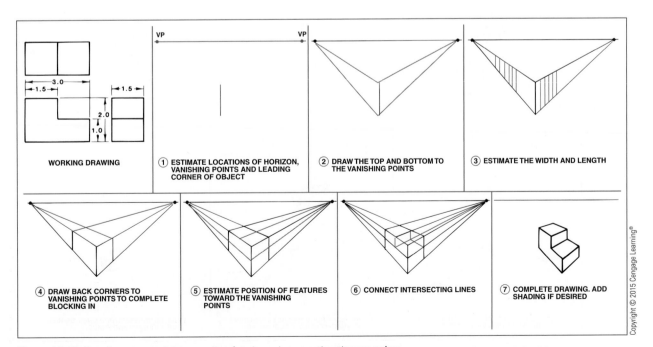

Figure 15-65 *Drawing a two-point perspective drawing using an estimating procedure.*

Step 3. Estimate the length of the object compared to the height of the front corner line. Mark the estimated length and depth with vertical lines drawn between the receding lines.

Step 4. Draw lines connecting the back receding corners to the opposite vanishing point.

Step 5. Draw details by estimating the position of features on the receding planes.

Step 6. Connect intersecting points with lines extending to the opposite vanishing point.

Step 7. Complete by erasing the construction lines, darkening the object lines, and shading.

CONSTRUCTION PROCEDURE

When a more dimensionally accurate two-point perspective drawing is required, the relationship of the station point, vanishing point, horizon, and picture plane must be precisely controlled. The following is the procedure for constructing a two-point perspective drawing that will closely approximate the true appearance of the object (**Figure 15-66**):

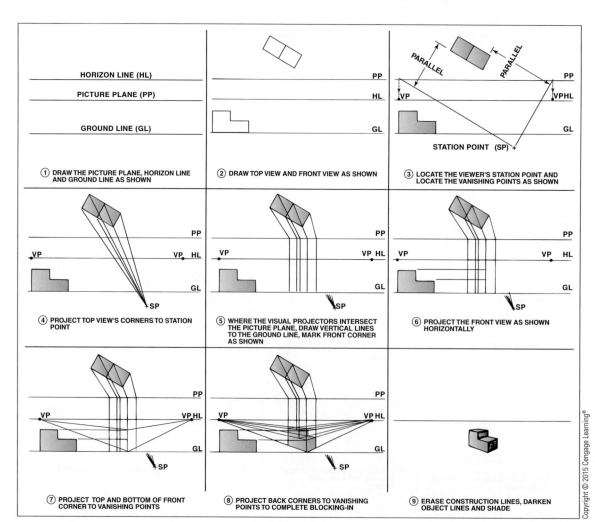

Figure 15-66 *Projection procedure for a two-point perspective drawing.*

Step 1. Draw horizontal lines representing the position of the picture plane, horizon, and ground line.

Step 2. Draw the orthographic top view of the object, rotated to the desired viewing angle above the picture plane line. Draw the orthographic front view of the object on the ground line.

Step 3. Locate the station point, and project lines of sight to the picture plane parallel to the sides of the top view. Draw vertical lines downward from the picture plane to intersect the horizon to locate the vanishing points.

Step 4. Draw lines connecting all corners of the object with the station point.

Step 5. From the points of intersection between the projected lines and the picture plane line, project vertical lines downward to intersect the ground line. Mark the line representing the front corner line.

Step 6: From the side view, project the object's height with horizontal lines to intersect the front corner line.

Step 7: Establish the height of the front corner line by marking the intersection between the vertical and horizontal projection lines. Project the lines from the bottom and top of the front corner line to each vanishing point. Draw vertical lines representing the width and depth of the object between the receding lines.

Step 8: Project lines from the back corners to the opposite vanishing points. Locate the position of depressions, cutaway areas, extensions, and holes on the planes. Project lines to the vanishing points to establish the final shape. Adjust the edges for any fillets and rounds.

Step 9: Darken the major object lines and erase all construction lines. Shade and render if desired.

PERSPECTIVE ELLIPSES

Circles on perspective drawings, as in axonometric drawings, are shown as ellipses. Since the receding lines of a perspective drawing are not parallel, ellipses are not symmetrical.

To be totally accurate, ellipses must be projected as shown in **Figure 15-67**. Small perspective ellipses can usually be drawn using an ellipse template. To use an ellipse template, follow the same procedure for using a template on axonometric drawings. The exception is that all centerlines and circle construction lines are

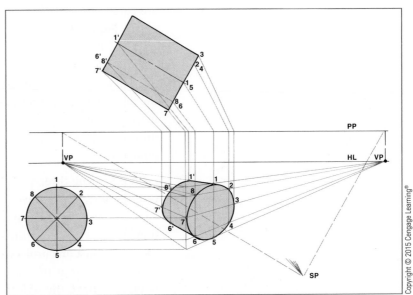

Figure 15-67 Projection procedure to draw circles in two-point perspective drawing.

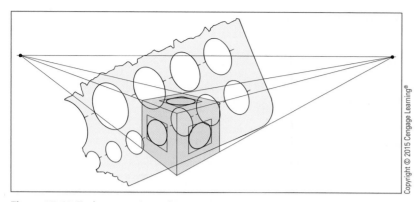

Figure 15-68 *To draw an estimated perspective circle, block in a perspective square and find the closest ellipse on an ellipse template.*

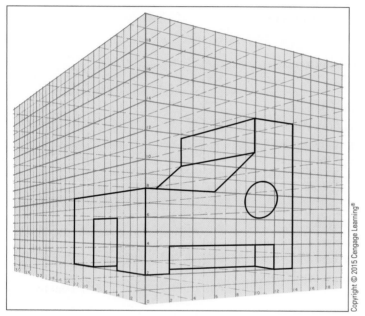

Figure 15-69 *Drawing on a perspective grid is fast and will help produce well-proportioned perspective drawings.*

projected to vanishing points. Once the outline of the circle is established, choose the appropriate ellipse template opening size. Position the opening as close as possible to align with the blocked-in area (**Figure 15-68**).

PERSPECTIVE GRIDS

Perspective grid paper, like isometric grid sheets, is often used as a guide in preparing perspective drawings. Grid sheets are available for one-, two-, and three-point perspective drawings, although two-point grids are the most popular. Grids are also available for either interior or exterior perspective drawings. Grids (**Figure 15-69**) are plotted on specific preset positions of the vanishing point, station point, and picture plane. This limits the options in selecting the exact vertical and horizontal position of the view and angles of the receding sides. Although they are excellent time-savers, the theory of perspective projection must be thoroughly understood to effectively use perspective grids.

THREE-POINT PERSPECTIVE

When a two-point perspective drawing contains long, narrow, vertical planes that extend far from the horizon, the object will appear distorted. This optical illusion is created because the vertical lines are parallel. This is not consistent with the normal way the eye sees the object. **Figure 15-70** shows a distorted two-point perspective drawing of this type. **Three-point perspective** drawings are used to correct this illusion by providing a third vanishing point (**Figure 15-71**). There are no parallel lines on a three-point perspective drawing since all lines connect to one of the three vanishing points. Lines representing horizontal surfaces are connected to vanishing points one and two, just as in two-point perspective drawings. All lines representing vertical planes are connected to vanishing point three.

In preparing three-point perspective drawings, follow the same procedures outlined in drawing two-point perspectives. Instead of drawing vertical parallel lines representing the width and depth of the object, draw vertical lines to a third vanishing point. This point is located below the object on a line projected vertically down from the front corner line. If the viewer is at the base of the building, the third vanishing point is placed above the object. This gives the added illusion of depth to the drawing.

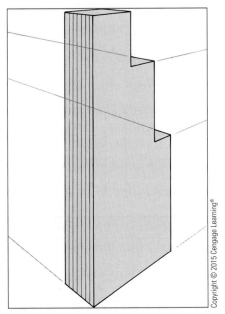

Figure 15-70 *Tall structures will appear distorted with a two-point perspective drawing.*

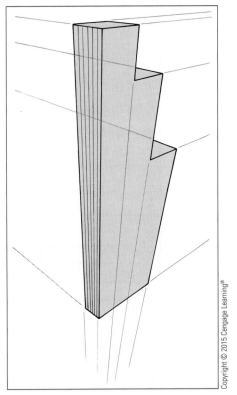

Figure 15-71 *Three-point perspective drawings for tall structures add realism.*

DRAFTING EXERCISES

1. Sketch the isometric drawings in **Figures 15-72** and **15-73**. Make your sketches approximately two times larger. If available, make your sketches on isometric grid paper. After completing the sketches, sketch the multiview drawing for each object.

2. With the manual drafting instruments or a CAD system, draw and dimension the multiview and isometric drawings, in **Figures 15-74** and **15-75**.

3. Draw and dimension a cabinet and a multiview drawing for each object in **Figures 15-76** and **15-77**.

4. Draw the isometric exploded views in **Figure 15-78**. Double the size from this page.

5. Draw the perspective **Figures 15-79** through **15-81**.

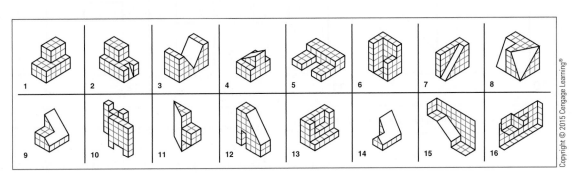

Figure 15-72 *Sketch each isometric exercise. Each square is 1/2".*

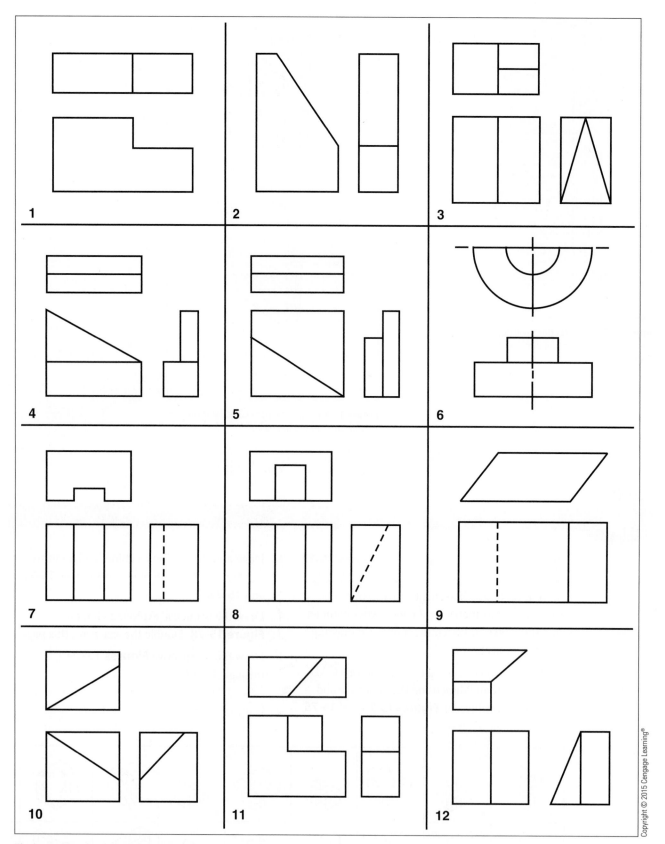

Figure 15-73 *Sketch the isometric for each multiview drawing.*

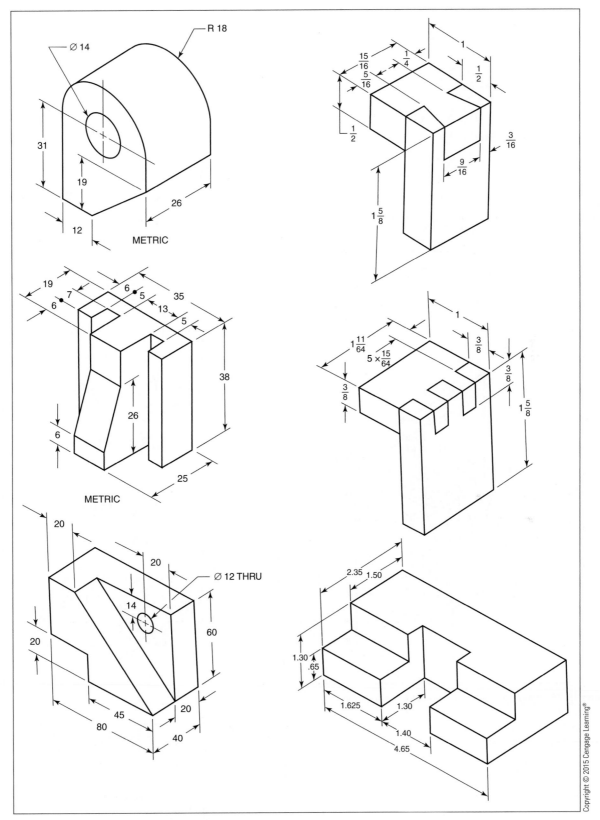

Figure 15-74 *Draw the multiview and isometric drawings for each object.*

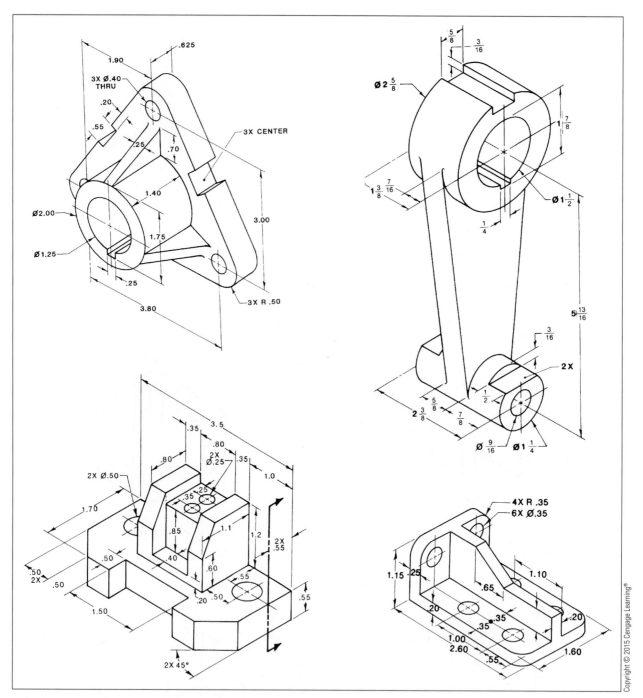

Figure 15-75 *Draw the multiview and isometric drawings for each object.*

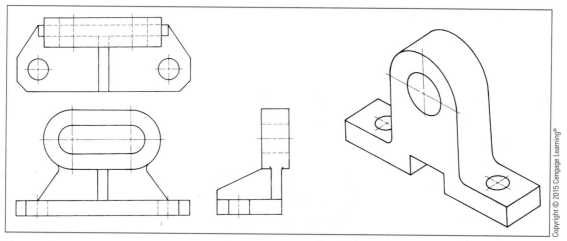

Figure 15-76 *Draw and dimension a dimetric and a cabinet drawing for each of the two figures. Double the size from the page.*

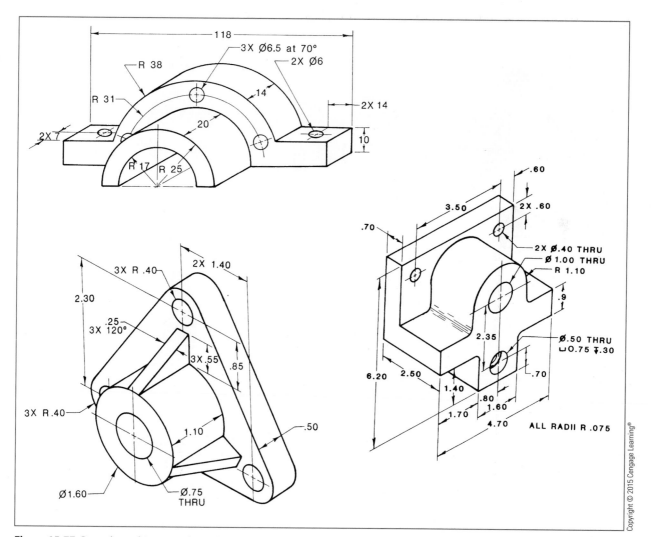

Figure 15-77 *Draw these objects as cabinet drawings.*

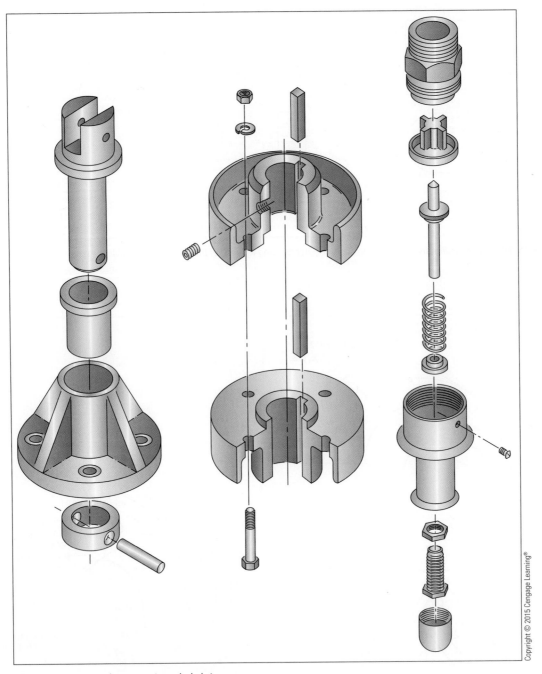

Figure 15-78 *Draw the isometric exploded views.*

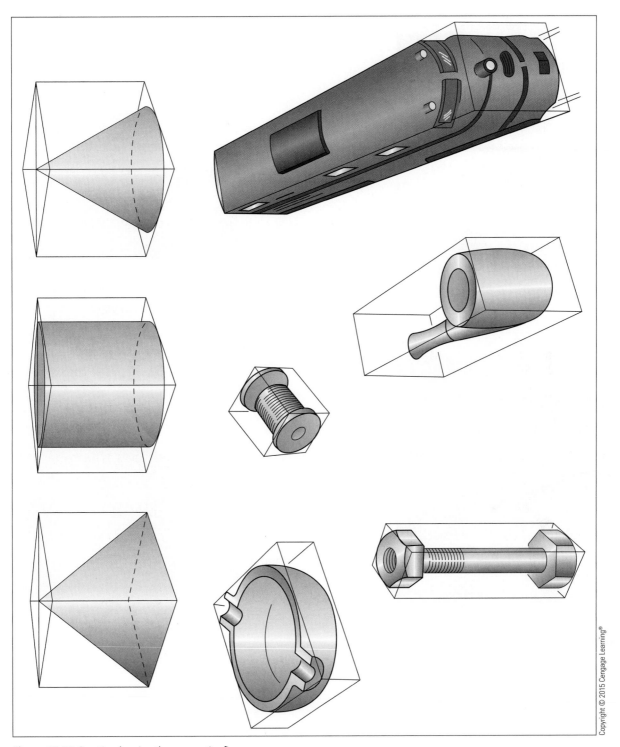

Figure 15-79 *Practice drawing the perspective figures.*

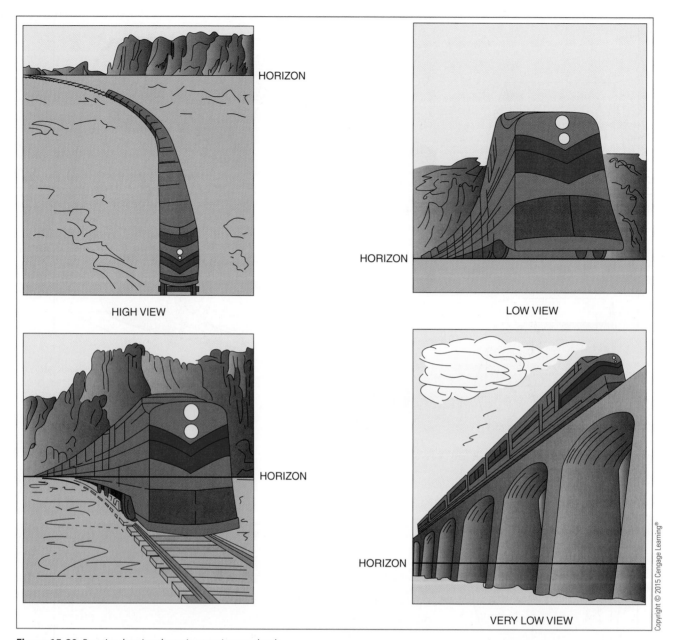

Figure 15-80 *Practice drawing the train at various eye levels.*

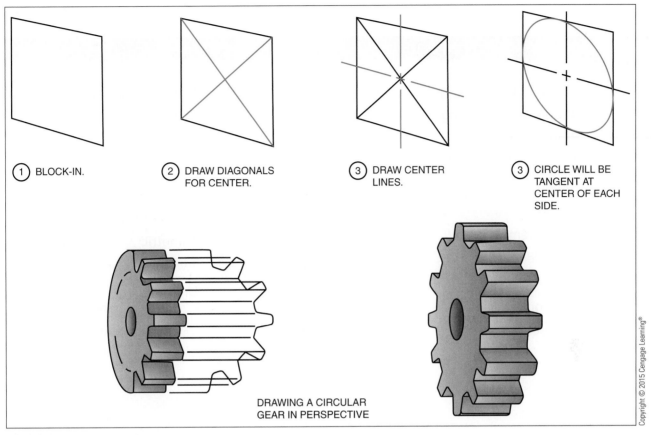

① BLOCK-IN.

② DRAW DIAGONALS FOR CENTER.

③ DRAW CENTER LINES.

③ CIRCLE WILL BE TANGENT AT CENTER OF EACH SIDE.

DRAWING A CIRCULAR GEAR IN PERSPECTIVE

Figure 15-81 *Practice drawing a gear in perspective.*

DESIGN EXERCISES

1. Redesign the freestanding base to hold and support a standard 35 mm camera as shown in **Figure 15-82**. Draw a pictorial drawing of your design.

2. Redesign the C-clamp in **Figure 15-83** so it will attach to a drafting table and support a tray for your drafting instruments and supplies.

3. Design a clipboard for 8.5" × 11" paper. Include in your design internal storage for paper and pencils. Draw a pictorial of your design.

4. Design a pull-toy for a three-year-old boy or girl. Draw a pictorial drawing of your design.

5. Design a standard checkerboard with a stylized design for the checkers. Draw a pictorial drawing of your design.

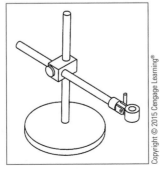

Figure 15-82 *Redesign the base to hold a camera.*

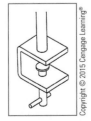

Figure 15-83 *Redesign the C-clamp to hold a supply of drafting instruments.*

KEY TERMS

Aligned dimensioning

Axonometric drawing

Broken-out section

Cutaway view

Dimetric

Ellipse

Ellipse template

Exploded views

Fillet

Full isometric section

Half isometric section

Horizon

Isometric drawing

Isometric axis

Isometric circles

Isometric lines

Lines of vision

Nonisometric lines

Oblique drawing

One-point
    perspective

Perspective drawing

Pictorial drawing

Picture plane

Round

Sectional view

Station point

Three-point perspective

Trimetric

Two-point perspective

Unidirectional
    dimensioning

Vanishing point

Fasteners

16

The student will be able to:

- Draw American standard internal and external thread symbols (simplified, schematic, and detailed)
- Draw ISO metric internal and external thread symbols (simplified, schematic, and detailed)
- Read and understand fastener tables
- Recognize the various types of fasteners used in industry

Introduction

A fastener is any device used to join separate parts or materials. There are two general types of fasteners—removable and permanent. The use of removable fasteners, such as screws, pins, keys, nuts, and bolts, allows attached parts or materials to be separated and reattached. Permanent fasteners, such as rivets, nails, and staples, and fastening methods, such as soldering, welding, and gluing, affix parts or materials that never need separating. The American Society of Mechanical Engineers (ASME), the International Standards Organization (ISO), and the **Industrial Fasteners Institute (IFI)** establish the standards for fastening methods and devices.

Threaded Fasteners

The functioning of all threaded fasteners is based on the principle of the inclined plane. An inclined plane is a simple machine that accomplishes work by reducing force by increasing distance. A screw thread is a spiral inclined plane. An evenly spaced spiral curve around a cylinder is called a **helix**. **Figure 16-1** shows how an inclined plane is applied to a cylinder to create a helix. The lead is the lateral distance a screw moves along the axis with one 360° revolution. The **lead per revolution** of the inclined plane is equal to the circumference of the cylinder.

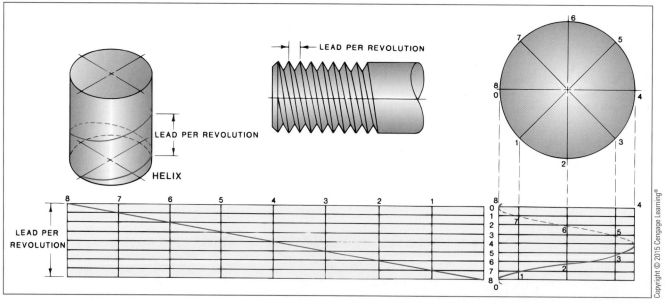

Figure 16-1 *The thread's form is a helix. One revolution equals the lead.*

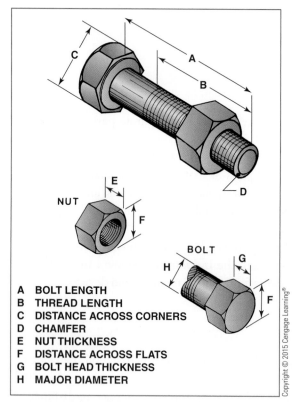

A BOLT LENGTH
B THREAD LENGTH
C DISTANCE ACROSS CORNERS
D CHAMFER
E NUT THICKNESS
F DISTANCE ACROSS FLATS
G BOLT HEAD THICKNESS
H MAJOR DIAMETER

Figure 16-2 *Nomenclature for nuts and bolts.*

THREADED FASTENER NOMENCLATURE

To work effectively with threaded fasteners using conventional drafting instruments or a CAD system, you must understand the following standard nomenclature (**Figures 16-2** and **16-3**).

- **Angle of thread**: The included angle formed between the major diameter and the depth of a thread.

- **Axis of fastener**: The cylindrical center of a fastener shaft.

- **Screw thread series**: Screws with similar combinations of pitches and diameters are divided into several related series. These are specified in the American National Standard Unified screw threads for screws, bolts, nuts, and other fasteners in ASME B1.1 (UN) and in ISO for metric threads. The following is the symbol designation for thread series:

  - Coarse thread series—UNC, NC

  - Fine thread series—UNF, NF

  - Extra-fine thread series—UNEF, NEF

- **Thread fit classes**: Threads are divided into the three tolerance fit classes, 1A, 2A, and 3A (for external threads) and 1B, 2B, and 3B (for external threads):

  - Class 1A and 1B—loose fits

  - Class 2A and 2B—general purpose fits

  - Class 3A and 3B—tight fits

- **Crest**: The point of widest diameter of a screw thread that lies on the major diameter plane.

- **Thread depth**: The distance from crest (major diameter) to the root (minor diameter) measured perpendicular to the axis

- **External thread:** Threads on the outside surface of a cylinder as on a bolt or screw.
- **Internal thread:** Threads on the inside surface of a hole as in a nut.
- **Lead:** The lateral distance the axis of a fastener moves during one 360° revolution of the fastener.
- **Major diameter:** The outside diameter of the threaded portion of a fastener measured from crest to crest.
- **Minor diameter:** The diameter of the thread measured from root to root perpendicular to the axis.
- **Root:** The deepest point or valley of a thread.
- **Thread forms:** The shape of the thread in profile section (**Figure 16-4**).
- **Right-hand thread:** As viewed from the head, a right-hand thread turns clockwise into an object or threaded opening.
- **Left-hand thread:** As viewed from the head, a left-hand thread turns counterclockwise into an object or threaded opening.
- **Single thread:** Threads with one continuous helical pattern in which the lead (one full turn) is equal to the single-thread pitch (**Figure 16-5**).
- **Multiple threads:** Threads that contain more than a single helical pattern side by side. For example, a double thread (**Figure 16-5**) includes two side-by-side helical threads; the lead is two times the pitch. In a triple-threaded fastener, the lead is three times the pitch. Multiple-threaded fasteners are used where few turns are desired to move the axis a maximum distance.
- **Pitch:** Pitch is the lateral distance between two related points on a thread (crest to crest or root to root). Pitch is expressed in the number of **threads per inch (TPI)**.
- **Pitch Diameter:** The pitch diameter is the diameter of an imaginary line drawn halfway between the major diameter and minor diameter (**Figure 16-6**).

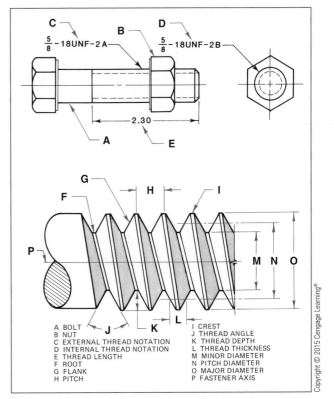

Figure 16-3 *Nomenclature for threaded nuts and bolts.*

THREAD SYMBOLISM

A true orthographic view or section of a threaded fastener is very time-consuming and complex to draw. Since most fasteners shown on working drawings are standard, a true orthographic or pictorial representation is not necessary. Simplified and schematic symbols are used on most working drawings. Simplified symbols show only the alignment of major and minor diameters, and the length of the threaded portion of the fastener. Schematic symbols represent crest and root lines with solid and hidden lines perpendicular to the axis.

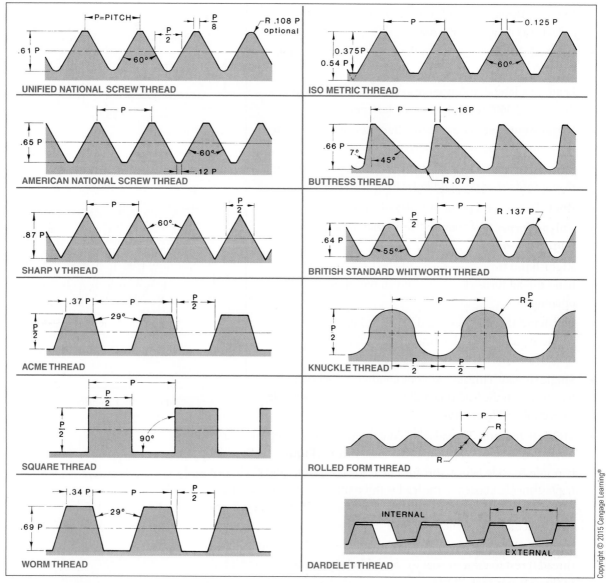

Figure 16-4 *Thread forms.*

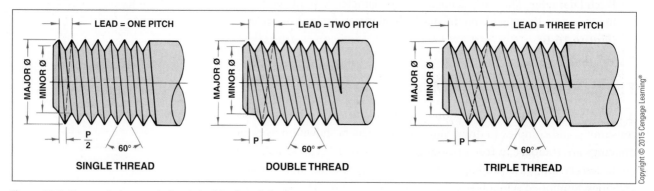

Figure 16-5 *One revolution equals the pitch of the threaded rod.*

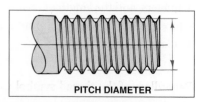

Figure 16-6 *The pitch diameter lies between the major and minor diameters.*

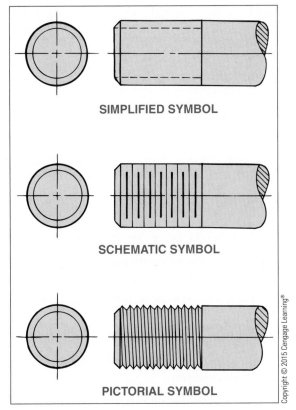

SIMPLIFIED SYMBOL

SCHEMATIC SYMBOL

PICTORIAL SYMBOL

Figure 16-7 *Symbols for drawing external threads. The simplified symbol is recommended for working drawings.*

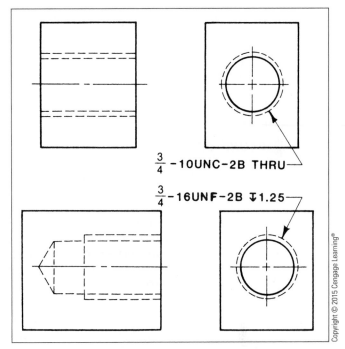

$\frac{3}{4}$ -10UNC-2B THRU

$\frac{3}{4}$ -16UNF-2B ⬎1.25

Figure 16-8 *Simplified symbols for drawing internal threads.*

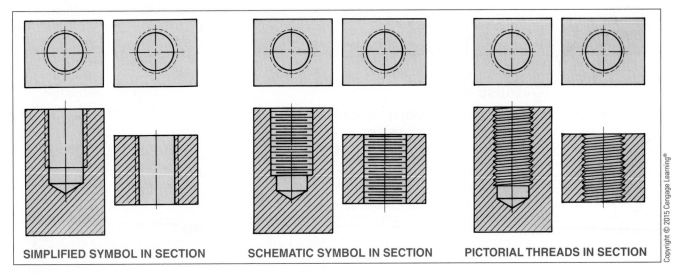

SIMPLIFIED SYMBOL IN SECTION   **SCHEMATIC SYMBOL IN SECTION**   **PICTORIAL THREADS IN SECTION**

Figure 16-9 *Drawing symbols for sectioned internal threads.*

Figure 16-7 shows a comparison of a simplified, a schematic, and a pictorial symbol for external threads. **Figure 16-8** shows simplified symbols for internal threads. When threads are shown in section, the three types of symbols are drawn (**Figure 16-9**); however, the simplified symbol is most widely used and recommended.

THREAD NOTATIONS

Regardless of the thread symbol used, all essential information cannot be effectively shown either graphically or with normal dimensioning methods. Notations are used to show the type, class, series, form, TPI, and

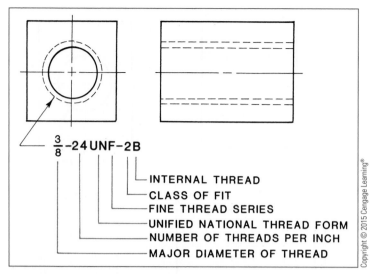

Figure 16-10 *Placement of internal thread notation's data.*

major diameter of fasteners on working drawings. **Figure 16-10** shows the notation method used to specify these features for internal threads. **Figure 16-11** shows the notation method used for external threads.

Drafting Procedures

A pictorial drawing of an external thread is drawn according to the following steps (**Figure 16-12**). (The major diameter should be drawn to scale, but all other measurements can be estimated.)

Step 1. Draw the major diameter and axis.

Step 2. Draw the minor diameter and the pitch diameter lines. Draw two 60° lines upward from the minor diameter line (root). As a check, the distance between the 60° lines on the pitch circle line should be about one-half the space of the 60° lines at the major diameter (crest).

Step 3. Continue to draw the 60° V threads for the full length of the threaded rod at the top and bottom.

Step 4. If the thread is a single lead, connect a line from a peak angle (vertex) on the major diameter to a peak angle on the opposite side, one thread over. Repeat for all the major (crest) and minor (root) diameters to complete the pictorial thread layout.

Step 5. Add a 45° chamfer and erase all construction lines. Darken to complete.

Simplified internal thread symbols are drawn according to the following steps (**Figure 16-13**):

Step 1. Draw the drill depth.

Step 2. Draw the major diameter and depth of the thread.

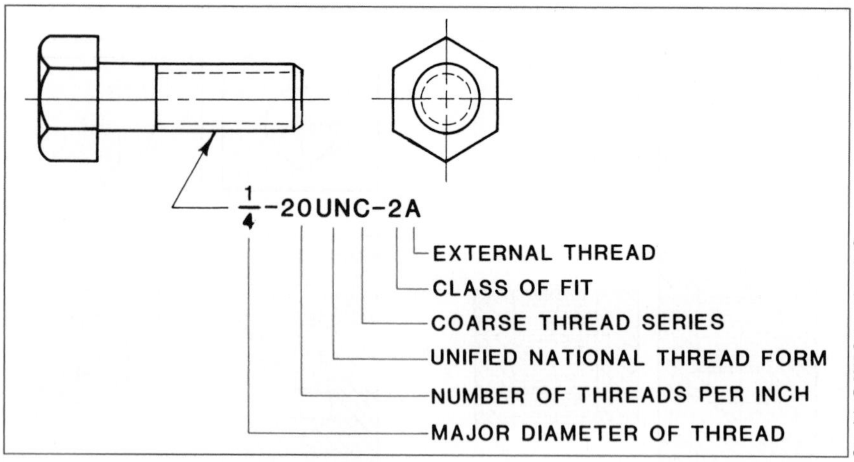

Figure 16-11 *Placement of external thread notation's data.*

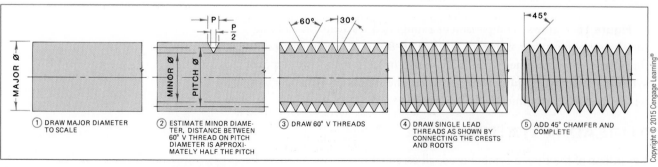

Figure 16-12 *Steps to draw external pictorial (detailed) V thread symbol.*

Simplified external thread symbols are drawn according to the following steps (**Figure 16-14**):

Step 1. Draw the outline and centerline of the fastener. The object lines, which parallel the axis centerline, represent the major diameter and are drawn to scale. Draw a line representing the threaded length.

Step 2. Draw a 45° chamfer on the opposite end. With construction lines, project a 60° included angle from the major diameter (crest) in order to determine the thread depth (root). At this point (thread depth), draw dashed lines parallel to the axis to represent the minor diameter.

Schematic internal thread symbols are drawn according to the following steps (**Figure 16-15**):

Step 1. Draw the outline of the tap drill hole's diameter to the required depth.

Step 2. Draw the outline of the thread's major diameter to the required depth.

Step 3. Estimate the pitch spacing and draw lines to the major diameter (crest to crest).

Step 4. Draw the lines for the minor diameter lines from root to root (use the tap drill line).

Step 5. Darken and add crosshatching as shown in **Figure 16-15**.

External schematic thread symbols are drawn following the five steps in **Figure 16-16**.

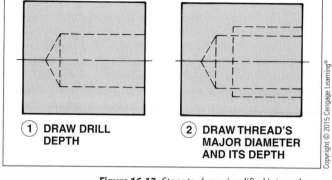

Figure 16-13 Steps to draw simplified internal thread symbols.

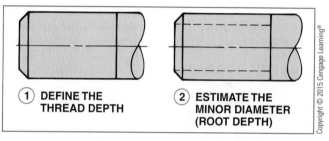

Figure 16-14 Steps to draw simplified external thread symbols.

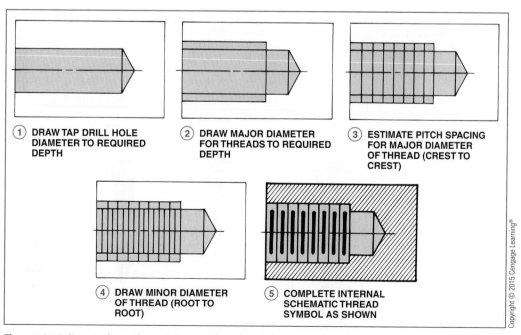

Figure 16-15 Steps to draw schematic internal thread symbols.

Types of Threaded Fastening Devices

There is a vast number of threaded fasteners, ranging from the most common (**Figure 16-17**) to very specialized. Threaded fasteners consist of a body, head, and point. **Figure 16-18** shows a variety of fastener heads, and **Figure 16-19** shows drive configurations used on fastener heads. Various types of fastener points are shown in **Figure 16-20**.

In designing the types and sizes of fasteners for a project, information concerning the class and dimensions is necessary to ensure that the fastener will fit and perform the desired function. To find this

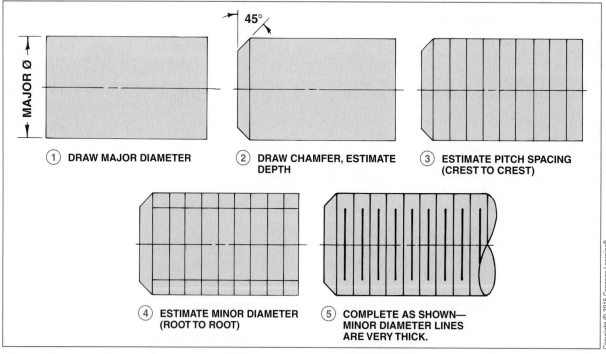

Figure 16-16 *Steps to draw schematic external thread symbols.*

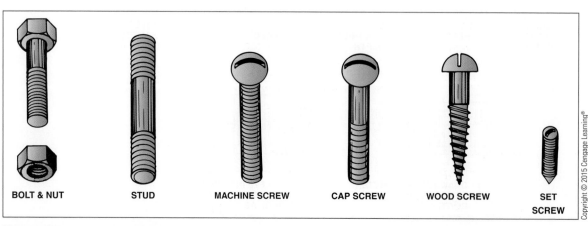

Figure 16-17 *Common thread fasteners.*

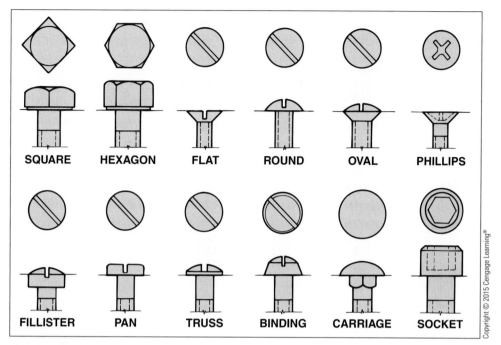

Figure 16-18 *Fastener heads.*

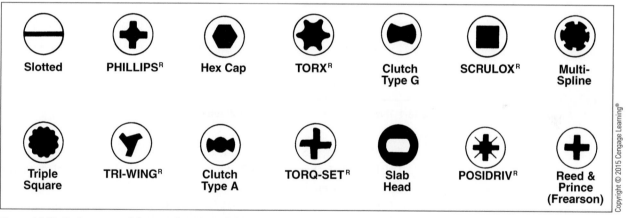

Figure 16-19 *Various types of drive configurations for threaded fasteners. (Reprinted from Technical Drawing and and Engineering Communcation by Goetsch, Chalk, Nelson, and Rickman, Delmar/Cengage Learning.)*

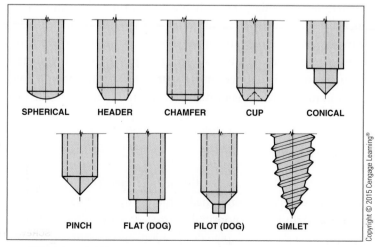

Figure 16-20 *Fastener points.*

Major Diameter	Number of Threads per Inch	
	Course UNC	Fine UNF
0 (.060)		80
1 (.073)	64	72
2 (.086)	56	64
3 (.099)	48	56
4 (.112)	40	48
5 (.125)	40	44
6 (.138)	32	40
8 (.164)	32	36
10 (.190)	24	32
12 (.216)	24	28
1/4	20	28
5/16	18	24
3/8	16	24
7/16	14	20
1/2	13	20
9/16	12	18
5/8	11	18
3/4	10	16
7/8	9	14
1	8	12
1 1/8	7	12
1 1/4	7	12
1 3/8	6	12
1 1/2	6	12
1 3/4	5	—
2	4 1/2	—
2 1/4	4 1/2	—
2 1/2	4	—
2 3/4	4	—
3	4	—
3 1/4	4	—
3 1/2	4	—

Table 16-1 *Thread table for commonly used ASME fasteners.*

detailed data, ASME, ISO, IFI, and manufacturer's charts are used. **Tables 16-1** through **16-3** show samples of this type of data for threaded fasteners.

METRIC THREAD SERIES

ASME conventions cover inch standards. They are not compatible with the ISO metric series. Today, the United States is the only country that uses an inch-based thread system, although the IFI has established U.S. metric standards for some fasteners. A comparison of U.S. standard (UN) threads and metric threads is shown in **Figure 16-21**. ISO metric notations differ from ASME notations (**Figure 16-22**). Metric notations include pitch in millimeters, major diameter in millimeters, and the ISO thread symbol.

PIPE THREADS

Pipes have either straight or tapered threads (**Figure 16-23**). They are drawn with either detailed or simplified symbols. In either case, the pipe thread notation includes the nominal diameter of the pipe in inches, threads per inch, standard used, pipe designation, and an entry indicating whether the pipe is straight or tapered.

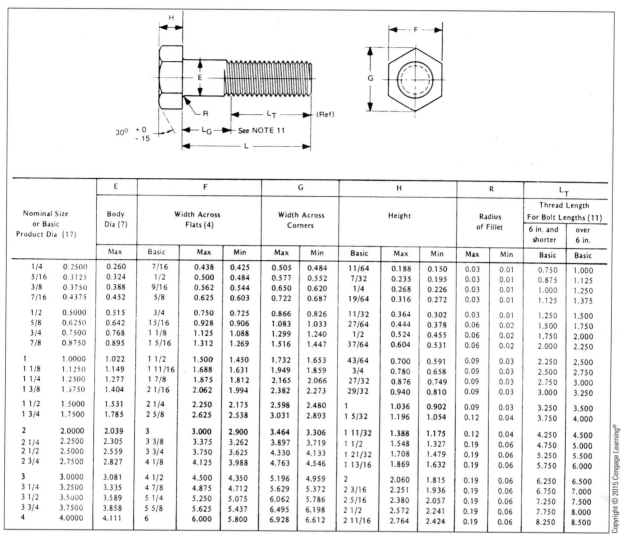

Table 16-2 Hex bolt table for commonly used ASME standard hex bolts.

Nominal Size or Basic Product Dia (17)		E Body Dia (7)	F Width Across Flats (4)			G Width Across Corners		H Height			R Radius of Fillet		L_T Thread Length For Bolt Lengths (11)	
													6 in. and shorter	over 6 in.
		Max	Basic	Max	Min	Max	Min	Basic	Max	Min	Max	Min	Basic	Basic
1/4	0.2500	0.260	7/16	0.438	0.425	0.505	0.484	11/64	0.188	0.150	0.03	0.01	0.750	1.000
5/16	0.3125	0.324	1/2	0.500	0.484	0.577	0.552	7/32	0.235	0.195	0.03	0.01	0.875	1.125
3/8	0.3750	0.388	9/16	0.562	0.544	0.650	0.620	1/4	0.268	0.226	0.03	0.01	1.000	1.250
7/16	0.4375	0.452	5/8	0.625	0.603	0.722	0.687	19/64	0.316	0.272	0.03	0.01	1.125	1.375
1/2	0.5000	0.515	3/4	0.750	0.725	0.866	0.826	11/32	0.364	0.302	0.03	0.01	1.250	1.500
5/8	0.6250	0.642	15/16	0.928	0.906	1.083	1.033	27/64	0.444	0.378	0.06	0.02	1.500	1.750
3/4	0.7500	0.768	1 1/8	1.125	1.088	1.299	1.240	1/2	0.524	0.455	0.06	0.02	1.750	2.000
7/8	0.8750	0.895	1 5/16	1.312	1.269	1.516	1.447	37/64	0.604	0.531	0.06	0.02	2.000	2.250
1	1.0000	1.022	1 1/2	1.500	1.450	1.732	1.653	43/64	0.700	0.591	0.09	0.03	2.250	2.500
1 1/8	1.1250	1.149	1 11/16	1.688	1.631	1.949	1.859	3/4	0.780	0.658	0.09	0.03	2.500	2.750
1 1/4	1.2500	1.277	1 7/8	1.875	1.812	2.165	2.066	27/32	0.876	0.749	0.09	0.03	2.750	3.000
1 3/8	1.3750	1.404	2 1/16	2.062	1.994	2.382	2.273	29/32	0.940	0.810	0.09	0.03	3.000	3.250
1 1/2	1.5000	1.531	2 1/4	2.250	2.175	2.598	2.480	1	1.036	0.902	0.09	0.03	3.250	3.500
1 3/4	1.7500	1.785	2 5/8	2.625	2.538	3.031	2.893	1 5/32	1.196	1.054	0.12	0.04	3.750	4.000
2	2.0000	2.039	3	3.000	2.900	3.464	3.306	1 11/32	1.388	1.175	0.12	0.04	4.250	4.500
2 1/4	2.2500	2.305	3 3/8	3.375	3.262	3.897	3.719	1 1/2	1.548	1.327	0.19	0.06	4.750	5.000
2 1/2	2.5000	2.559	3 3/4	3.750	3.625	4.330	4.133	1 21/32	1.708	1.479	0.19	0.06	5.250	5.500
2 3/4	2.7500	2.827	4 1/8	4.125	3.988	4.763	4.546	1 13/16	1.869	1.632	0.19	0.06	5.750	6.000
3	3.0000	3.081	4 1/2	4.500	4.350	5.196	4.959	2	2.060	1.815	0.19	0.06	6.250	6.500
3 1/4	3.2500	3.335	4 7/8	4.875	4.712	5.629	5.372	2 3/16	2.251	1.936	0.19	0.06	6.750	7.000
3 1/2	3.5000	3.589	5 1/4	5.250	5.075	6.062	5.786	2 5/16	2.380	2.057	0.19	0.06	7.250	7.500
3 3/4	3.7500	3.858	5 5/8	5.625	5.437	6.495	6.198	2 1/2	2.572	2.241	0.19	0.06	7.750	8.000
4	4.0000	4.111	6	6.000	5.800	6.928	6.612	2 11/16	2.764	2.424	0.19	0.06	8.250	8.500

Basic Major DIA & Pitch	Tap Drill DIA	INTERNAL THREADS Minor DIA MAX	Minor DIA MIN	EXTERNAL THREADS Major DIA MAX	Major DIA MIN	Clearance Hole
M1.6 × 0.35	1.25	1.321	1.221	1.576	1.491	1.9
M2 × 0.4	1.60	1.679	1.567	1.976	1.881	2.4
M2.5 × 0.45	2.05	2.138	2.013	2.476	2.013	2.9
M3 × 0.5	2.50	2.599	2.459	2.976	2.870	3.4
M3.5 × 0.6	2.90	3.010	2.850	3.476	3.351	4.0
M4 × 0.7	3.30	3.422	3.242	3.976	3.836	4.5
M5 × 0.8	4.20	4.334	4.134	4.976	-4.826	5.5
M6 × 1	5.00	5.153	4.917	5.974	5.794	6.6
M8 × 1.25	6.80	6.912	6.647	7.972	7.760	9.0
M10 × 1.5	8.50	8.676	8.376	9.968	9.732	11.0
M12 × 1.75	10.20	10.441	10.106	11.966	11.701	13.5
M14 × 2	12.00	12.210	11.835	13.962	13.682	15.5
M16 × 2	14.00	14.210	13.835	15.962	15.682	17.5
M20 × 2.5	17.50	17.744	17.294	19.958	19.623	22.0
M24 × 3	21.00	21.252	20.752	23.952	23.577	26.0
M30 × 3.5	26.50	26.771	26.211	29.947	29.522	33.0
M36 × 4	32.00	32.270	31.670	35.940	35.465	39.0
M42 × 4.5	37.50	37.799	37.129	41.937	41.437	45.0
M48 × 5	43.00	43.297	42.587	47.929	47.399	52.0
M56 × 5.5	50.50	50.796	50.046	55.925	55.365	62.0
M64 × 6	58.00	58.305	57.505	63.920	63.320	70.0
M72 × 6	66.00	66.305	65.505	71.920	71.320	78.0
M80 × 6	74.00	74.305	73.505	79.920	79.320	86.0
M90 × 6	84.00	84.305	83.505	89.920	89.320	96.0
M100 × 6	94.00	94.305	93.505	99.920	99.320	107.0

Table 16-3 Thread table for commonly used ISO metric fasteners.

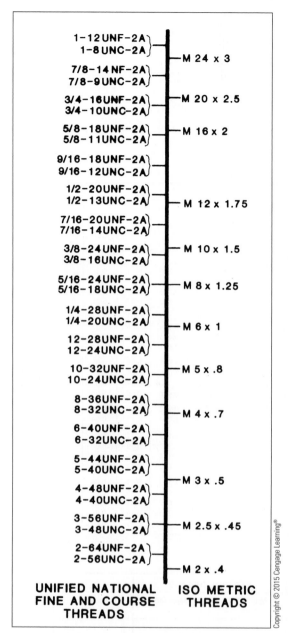

Figure 16-21 *Comparison of typical ASME's UN thread series to ISO metric threads.*

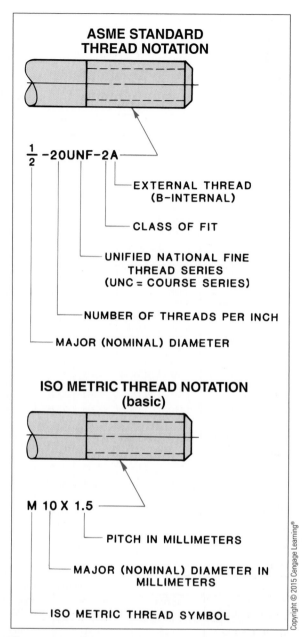

Figure 16-22 *Comparison of ASME's standard inch thread note to ISO metric thread note.*

Fastener Templates

The normal laying out and drawing of fasteners with instruments is a long and costly process. Consequently, most fasteners are drawn with templates using simplified fastener symbols. Templates provide outlines for two- and three-dimensional fastener drawings (**Figures 16-24**, **16-25a**, and **16-25b**).

Templates provide the outline of fastener components; however, combinations of template openings must be used to draw all the features of a fastener symbol. **Figure 16-26** shows the steps used to draw a bolt using a template. To use a template, first establish centerline (**Figure 16-27a**). In **Figure 16-27a**, a centerline positions the hexagon nut opening, providing a guide for the addition of concentric circles with a hole template. The use of some template openings may require reversing a template outline to draw identical sides of a symbol. This procedure is illustrated in the layout of the jamb nut shown in **Figure 16-27b**.

Templates do not always provide the exact outline of every fastener size and configuration. Some contain the most common sizes with the most common heads, bodies, and points. Most templates contain separate heads that can be combined with different lengths and points to produce the symbol needed. A machine screw template is used to draw the flat head and fillister machine screws in **Figure 16-28**.

Pictorial representations of fasteners are also accomplished with the use of templates (**Figure 16-29**). In using pictorial templates, the drafter must make sure that the template angle of projection is the same as the angle in the drawing. Most technical drawing pictorial templates include isometric angles, although ellipse templates cover a wide range of angles. Isometric fastener templates contain symbol outlines designed for the receding 30° planes and the top plane. These templates provide isometric views of unique features that, used with centerlines and isometric circles (35°), produce the final drawing.

Figures 16-29 and **16-30** show the procedure to draw an isometric hex nut and bolt. **Figure 16-31** shows the application of this method to drawing an isometric flat head machine screw. There are four isometric planes available on most templates: top, bottom, right side, and left side (**Figure 16-32**). Usually, the right-plane outlines are provided, and the template is reversed for the left receding plane. A wide variety of fasteners can be drawn pictorially with fastener and hole templates (**Figure 16-33**).

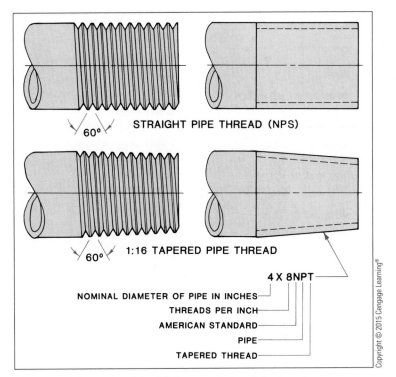

Figure 16-23 *Pictorial and simplified representation of pipe threads and notations.*

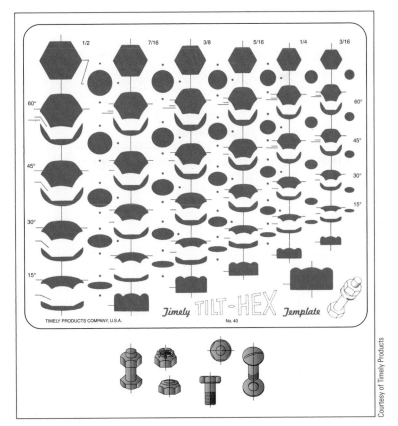

Figure 16-24 *Two- and three-dimensional bolt and nut drawing template.*

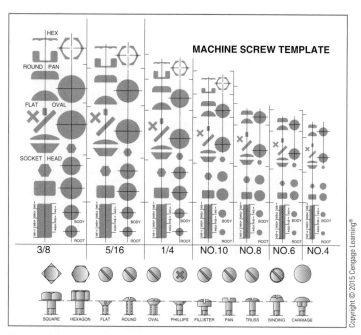

Figure 16-25a *2D machine screws template.*

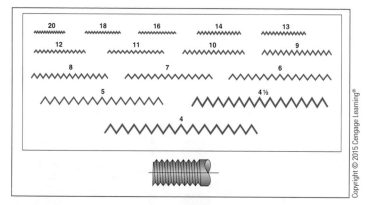

Figure 16-25b *Screw templates noting the threads per inch (TPI).*

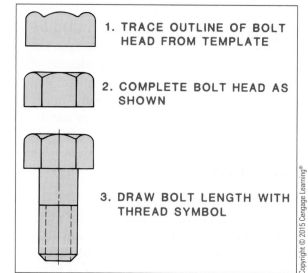

Figure 16-26 *Steps to draw a bolt with a template.*

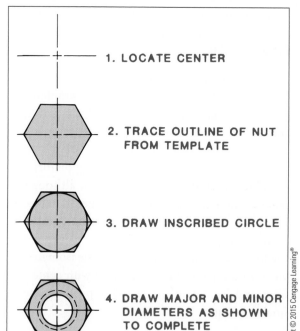

Figure 16-27a *Steps to draw the face view of a nut with a template.*

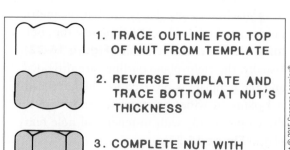

Figure 16-27b *Steps to draw a side view of a jamb nut with a template.*

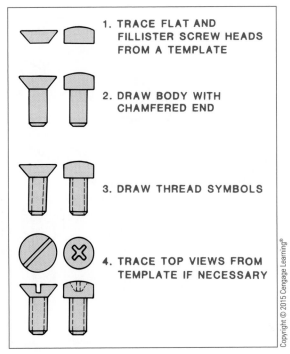

Figure 16-28 *Steps to draw a flat head and fillister head machine screw with a template.*

Threaded Permanent Fasteners

Most permanent fastening systems do not use fastening devices, but bond materials through heat, pressure, or chemical action. These include the use of adhesives, bonding agents, brazing, soldering, and welding (see Chapter 19). The major fastening devices used for permanent fastening are rivets.

Rivets are available in many different sizes and configurations. Large rivets are used primarily in heavy construction. Small rivets are used to join relatively thin materials, such as sheets and plates. Rivets

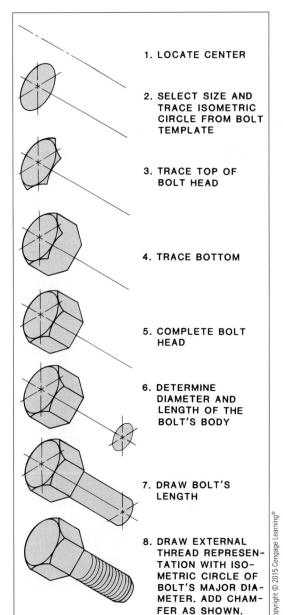

Figure 16-29 *Steps to draw an isometric hex bolt with a template.*

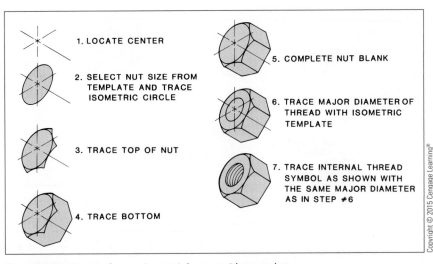

Figure 16-30 *Steps to draw an isometric hex nut with a template.*

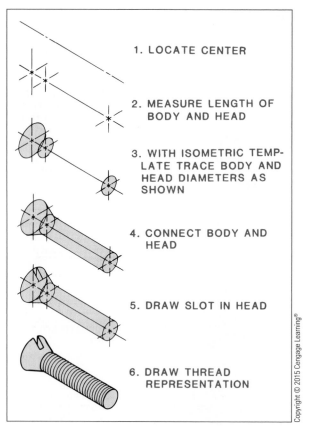

1. LOCATE CENTER

2. MEASURE LENGTH OF BODY AND HEAD

3. WITH ISOMETRIC TEMP- LATE TRACE BODY AND HEAD DIAMETERS AS SHOWN

4. CONNECT BODY AND HEAD

5. DRAW SLOT IN HEAD

6. DRAW THREAD REPRESENTATION

Figure 16-31 *Steps to draw an isometric flat head machine screw with a template.*

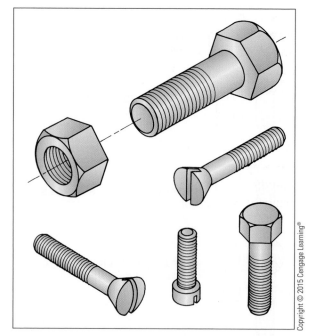

Figure 16-32 *Various positions of isometric planes.*

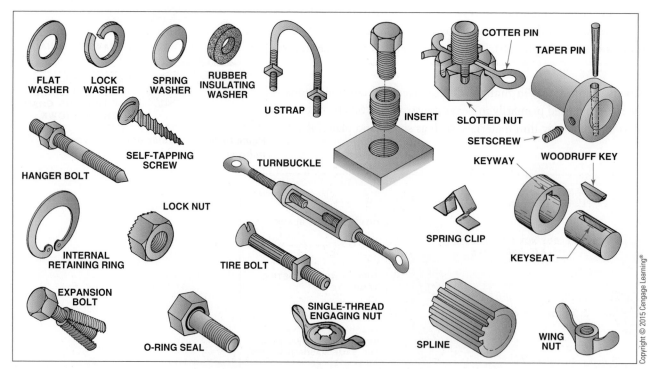

Figure 16-33 *Miscellaneous types of fasteners.*

have no threads but hold materials together when one head is flattened. Rivet specifications include the type of material, head shape, shaft diameter, and length (**Figure 16-34**). The symbols in **Figure 16-35** describe common rivet types.

Springs

A **spring** is a mechanical device, often in the form of a helical coil, that yields by expansion or contraction due to pressure, force, or stress applied. Springs are made to return to their normal form when the force or stress is removed. Springs are designed to store energy for the purpose of pushing

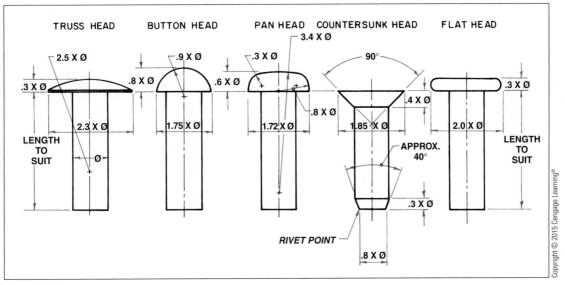

Figure 16-34 *American Standard small rivets. (Reprinted from Technical Drawing and Engineering Communication by Goetsch, Chalk, Nelson, and Rickman, Delmar/Cengage Learning.)*

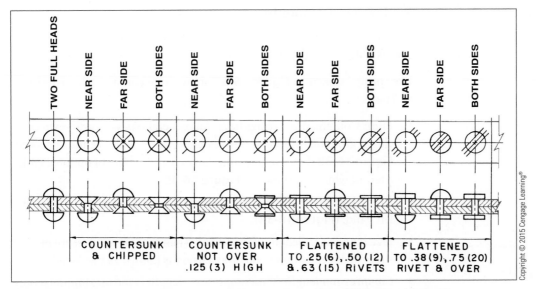

Figure 16-35 *Rivet symbols for working drawings. (Reprinted from Technical Drawing and Engineering Communication by Goetsch, Chalk, Nelson, and Rickman, Delmar/Cengage Learning.)*

or pulling machine parts by reflex action into certain desired positions. Improved spring technology provides springs with the ability to function a long time under high stresses. The effective use of springs in machine design depends upon five basic criteria, including material, application, functional stresses, use, and tolerances.

Continued research and development of spring materials have helped improve spring technology. The spring materials most commonly used include high-carbon spring steels, alloy spring steels, stainless spring steels, music wire, oil-tempered steel, copper-based alloys, and nickel-based alloys. Spring materials, depending upon use, may have to withstand high operating temperatures and high stresses under repeated loading.

Spring design criteria are generally based on material gauge, kind of material, spring index, direction of the helix, type of ends, and function. Spring-wire gauges are available from several different sources ranging in diameter from number 7/0 (.490 in.) to number 80 (.013 in.). The most commonly used spring gauges range from 4/0 to 40. There are a variety of spring materials available in round or square stock for use depending on spring function and design stresses. The spring index is a ratio of the average coil diameter to the wire diameter. The index is a factor in determining spring stress, deflection, and the evaluation of the number of coils needed and the spring diameter. Recommended index ratios range between 7 and 9, although other ratios commonly used a range from 4 to 16. The direction of the helix is a design factor when springs must operate in conjunction with threads or with one spring inside of another. In such situations, the helix of one feature should be in the opposite direction of the helix for the other feature. Compression springs are available with ground or unground ends. Unground, or rough, ends are less expensive than ground ends. If the spring is required to rest flat on its end, then ground ends should be used. Spring function depends upon two basic factors, compression and extension. Compression springs release their energy and return to the normal form when compressed. Extension springs release their energy and return to the normal form when extended (**Figure 16-36**).

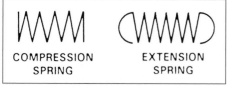

COMPRESSION SPRING   EXTENSION SPRING

Figure 16-36 *Compression and extension spring symbols.* Copyright © 2015 Cengage Learning®

SPRING TERMINOLOGY

The springs shown in **Figure 16-37** show some common characteristics, as follows.

- *Ends*: Compression springs have four general types of ends: open or closed grounded ends and open or closed ungrounded ends, shown in **Figure 16-38**. Extension springs have a large variety of optional ends, a few of which are shown in **Figure 16-39**.

- *Helix Direction*: The helix direction may be specified as right hand or left hand. (**Figure 16-38**).

- *Free length*: The length of the spring when there is no pressure or stress to affect compression or extension is known as free length.

- *Compression length*: The compression length is the maximum recommended design length for the spring when compressed.

- *Solid height*: The solid height is the maximum compression possible. The design function of the spring should not allow the spring to reach solid height when in operation unless this factor is a function of the machinery.

- *Loading extension*: The extended distance to which an extension spring is designed to operate is the loading extension length.

- *Pitch*: The pitch is one complete helical revolution, or the distance from a point on one coil to the same corresponding point on the next coil.

TORSION SPRINGS

Torsion springs are designed to transmit energy by a turning or twisting action. Torsion is defined as a twisting action that tends to turn one part or end around a longitudinal axis while the other part or end remains fixed. Torsion springs are often designed as anti-backlash devices or as self-closing or self-reversing units. A spiral torsion spring is constructed by winding a flat spring wire into a coil in the form of a spiral. The spiral torsion spring is designed to be wound up and exert force (**Figure 16-40**).

FLAT SPRINGS

Flat springs are arched or bent flat-metal shapes designed so that when placed in machinery, they cause tension on adjacent parts. The tension may be used to level parts, provide a cushion, or position the movement of one part relative to another. One of the most common examples of a flat spring is leaf springs on an automobile.

SPRING REPRESENTATIONS

There are three types of spring representations, detailed, schematic, and simplified as seen in **Figure 16-41**. Detailed spring drawings are used in situations that require a realistic representation, such as vendors' catalogs, assembly instructions, or detailed assemblies. Schematic spring representations are commonly used on drawings. The single-line schematic symbols are easy to draw and clearly represent springs without taking the additional time required to draw

Figure 16-37 *Spring characteristics.*

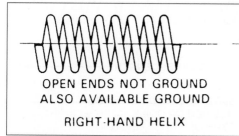

OPEN ENDS NOT GROUND
ALSO AVAILABLE GROUND

RIGHT-HAND HELIX

CLOSED ENDS GROUND
ALSO AVAILABLE NOT GROUND

LEFT-HAND HELIX

Figure 16-38 *Helix direction and compression spring end types.*

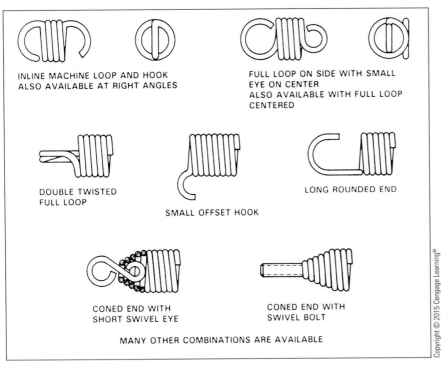

Figure 16-39 *Extension spring end types.*

INLINE MACHINE LOOP AND HOOK
ALSO AVAILABLE AT RIGHT ANGLES

FULL LOOP ON SIDE WITH SMALL
EYE ON CENTER
ALSO AVAILABLE WITH FULL LOOP
CENTERED

DOUBLE TWISTED
FULL LOOP

SMALL OFFSET HOOK

LONG ROUNDED END

CONED END WITH
SHORT SWIVEL EYE

CONED END WITH
SWIVEL BOLT

MANY OTHER COMBINATIONS ARE AVAILABLE

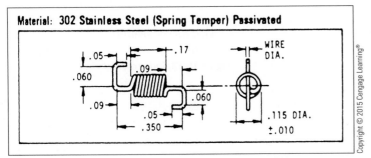

Figure 16-40 *The spiral torsion spring is designed to wind up and exert force in a rotating direction around the axis of the spring.*

Figure 16-41 *Spring representations.*

Copyright © 2015 Cengage Learning®

a detailed spring. The use of simplified spring drawings is limited to situations where the clear resemblance of a spring is not necessary. While very easy to draw, the simplified spring symbol must be accompanied by clearly written spring specifications. The simplified spring representation is not very useful in assembly drawings or other situations that require a visual comparison of features.

SPRING SPECIFICATIONS

No matter which representation is used, there are several important specifications that must accompany the spring symbol. Spring information is generally lettered in the form of a specific or general note.

Spring specifications include outside or inside diameter, wire gauge, kind of material, type of ends, surface finish, free and compressed length, and number of coils. Other information, when required, may include spring design criteria and heat treatment specifications. The information is often provided on a drawing as shown in **Figure 16-42**. The material note is usually found in the title block.

Since fasteners are standard items, they are manufactured according to standard specifications. The drafter or designer selects the appropriate fasteners for the design need. These standard fasteners include the full range of nuts, bolts, washers, cotter pins, rivets, and other fastening devices.

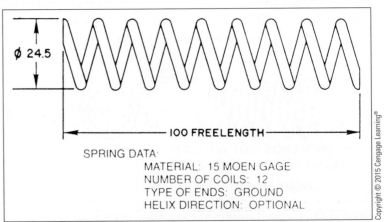

Figure 16-42 *Spring drawing with spring data chart.*

DRAFTING EXERCISES

1. Sketch **Figures 16-43** through **16-45**.

2. With an isometric fastener template, draw the various nuts in **Figure 16-46**.

3. With drawing instruments or a CAD system, draw the fasteners in **Figures 16-47** through **16-51**.

4. With drawing instruments or a CAD system, draw the metric fasteners in **Figures 16-52** and **16-53**.

5. Draw and label the twenty-three fastener symbols in **Figure 16-54**.

6. Draw with instruments or a CAD system the threaded machine parts in **Figures 16-55** through **16-57**.

7. Perform the following ECOs for **Figure 16-55**:

   - Change all UNF threads to UNC.

   - Add a fifth threaded fastener to the flange.

   - Add an external UNC thread 1.5" from the top of the shaft.

   - Add an internal UNF thread 2.0" from the top of the shaft's hole.

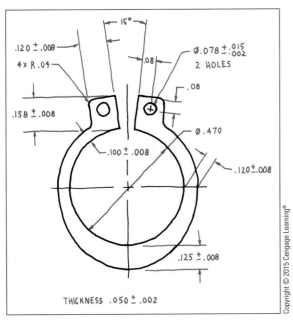

Figure 16-43 *Draw the working drawings of the steel retaining ring.*

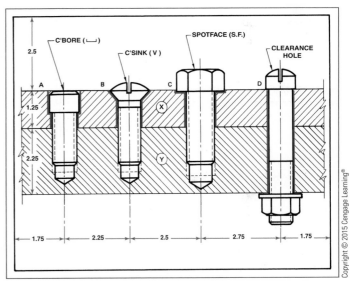

Figure 16-45 *Draw or sketch each fastener and add a thread notation. (Reprinted from Technical Drawing and Engineering Communication by Goetsch, Chalk, Nelson, and Rickman, Delmar/Cengage Learning.)*

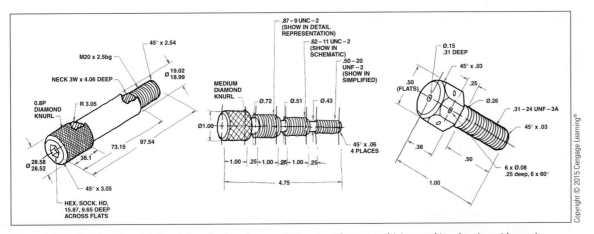

Figure 16-44 *Sketch a pictorial of each item freehand or on a CAD system, Draw a multiview working drawing with metric or inch/decimal dimensions. (Reprinted from Mechanical Drafting by Madsen, Schumaker, and Stewart, Delmar Learning.)*

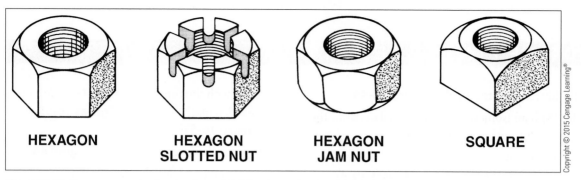

Figure 16-46 *Draw an example of each nut as a detail drawing and a pictorial drawing.*

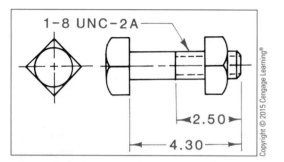

Figure 16-47 *Draw with instruments or with a CAD system a working drawing and isometric view of the square head bolt and nut.*

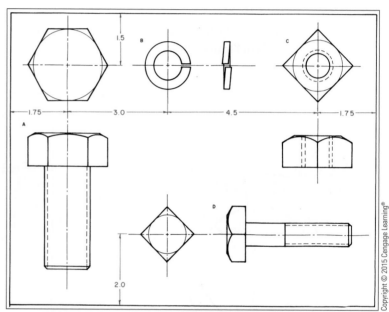

Figure 16-48 *On an A-size format lay out the fasteners as shown. Fastener data is: Hex bolt 1-3/8-UNC x 3.0 long; Lock washer fits a 7/8" diameter cap screw. Square nut 1-12 UNF 2B; Square head cap screw 3/4-10 UNC x 3.0 long. (Reprinted from Technical Drawing and Engineering Communication by Goetsch, Chalk, Nelson, and Rickman, Delmar/Cengage Learning.)*

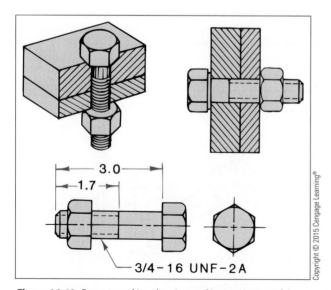

Figure 16-49 *Draw a working drawing and isometric view of the hex bolt and nut with instruments or a CAD system.*

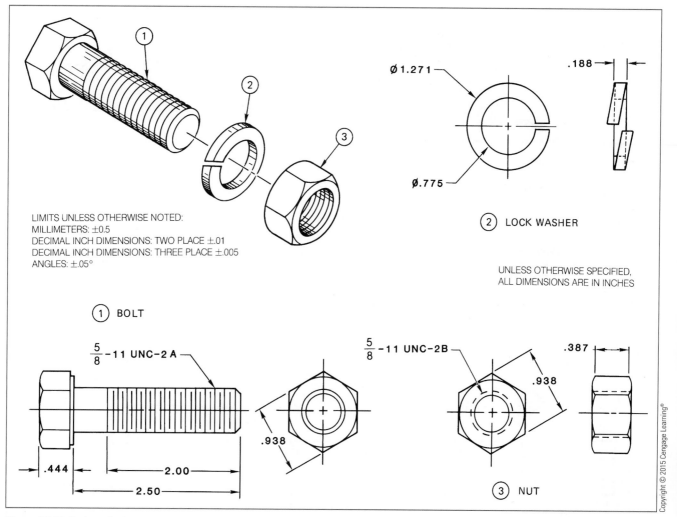

LIMITS UNLESS OTHERWISE NOTED:
MILLIMETERS: ±0.5
DECIMAL INCH DIMENSIONS: TWO PLACE ±.01
DECIMAL INCH DIMENSIONS: THREE PLACE ±.005
ANGLES: ±.05°

Ø1.271

Ø.775

.188

② LOCK WASHER

UNLESS OTHERWISE SPECIFIED,
ALL DIMENSIONS ARE IN INCHES

① BOLT

$\frac{5}{8}$ -11 UNC-2 A

$\frac{5}{8}$ -11 UNC-2B

.938

.387

.938

.444   2.00

2.50

③ NUT

Figure 16-50 *Redraw the fasteners.*

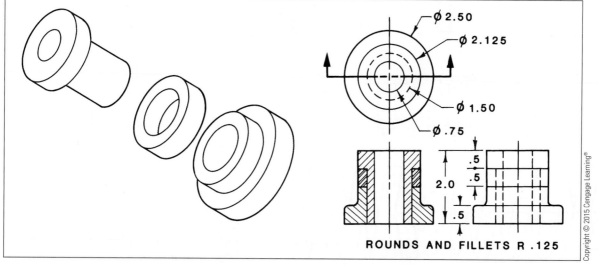

Ø 2.50
Ø 2.125
Ø 1.50
Ø .75

.5
.5
2.0
.5

ROUNDS AND FILLETS R .125

Figure 16-51 *Draw the working drawing of the bearing assembly. Plan a thread system and add all thread notes and thread symbols.*

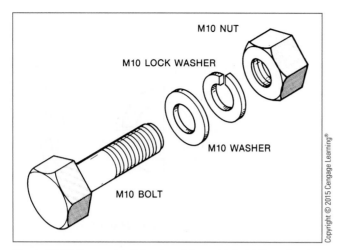

Figure 16-52 *Draw a detail drawing of the metric bolt (45 mm long) and the washer, lock washer, and nut.*

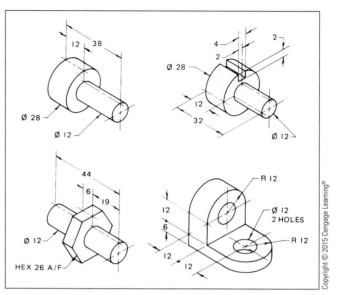

Figure 16-53 *Draw a multiview drawing and isometric view for each item. Dimension and add metric thread notations for all holes and shafts.*

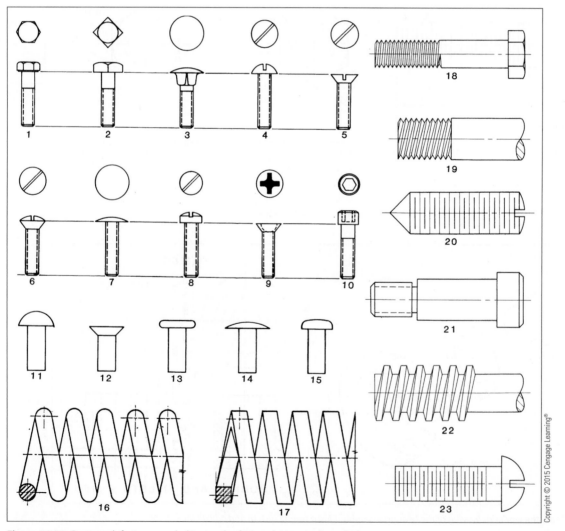

Figure 16-54 *Draw each fastener symbol to a scale of 2:1 and name each symbol.*

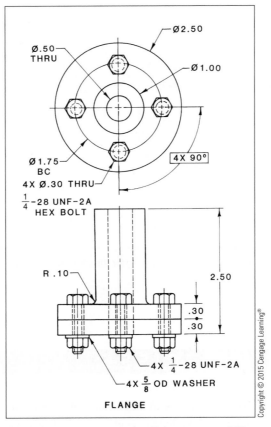

Figure 16-55 *Draw accurately with instruments or CAD.*

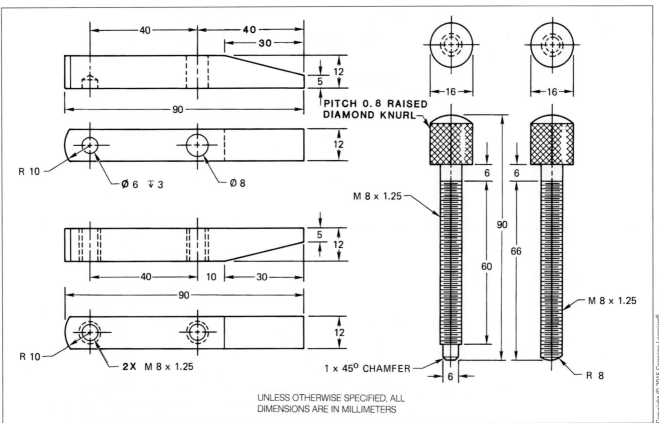

Figure 16-56 *Draw the detail drawings of the jeweler's clamp. Then complete the assembly drawing, noting all metric threads.*

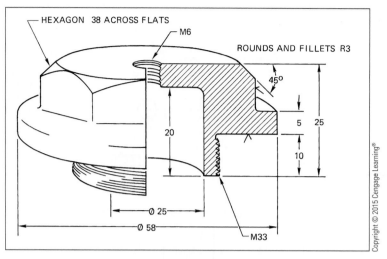

Figure 16-57 *Draw a multiview drawing for the metric hexagon cap.*

DESIGN EXERCISES

1. Redesign and complete the plate with a hook or holder to be fastened with threaded fasteners to the guide in **Figure 16-58**. A .75" diameter rope supporting several hundred pounds will pass through the hook or holder unit.

2. Design a threaded fastener to connect the two spools in **Figure 16-59**.

3. Study the old and new designs in **Figure 16-60**. See if you can improve on the new design.

4. Using any type of mass-produced fasteners or brackets, design an anti-break-in device for a sliding aluminum window.

5. Design an anti-break-in device for the front door of a residence.

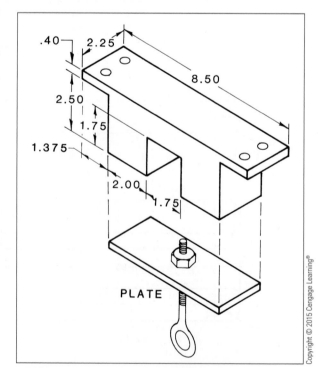

Figure 16-58 *Redesign this holder unit following the instructions in Exercise 1.*

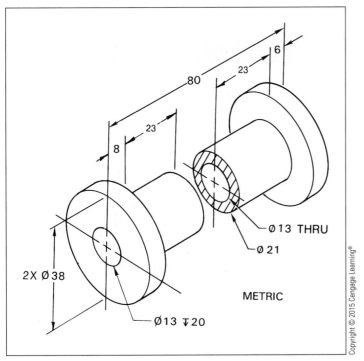

Figure 16-59 *Design a threaded fastener to hold the two spool forms together. Draw a multiview drawing of each part.*

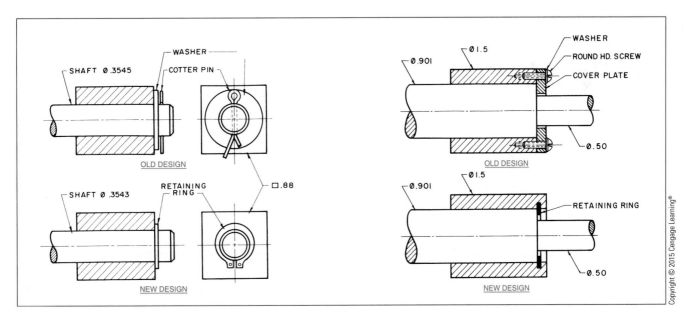

Figure 16-60 *Sketch the multiview drawings and isometric views of the fastening system. (Reprinted from Technical Drawing and Engineering Communication by Goetsch, Chalk, Nelson, and Rickman, Delmar/Cengage Learning.)*

KEY TERMS

Angle of thread

Axis of fastener

Compression length

Crest

External thread

Flat springs

Free length spring

Helix

Industrial Fasteners
  Institute (IFI)

Internal thread

Lead per revolution

Left-hand thread

Major diameter

Minor diameter

Multiple threads

Pitch

Pitch diameter

Right-hand thread

Rivets

Root

Screw thread series

Single thread

Spring

Thread depth

Thread fit classes

Thread forms

Torsion springs

Green Design in Industry

The student will be able to:

- Understand how toxins will affect the environment
- Understand how sustainability of the earth's resources will help the environment
- Develop the steps to design a green product
- Recognize toxic materials
- Conserve energy
- Understand the processes for the development of green energy

Introduction

A definition of green design is the development of the industrial processes and their products and minimizing the pollution that effects human health and the environment. Green materials and products are defined by being nontoxic, and recyclable, and have a long life cycle. Early industrial methods were powered by steam or water, and later by coal. Our industries over the past 100 years have rapidly expanded and new production technologies methods have been invented. With the modern expansion of the industrial revolution, there many beneficial products have been developed for convenience and health. There is also a darker side to industrialization, however. Everything that industry produces creates some sort of an environment impact, causing industry to focus on the issue of green design to answer the questions of "How can we make our products greener?" (**Figure 17-1**). The problem today is that many engineers and designers are not trained to understand how the production and the finished product will affect humans and the environment.

Sustainability

A major problem that industry has today is the **sustainability** of the earth's resources. Are they renewable, such as the forests? Are they recyclable? Will the earth's resources be depleted? Every industrial

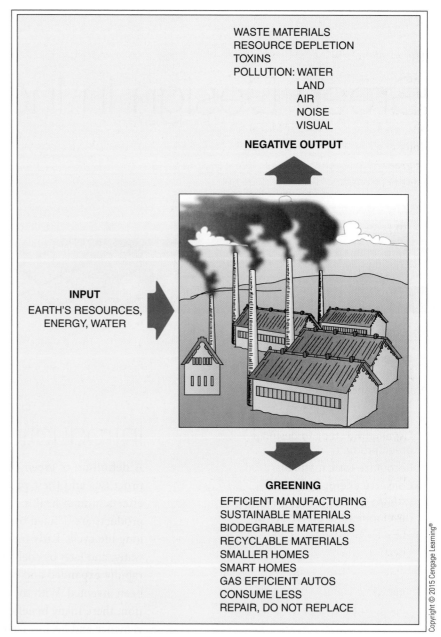

WASTE MATERIALS
RESOURCE DEPLETION
TOXINS
POLLUTION: WATER
LAND
AIR
NOISE
VISUAL
NEGATIVE OUTPUT

INPUT
EARTH'S RESOURCES,
ENERGY, WATER

GREENING
EFFICIENT MANUFACTURING
SUSTAINABLE MATERIALS
BIODEGRABLE MATERIALS
RECYCLABLE MATERIALS
SMALLER HOMES
SMART HOMES
GAS EFFICIENT AUTOS
CONSUME LESS
REPAIR, DO NOT REPLACE

Figure 17-1 *Manufacturing effects.*

product has its own materials used for production. However, materials will vary, and some of them may be toxic and they will have a negative effect on the health of the earth and its occupants (**Figure 17-2**).

Worldwide, industry consumes more than 10 billion tons of raw materials that were taken from the earth each year. When using renewable and local resources, whenever possible, it will help to conserve the nonrenewable resources, save energy in production, reduce transportation costs, and improve the planets health.

Energy

The usage of energy for industrial and residential production has very large cumulative impact on the environment. Even the use of the small LEDs in home electronics and appliances for stand-by power accumulates a large amount of electricity (**vampire energy**). The use of coal-, oil-, and gas-driven generators for electrical power has had a tremendous negative effect in our society (**Figure 17-3**). The following types of green power will help to eliminate many of our environmental problems (**Figure 17-4**):

- **Geothermal energy** uses hot water and steam from deep underground to generate power.

- **Photovoltaic panels** directly convert sunlight into **direct current (DC)**.

- **Hydroelectricity** uses falling water from a damn, ocean tides, or waves to generate electricity.

- **Biomass** is organic matter or surplus agricultural products burned to generate electricity. Biomass is also fermented into liquid fuels.

- **Windmills/wind turbines** use the force of winds to generate electricity.

- **Passive solar** energy uses design features of a structure to capture the sun's heat.

- **Active solar** systems use panels of water, oil, or air to capture the sun's heat for home heating and/or heating water.

- **Combustible hydrogen** is separated from natural gas and water and used as an energy supply.

- **Microbial fuel cells** are powered by bacteria growing on a biofilm, which produces an electric current.

The following suggestions will significantly help with the conservation of energy and our natural resources (**Figure 17-5**):

- Use **smart structures**, which do not generate power but reduce energy consumption by controlling electrical items that are not in use.

- Encouraging the government and utility companies to develop nonpolluting energy sources.

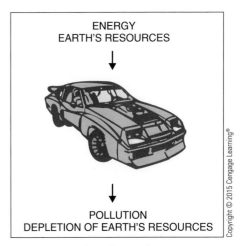

Figure 17-2 *Failure of green planning.*

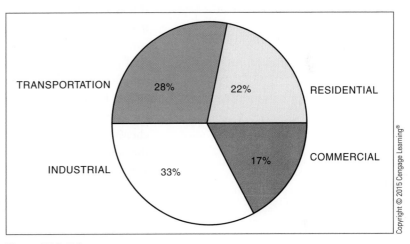

Figure 17-3 *U.S. energy usage.*

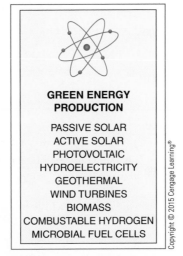

Figure 17-4 *The various methods used to produce green energy.*

Figure 17-5 *Efficient green planning.*

- Continue development of fuel-efficient automobiles.
- Encourage the sale of compressed natural gas (CNG) automobiles.
- Fully insulate and weatherstrip all openings in all buildings.

Ecology and the Environment

The terms "ecology and environment" are used to describe the physical conditions that influence the life of all people, animals, and plants, plus the condition of the earth.

With the arrival of the Industrial Age and the population explosion, large-scale production of machine-made goods introduced thousands of pollutants into the air, water, and soil on an unprecedented scale. Today, there are over two million synthetic materials and chemicals. These marvels of industrial sciences have produced many chemical compounds (**Figure 17-6**) that enhance, clean, preserve, speed up, and beautify our lives. Most of these chemical products are safe, but some are irritants, some are dangerous, and some are lethal. When a new or improved industrial product is released, the long-term effect on our health and environment is rarely studied. The development of each of these chemicals should be subject to final approval for safety by the government. However the government cannot keep pace with the deluge. Even when a toxic chemical is banned, it may be exported to another country. The questions that industry and government must answer about these new products are: What is the quantity? How is it stored? How is it diluted or mixed? Will it become harmless with time? The personnel involved in the fields of **architecture, engineering,** and **construction (AEC)** must address these questions.

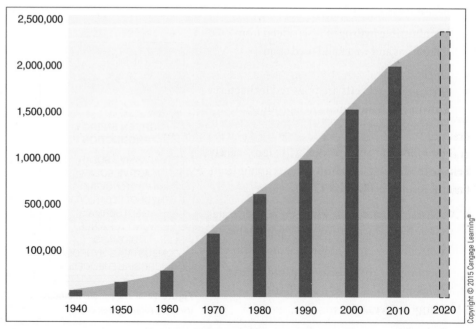

Figure 17-6 *Number of synthetic chemicals developed.*

An example a few of the most common toxic chemicals are:

- **Volatile organic compounds (VOC)** are a category of thousands of different chemicals that are in many types of newly developed materials. They may cause cancer or many other internal damages.

- **Phthalates** are used in many plastic products to soften the plastic. They cause bronchial illnesses.

- **Halogenated flame retardants (HFR)** have saved lives, but they may cause brain and thyroid damage.

- **Perfluorocarbons (PFC)** are used in the treatment of fabrics and Teflon products. It may cause development problems with children.

- **Lead** is found in paints, drinking water, toys, and lead-glazed pottery. It may cause mental impairment and developmental difficulties in children.

- **Formaldehyde** is found in many products such as in treated wood, carpets, and fabrics. It causes fatigue and possibility cancer.

- **Asbestos** is used for many types of insulation. It causes lung disease and cancer.

- **Tobacco** smoke has over 2000 chemicals of which 40 are cancerous.

Power plants that burn fossil fuels, such as coal, oil, or natural gas, create the pollutants carbon dioxide (CO_2), sulfur dioxide (SO_2), sulfur trioxide (SO_3), and nitrogen dioxide (NO_2). Also released are many trace metals, all of which cause many types of illnesses.

Planning Green

A continuing educational green planning program for the new and practicing engineers and designers will help to develop an informed green industry (**Figure 17-7**). The program should cover:

- Green design
- Green products
- The long-term effects of materials and chemicals on society and the environment.

Leadership in Energy and Environmental Design (LEED) assists in the creation of industrial developments that are environmentally friendly, energy efficient, and durable. They will perform a study of an industrial product or project and offer four different levels of certificates, depending how environmental a project may be.

The basic principles of green design are:

- Minimize the depletion of natural resources.
- Use renewable energy.
- Reduce CO_2 during manufacturing.
- Eliminate the use of toxic materials.
- Reduce, as much as possible, the pollution of the air, water, and land.

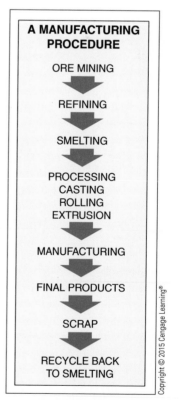

A MANUFACTURING PROCEDURE

ORE MINING

REFINING

SMELTING

PROCESSING
CASTING
ROLLING
EXTRUSION

MANUFACTURING

FINAL PRODUCTS

SCRAP

RECYCLE BACK
TO SMELTING

Figure 17-7 *Planning for a green manufacturing procedure.*

- Prevent waste.
- Study the effects of industrial materials over the short and long term.
- Maximize the efficiency of manufacturing.
- Develop long-term durable products.
- Use water efficiently during manufacturing.
- Use as many sustainable materials as possible.
- Use as many recyclable materials as possible.
- Recycle as many materials as possible.
- Do not replace items that can be repaired.

Green Setbacks

Going green often becomes a marketing gimmick as a way to increase profits. This is the result of greed by corporations and individuals. They want us to start replacing autos, appliances, housing, and the like. and buy eco-friendly products. What they do not say is that a lot of energy and resources are used to manufacture and transport these new green products. What's more, the items we discard will end up in a landfill. Many causes of the green setbacks are:

- Building cheap and more quickly made products for more profit.
- Using too many dangerous synthetic chemicals in manufacturing.
- The billions of tons of waste products produced each year.
- Government officials being reluctant to offend powerful corporations if this interferes with their contributions.

Conclusion

Green planning embraces the concept that decisions to protect human health and the environment can have the greatest impact when applied early to the design, engineering, testing, and development of industrial processes and products.

Our growing concern is not new. More than 100 years ago John Muir, a well-known naturalist, strove to improve the environment. He said "When we try to pick out anything by itself, we find it hitched to everything in the universe."

DESIGN EXERCISES

1. Go online to LEED (www.usgbc.org) and find out about their point system and certificates.

2. Design a curbside mailbox with sustainable materials.

3. Make a list of sustainable materials that could go into the manufacturing of an automobile.

4. Make a list of toxic materials that may go into the manufacturing of an automobile.

5. Make a list of sustainable materials that could go into the construction of a home.

6. Make a list of toxic materials that may go into the construction of a house.

7. Make a list of possible toxic materials in your classroom.

8. Study the green energy sources in **Figure 17-8**. Write a paper on how you may make your home more energy efficient.

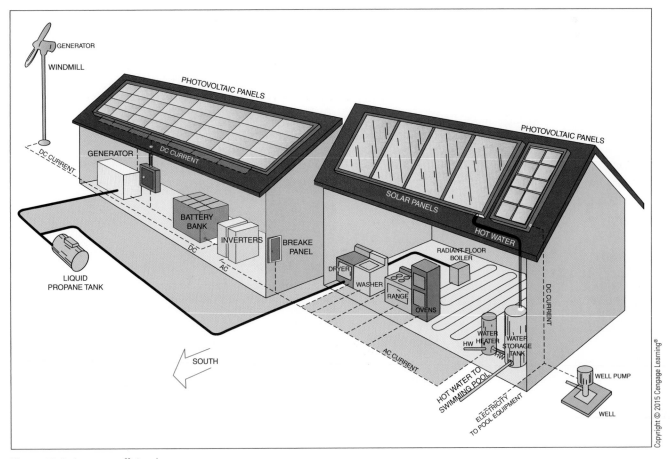

Figure 17-8 *An energy efficient home.*

KEY TERMS

Active solar

Architecture, engineering,
   and construction (AEC)

biomass

Direct current (DC)

Geothermal energy

Hydroelectricity

Leadership in Energy and
   Environmental Design
   (LEED)

Microbial fuel cells

Passive solar

Perfluorocarbons
   (PFC)Photovoltaic
   panel

Smart structure

sustainability

Vampire energy

Volatile organic
   compounds (VOC)

Windmills/wind turbines

Working Drawings

Introduction

Working drawings provide specific information and instructions needed to manufacture or construct products and structures. This includes detailed graphics of precise shapes and dimensions of exact sizes. Working drawings also contain other notations for manufacturing, such as material descriptions, finishing methods, and tolerances. Working drawings are used in all phases of manufacturing, including fabrication and assembly. All information needed to manufacture a product must be included in a working drawing. The manufacturer must not be forced to guess about any detail to produce a product as designed.

Working Drawings

There are two general categories of working drawings—detail drawings and assembly drawings. **Working drawings** usually include fully dimensioned multiview drawings of the assembled product, plus detail drawings for each part and component (**Figure 18-1**). **Detail drawings** contain information relating to the manufacture of parts. **Assembly drawings** describe the relationship of parts and the product assembly.

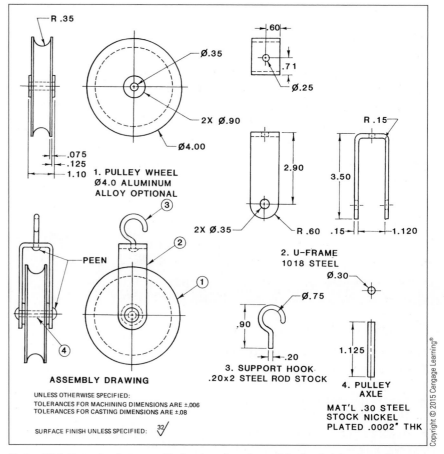

Figure 18-1 *Examples of an assembly drawing of a pulley and the detailed drawings of its parts.*

Pictorial drawings may be used as working drawings if the exact shape and necessary dimensions can be shown with clarity (**Figure 18-2**). Pictorial drawings are usually used to illustrate product assembly, whereas multiview drawings are used to convey manufacturing details.

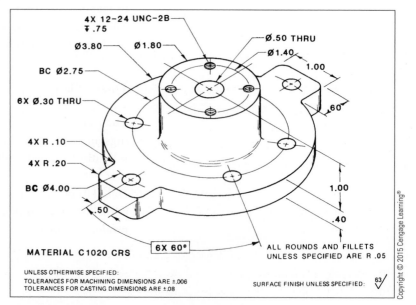

Figure 18-2 *Pictorial working drawing of a plate cover.*

DETAIL DRAWINGS

A product may consist of a single part or contain hundreds or thousands of parts (**Figure 18-3**). The manufacture of each part requires a working drawing containing specifications for its manufacture. These include dimensions, tolerances, scale, material, machining and casting details, fastening methods, finishing instructions, identification numbers, parts and standard purchase lists, and indexes to related part drawings. A typical working drawing of a single part is shown in **Figure 18-4**.

Since different parts may be manufactured and/or assembled in different locations, it is imperative that uniform standards, such as ASME, be used in the preparation of all detail drawings. Unless the entire manufacturing operation is contained in one location, detail drawings for one part should not be placed on the same sheet as drawings for another part. Even then, careful indexing and cross-indexing are necessary to avoid confusion. Drawing sheet layout standards, which outline indexing methods, are covered in Chapter 6.

Specialized detail drawings are often prepared in separate sets for different manufacturing operations. Specialized sets may be prepared for foundry use, rough machining, finish machining, and surface finishing. The detail drawings show only the information needed for the particular operation. When this is done, great care must be taken to ensure that the individual sets are totally compatible, especially in the assignment of tolerances.

ASSEMBLY DRAWINGS

A drawing that shows the relationship of two or more parts of a product is an assembly drawing (**Figure 18-5**). There are many different types of assembly drawings. The amount of detail and drawing type depend on its use during manufacturing. Assembly drawings do not contain as much detail as detail drawings. Most are prepared without hidden lines to promote clarity. The following are the most common types of assembly drawings.

- **General assembly drawings** include orthographic drawings of assembled products, parts lists, and related data needed in the manufacturing process (**Figure 18-6**).

- **Layout assembly drawings** show rough preliminary designs, design ideas, sometimes as a freehand sketch, and/or pictorial forms. They show overall size, form, location, and relationship of only major components (**Figure 18-7**).

Figure 18-3 *Space shuttle.*

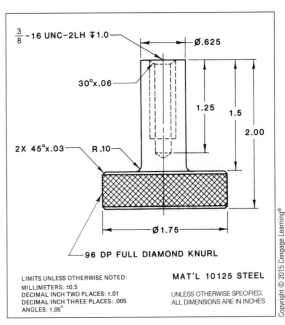

Figure 18-4 *Detail working drawing for an adjustment knob from a lathe assembly.*

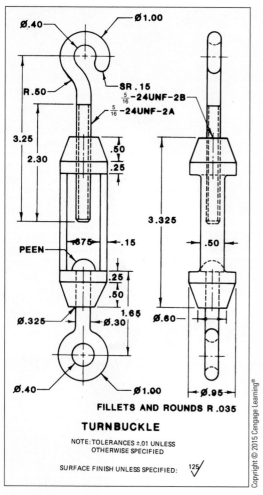

FILLETS AND ROUNDS R .035

TURNBUCKLE

NOTE: TOLERANCES ±.01 UNLESS
OTHERWISE SPECIFIED

SURFACE FINISH UNLESS SPECIFIED: 125

Figure 18-5 *Assembly drawing.*

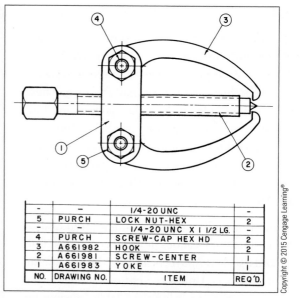

–	–	1/4-20 UNC	–
5	PURCH	LOCK NUT-HEX	2
–	–	1/4-20 UNC X I 1/2 LG.	–
4	PURCH	SCREW-CAP HEX HD	2
3	A661982	HOOK	2
2	A661981	SCREW-CENTER	I
I	A661983	YOKE	I
NO.	DRAWING NO.	ITEM	REQ'D.

Figure 18-6 *General assembly drawing and parts list for a gear puller. (Reprinted from Technical Drawing and Engineering Communication by Goetsch, Chalk, Nelson, and Rickman, Delmar/Cengage Learning.)*

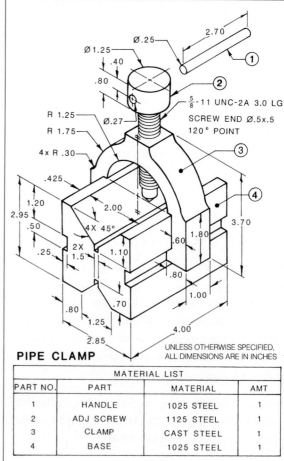

PIPE CLAMP

UNLESS OTHERWISE SPECIFIED,
ALL DIMENSIONS ARE IN INCHES

MATERIAL LIST			
PART NO.	PART	MATERIAL	AMT
1	HANDLE	1025 STEEL	1
2	ADJ SCREW	1125 STEEL	1
3	CLAMP	CAST STEEL	1
4	BASE	1025 STEEL	1

LIMITS UNLESS OTHERWISE NOTED:
MILLIMETERS: ±0.5
DECIMAL INCH DIMENSIONS: TWO PLACE ±.01
DECIMAL INCH DIMENSIONS: THREE PLACE ±.005
ANGLES: ±.05°

Figure 18-7 *Pictorial assembly.*

- **Design assembly drawings** are refined versions of layout assembly drawings with more accurate and refined dimensions, linework, and manufacturing data (**Figure 18-8**).

- **Working assembly drawings** are fully dimensioned and noted drawings that are used directly in the manufacturing process (**Figure 18-9**).

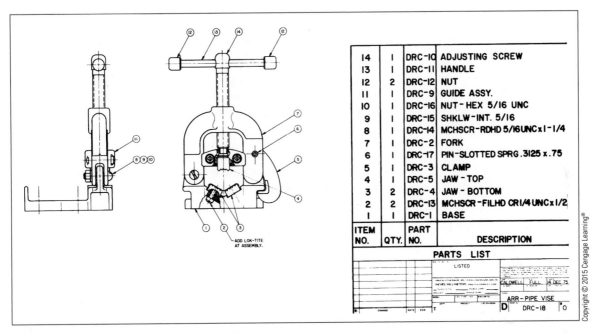

Figure 18-8 *Design assembly drawing for a pipe vise.*

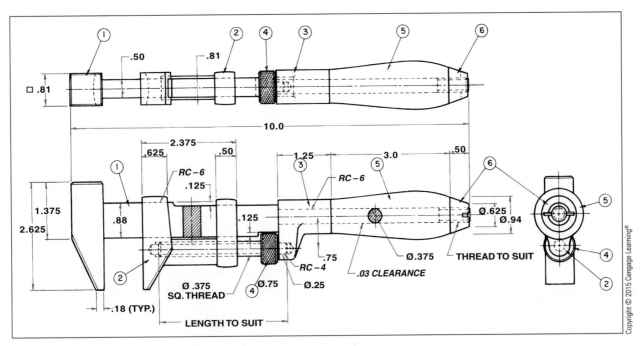

Figure 18-9 *Working assembly drawing for a wrench.*

- **Erection assembly drawings** are used in the construction industry and include all information necessary for the erection of structural framework. Detailed information relating to structural members is omitted (**Figure 18-10**).

- **Subassembly drawings** show the smallest assembled components that are parts of a larger component. Working drawings may include many layers of subassembly drawings, depending on the size and complexity of the finished product (**Figure 18-11**).

- **Pictorial assembly drawings** are substitutes for multiview assembly drawings if the product is not complex or if it is intended for nontechnical interpretation (**Figure 18-12**). Many pictorial assembly drawings are sectional views designed to show the interior structure and relationship of static and moving parts and components (**Figure 18-13**). Pictorial assembly drawings are often drawn in exploded-view form (see Chapter 15) to show the relationship of parts before assembly.

- **Outline assembly drawings** show only the major outline of an assembly or subassembly and include only a few basic notations. Symmetrical objects are often sparsely detailed on one half with only major outlines shown on the other half (**Figure 18-14**). They are also used to show only the basic positioning of subassemblies. Catalog publications often use outline assembly drawings.

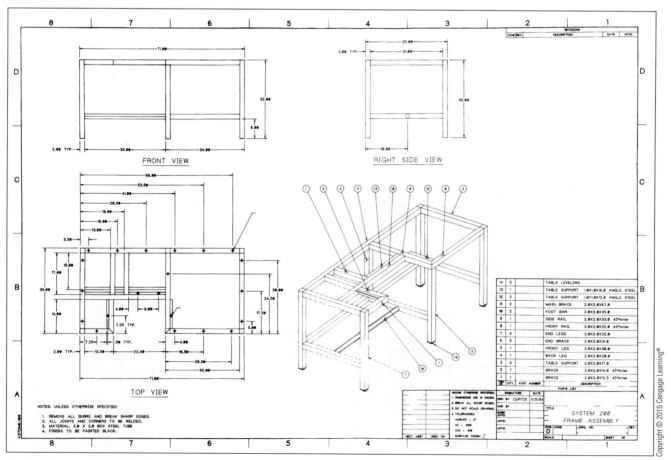

Figure 18-10 *Erection assembly drawing of a bench frame.*

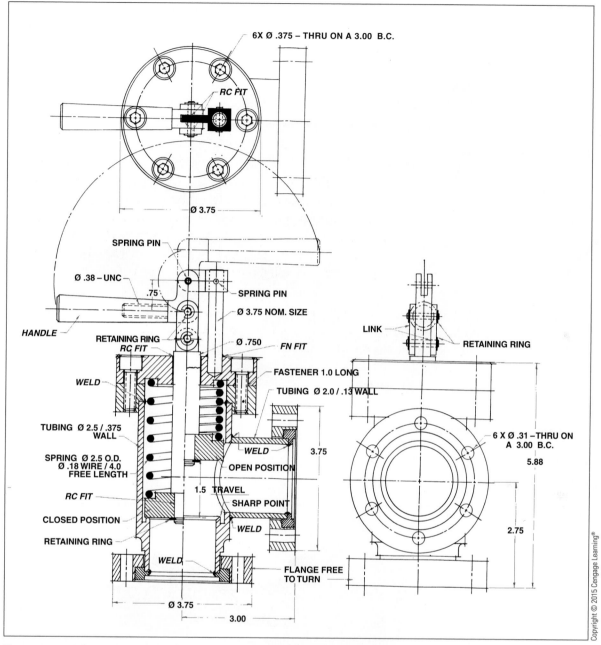

Figure 18-11 *Subassembly drawing. (Reprinted from Technical Drawing and Engineering Communication by Goetsch, Chalk, Nelson, and Rickman, Delmar/Cengage Learning.)*

- **Operation assembly drawings** illustrate the mechanical operation. Multiple positions of moving parts are often shown with phantom lines, sections, and enlarged details (**Figure 18-15**).

- **Diagram assembly drawings** show the form and location of all subassemblies with multiview drawings. Partial cutaway sections and limited hidden lines are often used to show only the basic outline of hidden parts. Phantom lines are used to depict adjacent parts (**Figure 18-16**).

- **Installation assembly drawings** show how to assemble products. Only the details and dimensions needed for assembly are included (**Figure 18-17**). These often contain multiview drawings and a pictorial drawing as an aid to interpretation.

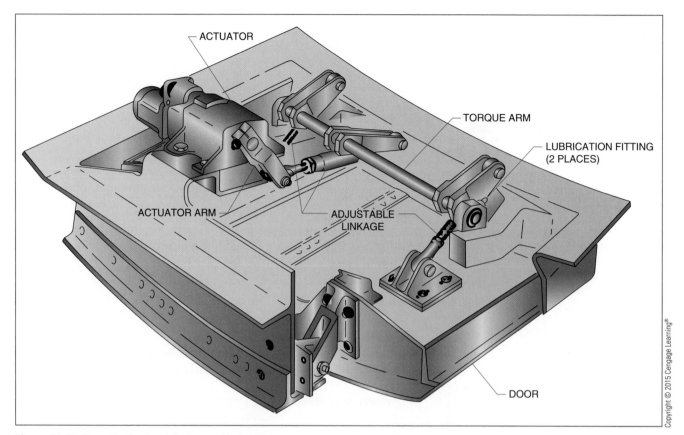

Figure 18-12 *Example of a pictorial subassembly drawing.*

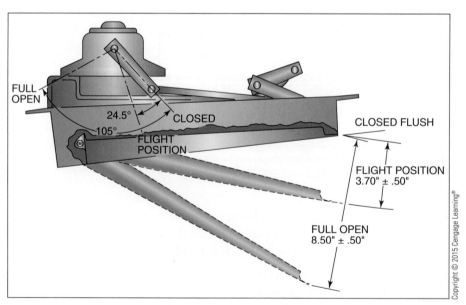

Figure 18-13 *An interior subassembly drawing showing components and moving parts.*

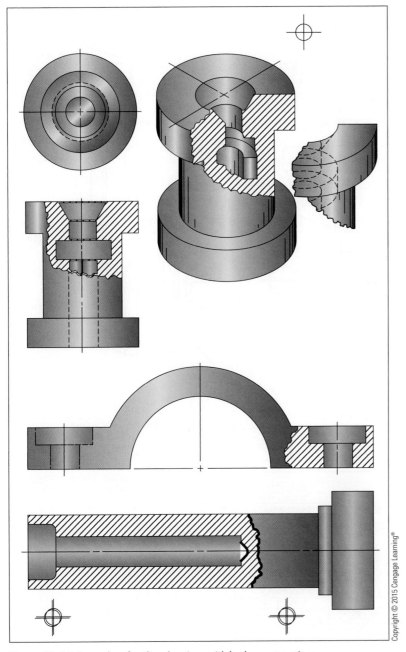

Copyright © 2015 Cengage Learning®

Figure 18-14 *Examples of outline drawings with broken-out sections.*

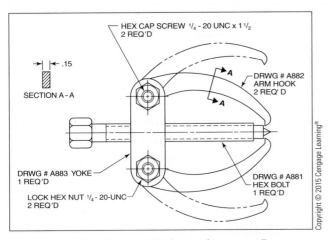

Copyright © 2015 Cengage Learning®

Figure 18-15 *Operation assembly drawing for a gear puller.*

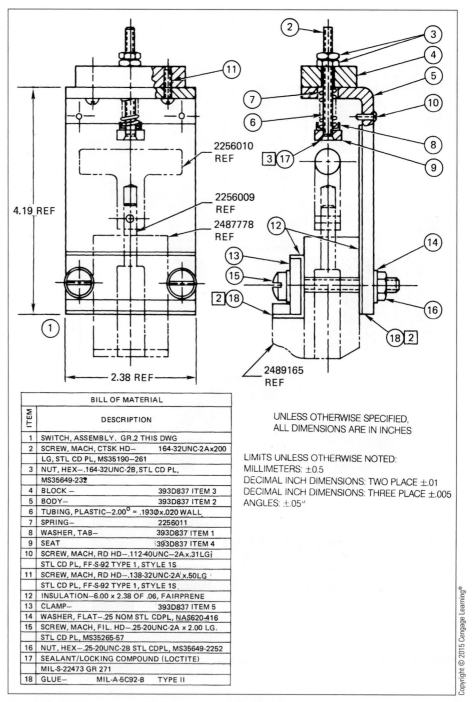

Figure 18-16 *Diagram assembly drawing of a valve adjuster.*

BILL OF MATERIAL	
ITEM	DESCRIPTION
1	SWITCH, ASSEMBLY. GR.2 THIS DWG
2	SCREW, MACH, CTSK HD— .164-32UNC-2A x200 LG, STL CD PL, MS35190—261
3	NUT, HEX—.164-32UNC-2B,STL CD PL, MS35649-232
4	BLOCK— 393D837 ITEM 3
5	BODY— 393D837 ITEM 2
6	TUBING, PLASTIC—2.00° = .1930 x.020 WALL
7	SPRING— 2256011
8	WASHER, TAB— 393D837 ITEM 1
9	SEAT 393D837 ITEM 4
10	SCREW, MACH, RD HD—.112-40UNC—2A x.31LG STL CD PL, FF-S-92 TYPE 1, STYLE 1S
11	SCREW, MACH, RD HD—.138-32UNC-2A x.50LG STL CD PL, FF-S-92 TYPE 1, STYLE 1S
12	INSULATION—6.00 x 2.38 OF .06, FAIRPRENE
13	CLAMP— 393D837 ITEM 5
14	WASHER, FLAT—.25 NOM STL CDPL, NAS620-416
15	SCREW, MACH, FIL. HD—.25-20UNC-2A x 2.00 LG. STL CD PL, MS35265-57
16	NUT, HEX—.25-20UNC-2B STL CDPL, MS35649-2252
17	SEALANT/LOCKING COMPOUND (LOCTITE) MIL-S-22473 GR 271
18	GLUE— MIL-A-5C92-B TYPE II

UNLESS OTHERWISE SPECIFIED,
ALL DIMENSIONS ARE IN INCHES

LIMITS UNLESS OTHERWISE NOTED:
MILLIMETERS: ±0.5
DECIMAL INCH DIMENSIONS: TWO PLACE ±.01
DECIMAL INCH DIMENSIONS: THREE PLACE ±.005
ANGLES: ±.05°

Working Drawing Dimensions

Dimensions on working drawings are used when manufacturing products to an exact specified size. Dimensions must be selected and placed to insure that no further calculations are required during any stage of the manufacturing process. All related drawings must use the same standards for tolerancing to insure proper fitting in the final assembly. **Figure 18-18** shows a summary of the use of the U.S. customary system of measure. **Figures 18-19** through **18-21** show the metric system of measure.

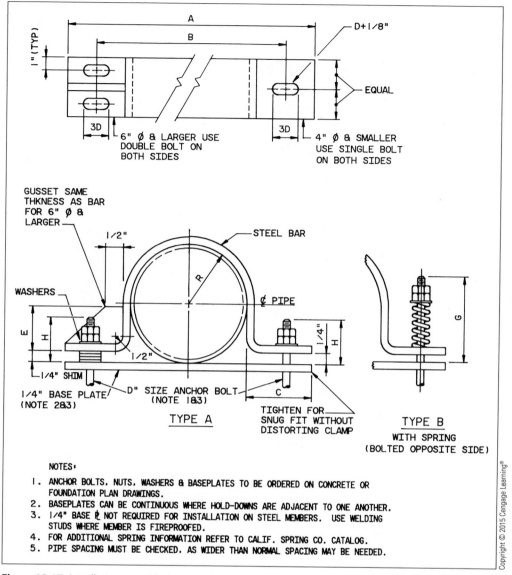

NOTES:

1. ANCHOR BOLTS, NUTS, WASHERS & BASEPLATES TO BE ORDERED ON CONCRETE OR FOUNDATION PLAN DRAWINGS.
2. BASEPLATES CAN BE CONTINUOUS WHERE HOLD-DOWNS ARE ADJACENT TO ONE ANOTHER.
3. 1/4" BASE ℄ NOT REQUIRED FOR INSTALLATION ON STEEL MEMBERS. USE WELDING STUDS WHERE MEMBER IS FIREPROOFED.
4. FOR ADDITIONAL SPRING INFORMATION REFER TO CALIF. SPRING CO. CATALOG.
5. PIPE SPACING MUST BE CHECKED, AS WIDER THAN NORMAL SPACING MAY BE NEEDED.

Figure 18-17 *Installation assembly drawing of a base plate for pipes.*

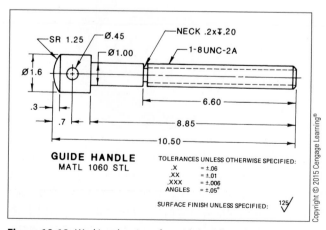

Figure 18-18 *Working drawing of a guide handle with U.S. customary dimensions (ASME).*

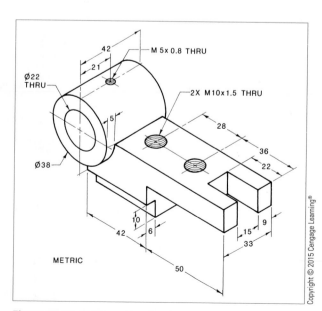

Figure 18-19 *Metric working drawing.*

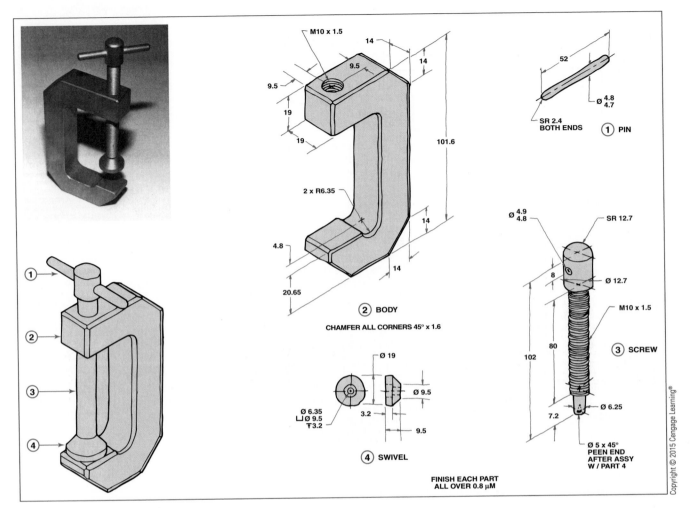

Figure 18-20 *Metric working drawing.*

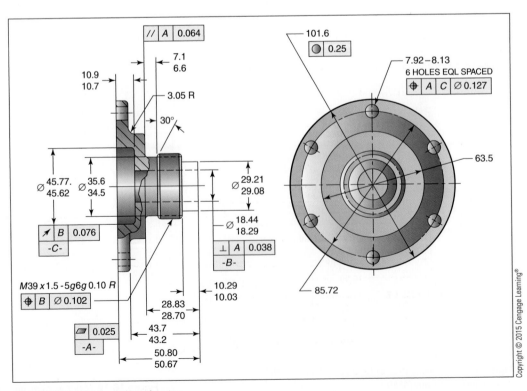

Figure 18-21 *Metric working drawing.*

Although working drawing dimensions must be extremely accurate and consistent, specific manufacturing processes should not be specified. Exactly how the product is manufactured is the responsibility of the manufacturer, not the engineer, designer, or drafter. Only the final desired result is specified on working drawings. Outcomes may be specified by machine process notations on working drawings. In these cases, the end result is also the goal, not the process (**Figure 18-22**). Since there are other manufacturing methods that can produce the same result, only the size and form of the final product are interpreted from the drawing—not the specific process.

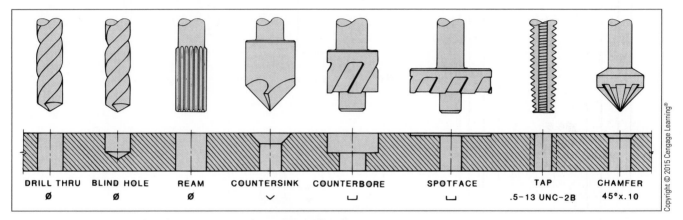

Figure 18-22 *Typical types of machining operations depicted on working drawings.*

DRAFTING EXERCISES

The following working drawing exercises are arranged from elementary to complex. All exercises may be drawn freehand, manual drafting, and/or with computer-aided drafting. Following your instructor's instructions, draw the multiview and/or the isometric for **Figures 18-23** through **18-62**.

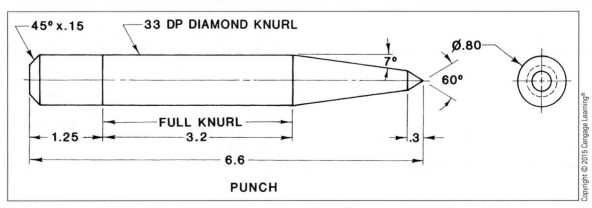

Figure 18-23 *Punch.*

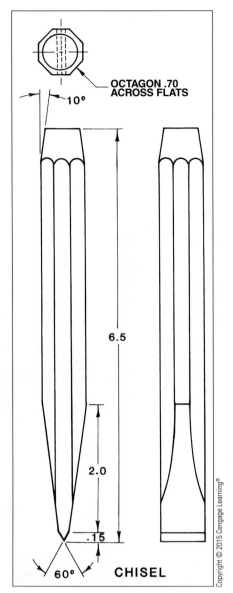

Figure 18-24 *Punch.*

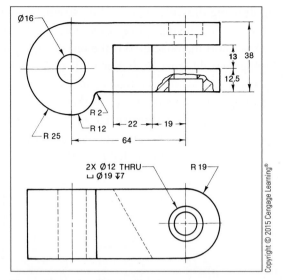

Figure 18-25 *Landing gear bearing.*

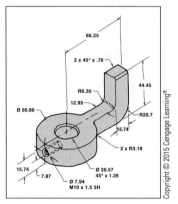

Figure 18-26 *Lathe dog.*

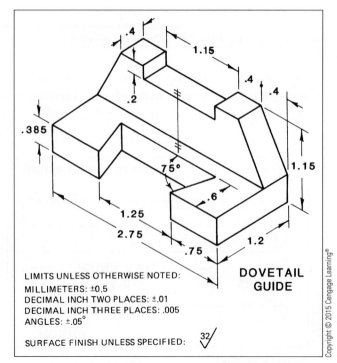

LIMITS UNLESS OTHERWISE NOTED:

MILLIMETERS: ±0.5
DECIMAL INCH TWO PLACES: ±.01
DECIMAL INCH THREE PLACES: .005
ANGLES: ±.05°

SURFACE FINISH UNLESS SPECIFIED:

Figure 18-27 *Dovetail guide.*

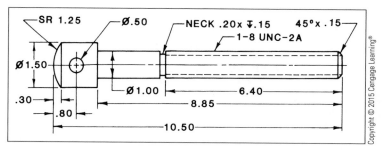

Figure 18-28 *Threader arbor.*

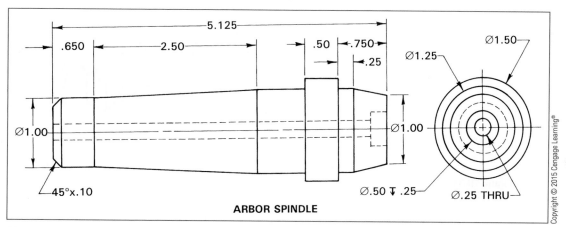

Figure 18-29 *Arbor spindle.*

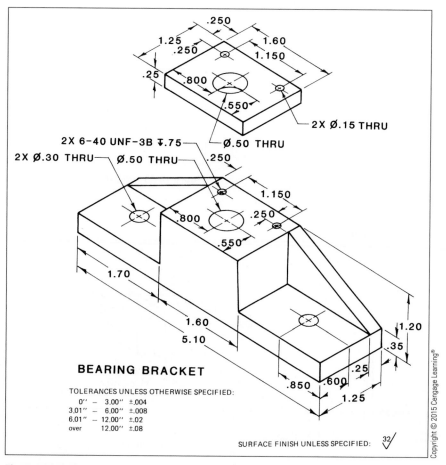

Figure 18-30 *Bearing bracket.*

Ø1.00 ⌁.175   .75

□.40 THRU

R .80
2X

Ø1.50

4 X
R .10

4.4

.45

.35

CRS

Figure 18-31 *Locking Handle.*

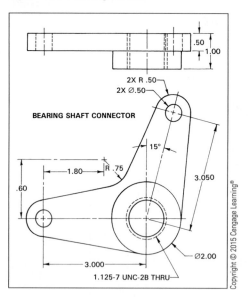

BEARING SHAFT CONNECTOR

2X R .50
2X Ø.50

15°

R .75

1.80

.60

3.050

3.000

Ø2.00

1.125-7 UNC-2B THRU

Figure 18-32 *Bearing shaft connector.*

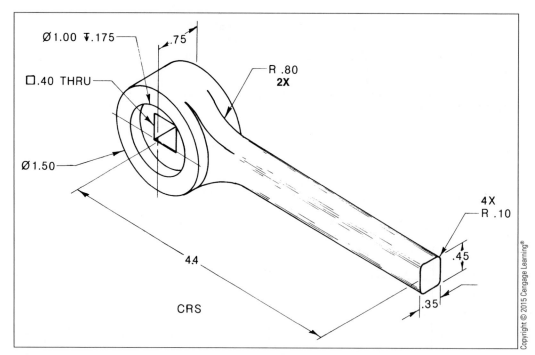

Figure 18-33 *Bracket support.*

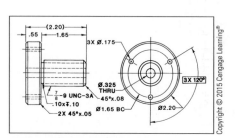

(2.20)
.55   1.65

3X Ø.175

Ø.325
THRU
45°x.08

7/8 - 9 UNC-3A
.10x⌁.10
2X 45°x.05

Ø1.65 BC

Ø2.20

3X 120°

Figure 18-34 *Threaded spindle.*

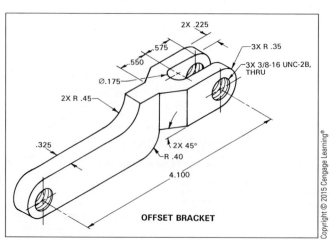

2X .225
.575
.550
Ø.175

3X R .35

3X 3/8-16 UNC-2B,
THRU

2X R .45

.325

R .40

2X 45°

4.100

OFFSET BRACKET

Figure 18-35 *Offset bracket.*

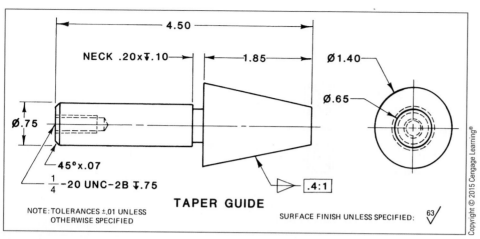

NECK .20x⍶.10

Ø.75

45°x.07

¼ –20 UNC–2B ⍶.75

4.50

1.85

Ø1.40

Ø.65

.4:1

TAPER GUIDE

NOTE: TOLERANCES ±.01 UNLESS OTHERWISE SPECIFIED

SURFACE FINISH UNLESS SPECIFIED: 63

Figure 18-36 *Taper guide.*

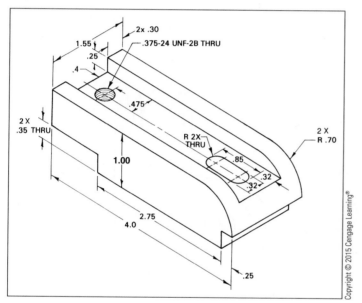

2x .30

.375-24 UNF-2B THRU

1.55

.25

.4

.475

R 2X THRU

2 X R .70

.85

.32

.32

2 X .35 THRU

1.00

2.75

4.0

.25

Figure 18-37 *Slip guide.*

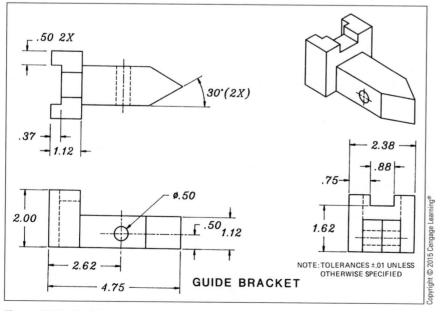

.50 2X

30°(2X)

.37

1.12

2.00

Ø.50

.50 1.12

2.62

4.75

GUIDE BRACKET

2.38

.88

.75

1.62

NOTE: TOLERANCES ±.01 UNLESS OTHERWISE SPECIFIED

Figure 18-38 *Guide Bracket.*

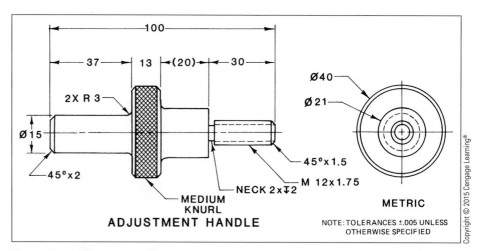

Figure 18-39 *Adjustment handle.*

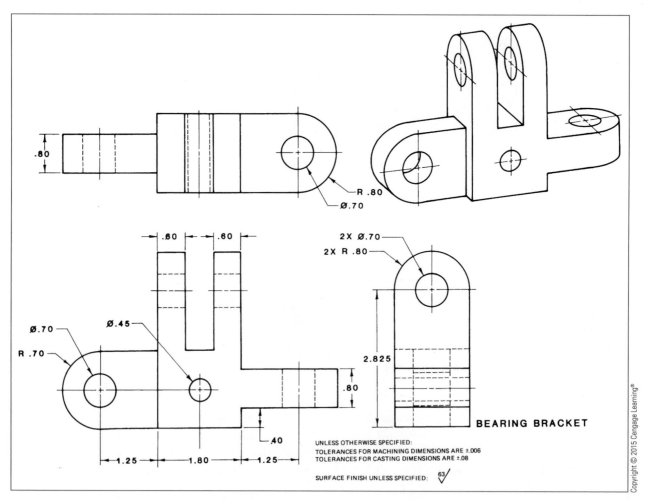

Figure 18-40 *Bearing bracket.*

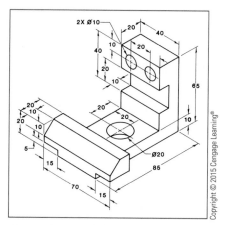

Figure 18-41 *Modular guide.*

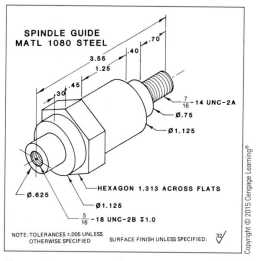

Figure 18-42 *Spindle guide.*

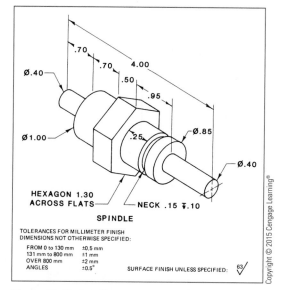

Figure 18-43 *Spindle.*

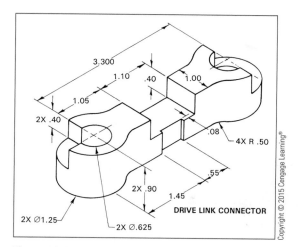

Figure 18-44 *Drive link connector.*

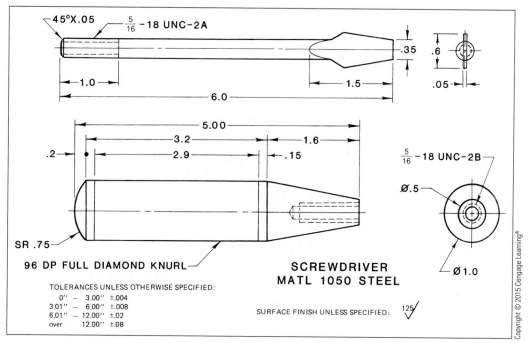

Figure 18-45 *Screwdriver.*

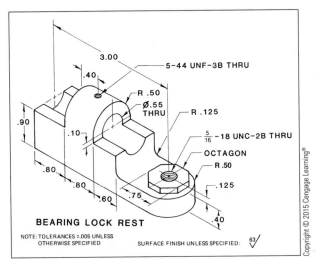

Figure 18-46 *Bearing lock rest.*

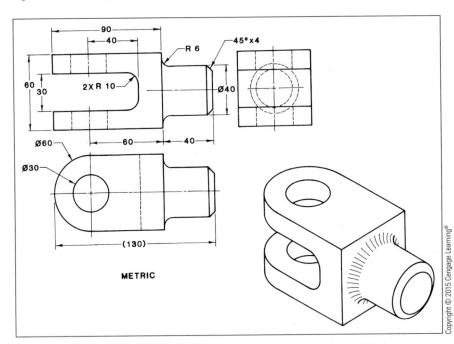

Figure 18-47 *Metric harness.*

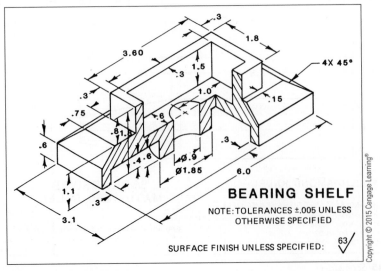

Figure 18-48 *Bearing shelf.*

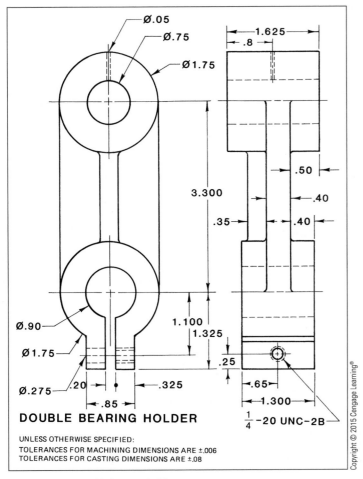

DOUBLE BEARING HOLDER

UNLESS OTHERWISE SPECIFIED:
TOLERANCES FOR MACHINING DIMENSIONS ARE ±.006
TOLERANCES FOR CASTING DIMENSIONS ARE ±.08

Figure 18-49 *Double bearing holder.*

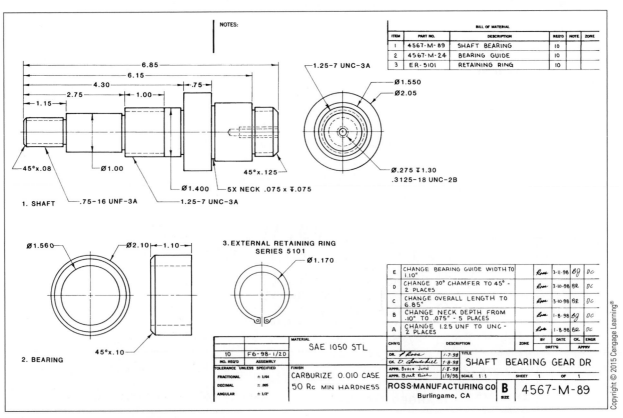

Figure 18-50 *Shaft bearing gear drive.*

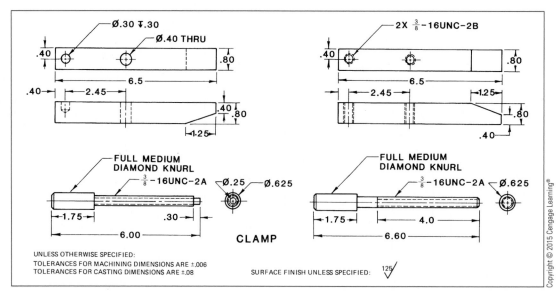

Figure 18-51 *Clamp.*

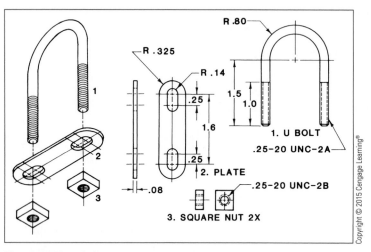

Figure 18-52 *U-bolt.*

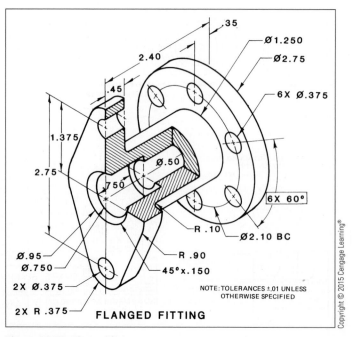

Figure 18-53 *Flanged fitting.*

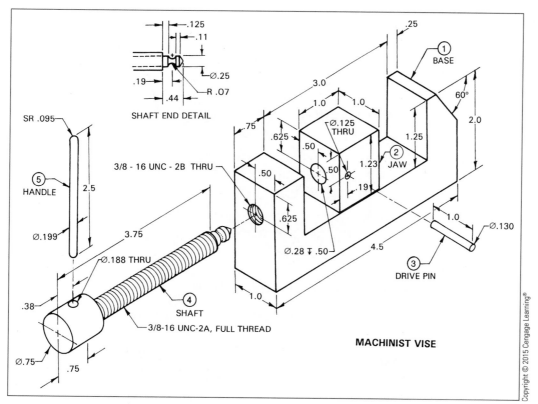

Figure 18-54 *Machinist vise.*

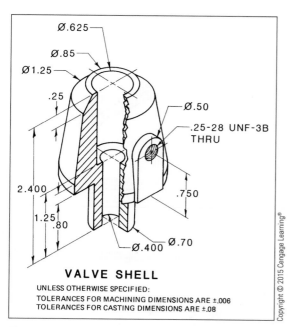

Figure 18-55 *Valve shell.*

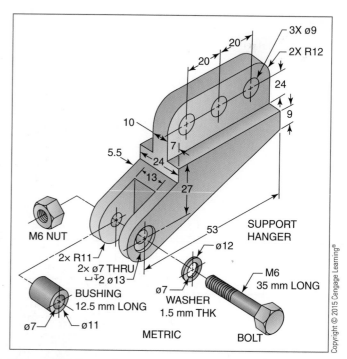

Figure 18-56 *Support hanger.*

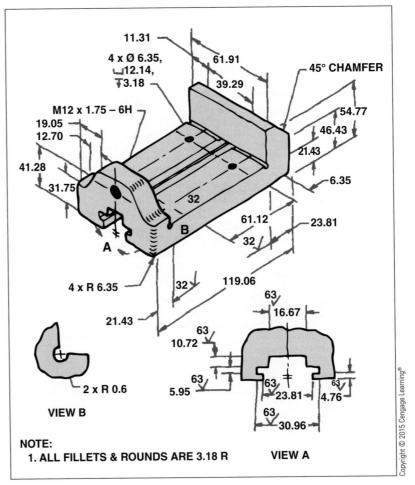

Figure 18-57 *Vise bed.*

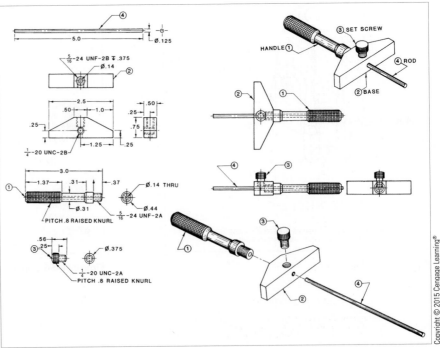

Figure 18-58 *Depth gage.*

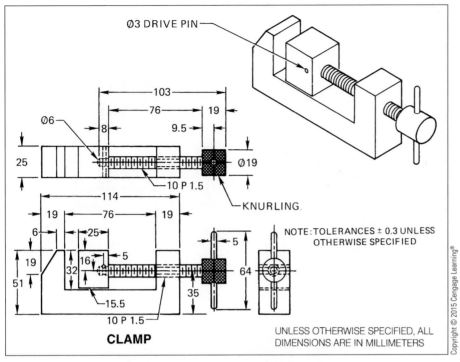

Figure 18-59 *Machinist clamp.*

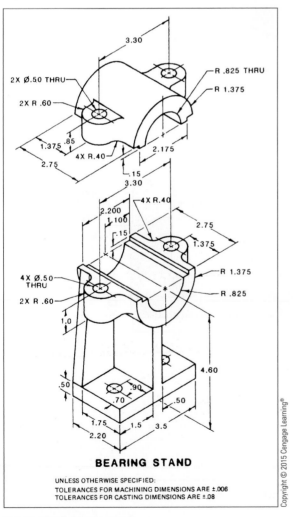

BEARING STAND

UNLESS OTHERWISE SPECIFIED:
TOLERANCES FOR MACHINING DIMENSIONS ARE ±.006
TOLERANCES FOR CASTING DIMENSIONS ARE ±.08

Figure 18-60 *Bearing stand.*

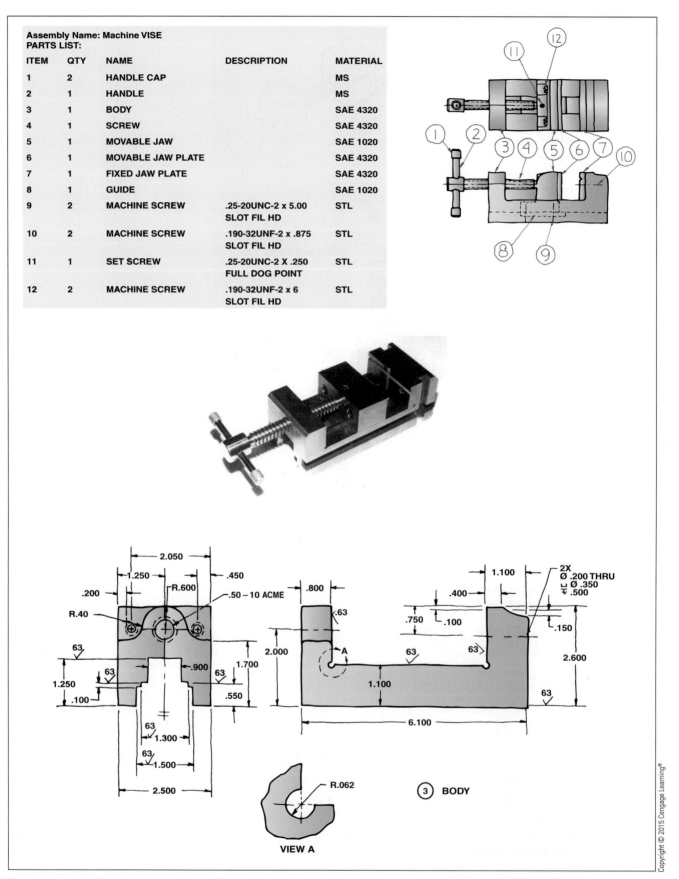

Assembly Name: Machine VISE
PARTS LIST:

ITEM	QTY	NAME	DESCRIPTION	MATERIAL
1	2	HANDLE CAP		MS
2	1	HANDLE		MS
3	1	BODY		SAE 4320
4	1	SCREW		SAE 4320
5	1	MOVABLE JAW		SAE 1020
6	1	MOVABLE JAW PLATE		SAE 4320
7	1	FIXED JAW PLATE		SAE 4320
8	1	GUIDE		SAE 1020
9	2	MACHINE SCREW	.25-20UNC-2 x 5.00 SLOT FIL HD	STL
10	2	MACHINE SCREW	.190-32UNF-2 x .875 SLOT FIL HD	STL
11	1	SET SCREW	.25-20UNC-2 X .250 FULL DOG POINT	STL
12	2	MACHINE SCREW	.190-32UNF-2 x 6 SLOT FIL HD	STL

Figure 18-61a *Machine vise. Part 1. (Reprinted from Mechanical Drafting by Madsen, Schumaker, and Stewart, Delmar Learning.)*

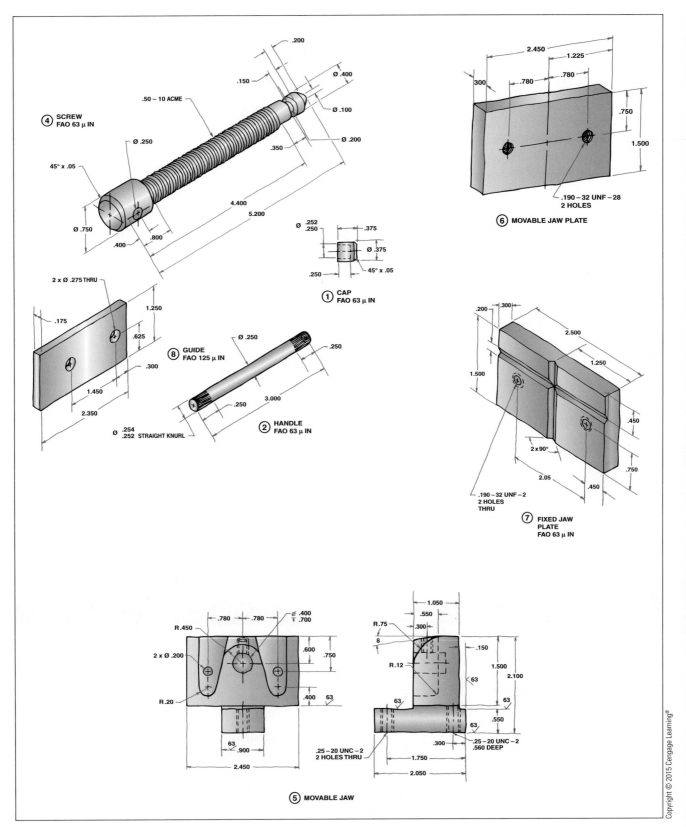

Figure 18-61b *Machine vise. Part 2. (Reprinted from Mechanical Drafting by Madsen, Schumaker, and Stewart, Delmar Learning.)*

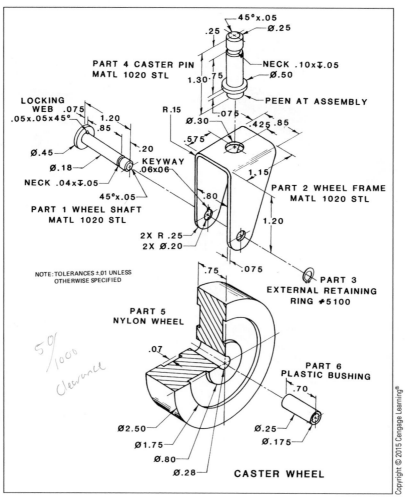

Figure 18-62 *Castor wheel.*

DESIGN EXERCISES

1. Redesign the items in **Figures 18-63** through **18-66** so that they will perform their tasks more efficiently.

2. Design a more efficient type of kitchen cabinet door clasp.

3. Design an esthetic drawer pull for kitchen cabinet drawers.

4. Design a childproof latch for bathroom medicine cabinets.

5. Design an esthetic hubcap for an automobile. Use standard dimensions.

6. Design a furniture module to hold a CAD system's hardware.

7. Design a removable clasp that may be used to lift heavy 2' × 2' × 3' boxes of books.

8. Redesign the caster wheel in **Figure 18-62**.

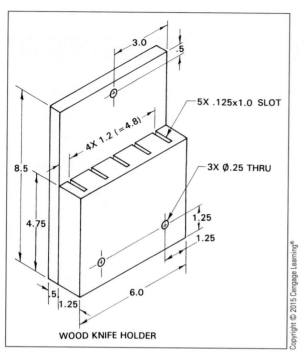

WOOD KNIFE HOLDER

Figure 18-63 *Redesign the knife holder to hold three more large knives (maximum blade width of 1.75" and a sharpening steel of .5" diameter).*

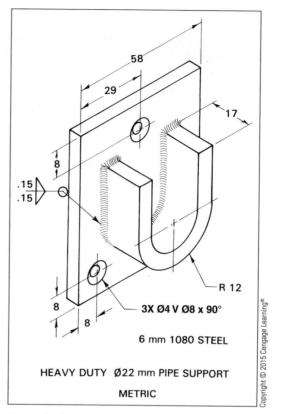

HEAVY DUTY Ø22 mm PIPE SUPPORT

METRIC

Figure 18-64 *Design a locking mechanism for the steel pipe's installation.*

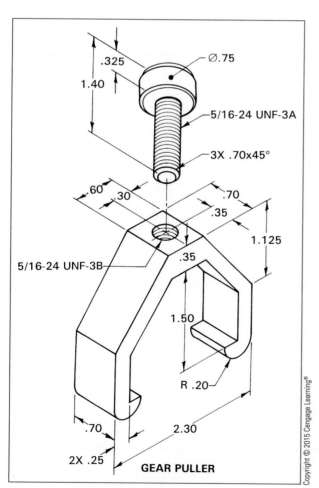

GEAR PULLER

Figure 18-65 *Redesign the gear puller for a 4" diameter gear removal.*

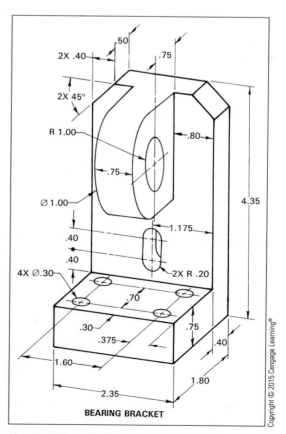

BEARING BRACKET

Figure 18-66 *Redesign the bearing bracket holder to support a 1.25" diameter bearing with a total weight of 1,500 lbs. (double the weight of the original design).*

KEY TERMS

Assembly drawing

Design assembly drawing

Detail drawing

Diagram assembly drawing

Erection assembly drawing

General assembly drawing

Installation assembly drawing

Layout assembly drawing

Operation assembly drawing

Outline assembly drawing

Pictorial assembly drawing

Subassembly drawing

Working assembly drawing

Working drawing

Welding Drawings

The student will be able to:

- Read welding abbreviations
- Select the proper type of weld
- Read welding symbols
- Draw a weld reference symbol
- Select the proper weld finish
- Specify contour symbols
- Specify the size and strength of a weld
- Specify the length of a weld
- Specify the pitch of a weld
- Draw a weld symbol with a CAD system

Introduction

When designing products with multiple parts, designers have several options for joining parts, including mechanical devices, adhesive bonding, and welding. The joining method depends on the material to be joined, the strength of the joint required, and the need for disassembly. Mechanical devices include nails, screws, bolts, rivets, keys, and splines. Adhesive bonding methods range from cement and epoxy glues to plastic bonding solutions. Welding is a method of permanently joining metals and plastics through the application of heat. Welding is usually chosen when great strength, rigidity, permanence, or liquid-proofing is required.

Welding began over 4,000 years ago after the discovery of iron, but the science and technology of welding are relatively new. Early welders simply heated two pieces of iron in a forge, dipped them into a flux, and then pounded the pieces together with a hammer. Later, oxygen and acetylene gas were added to increase the amount of heat produced. This enabled harder metals to be welded. Nevertheless, welding was restricted to iron and iron alloys until the twentieth century. Today, hundreds of metallurgical processes are used to weld thousands of different metals and alloys. Vehicles, aircraft, ships, and machinery contain thousands of welds of different types and categories. These modern welds are usually stronger, lighter, more durable, and more economical than other methods of assembly (**Figure 19-1**).

Figure 19-1 *Welding is a major procedure used to bond metals and plastics.*

Welding Processes

The specific metallurgic combination of heat, energy, and materials used to create a weld between two metals is called a **welding process**. There are hundreds of processes and combinations of processes now available. **Figure 19-2** shows the most commonly used processes organized under the major categories of welding. The abbreviations shown next to each process are used to specify the welding process on working drawings. All welding processes join metals together using either solid-state welding, fusion welding, brazing, or soldering.

SOLID-STATE WELDING

Solid-state welding involves heating two workpieces and forcing them together until the hot material on the adjoining surfaces becomes mixed. Solid-state welding processes include forge welding, friction welding, explosive welding, and roll welding. Thin, soft, malleable metals such as wrought iron, copper, and brass are well suited to solid-state welding processes.

FUSION WELDING

Fusion welding involves the common edge of two metal parts being melted to a molten state. The molten parts, combined with a filler material from a welding rod, form a molten puddle. When the puddle cools and solidifies, the separate pieces are permanently fused as one. The most common fusion welding processes include **arc welding**, **gas welding**, and **resistance welding**.

Arc Welding. This process uses an electric arc struck between a workpiece and an electrode. The extreme heat from the arc fuses the metals together with material from the consumable electrode. Metals well suited to arc welding include copper, brass, soft iron, aluminum, low-carbon steel, cast steel, stainless steel, and nickel alloys.

Gas Welding. The combination of gas (usually acetylene or hydrogen) with air (or oxygen) is known as **oxyacetylene welding**. Oxyacetylene welding can produce temperatures that will weld most metals, except very high carbon steel and chromium.

Resistance Welding. Both heat and pressure are used in resistance welding. A concentrated electrical current is passed through adjoining metals. The resistance to the electrical charge melts the metal while pressure is applied to the areas to be joined. When the metal cools and the pressure is removed, the separate pieces are permanently joined. When the electrical charge is concentrated on a very small area, the type of welding is spot welding. When the charge is moved along the edge of a joint, the type of welding is seam welding. Resistance welding is widely used to spot or seam weld sheet metal parts together.

WELDING PROCESS	LETTER DESIGNATION
ARC WELDING	
• Bare metal arc welding	BMAW
• Stud welding	SW
• Gas shielded stud welding	GSSW
• Submerged arc welding	SAW
• Gas tungsten arc welding	GTAW
• Gas metal arc welding	GMAW
• Atomic hydrogen welding	AHW
• Shielded metal arc welding	SMAW
• Twin carbon arc welding	TCAW
• Gas carbon arc welding	GCAW
• Flux cored arc welding	FCAW
RESISTANCE WELDING	
• Flash welding	FW
• Upset welding	UW
• Percussion welding	PEW
THERMIT WELDING	
• Nonpressure thermit welding	NTW
• Pressure thermit welding	PTW
GAS WELDING	
• Pressure gas welding	PGW
• Oxyhydrogen welding	OHW
• Oxyacetylene welding	OAW
• Air-acetylene welding	AAW
FORGE WELDING	
• Roll welding	RW
• Die welding	DW
• Hammer welding	HW
FLOW WELDING	FLOW

INDUCTION WELDING	IW
BRAZING	
• Torch brazing	TB
• Twin carbon arc brazing	TCAB
• Furnace brazing	FB
• Induction brazing	IB
• Resistance brazing	RB
• Dip brazing	DB
• Block brazing	BB
• Flow brazing	FB
CUTTING PROCESS	
• Arc cutting	AC
• Air-carbon arc cutting	AAC
• Carbon arc cutting	CAC
• Metal arc cutting	MAC
• Oxygen cutting	OC
• Chemical flux cutting	FOC
• Metal powder cutting	POC
• Arc oxygen cutting	AOC

Figure 19-2 *Categories of welding processes.*

BRAZING AND SOLDERING

These processes use a filler rod that melts at a lower temperature, leaving the workpiece metals solid. If the filler rod melts at less than 840°F (450°C), the process is called **soldering**. If the filler rod melts at a temperature higher than 840°F, but lower than the melting temperature of the base metal, the process is called **brazing**. When the molten filler rod cools and hardens, it adheres to both parts, forming a metallurgical bond. Since brazing or soldering temperatures are not hot enough to melt and mix the two workpiece metals, the joints are not as strong as those created by fusion or solid-state welding

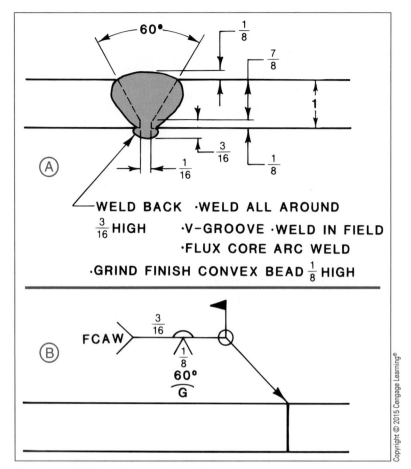

WELD BACK   ·WELD ALL AROUND
$\frac{3}{16}$ HIGH   ·V-GROOVE ·WELD IN FIELD
                ·FLUX CORE ARC WELD
·GRIND FINISH CONVEX BEAD $\frac{1}{8}$ HIGH

Figure 19-3 *All the weld data in drawing A is included in the weld symbol in drawing B.*

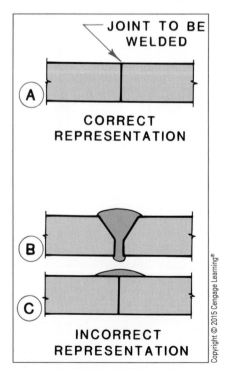

Figure 19-4 *The actual weld are is not shown on the working drawing.*

processes. Therefore, brazed or soldered joints are not used in areas of severe impact, high heat, shear, torsion, or tension stress.

Welding Symbols

When welders read working drawings, they need to know which welding process should be used. They also need to know the size and type of each weld, the location, the type of joint, and the finish to be used. It is impossible or impractical to draw, dimension, or label all of this information on each weld location on complex working drawings. **Figure 19-3** shows a drawing of a weld joint fully dimensioned at **A**. A weld symbol containing the same information is shown at **B**. Notice how much less space the symbol information requires.

ANATOMY OF A WELD SYMBOL

Because of limited dimensioning space on most drawings, the actual appearance or dimensions of a weld are not drawn. The parts are drawn as they appear after welding without the weld bead or removed metal shown. For example, the material to be welded is shown at A in **Figure 19-4** as being butted together. Examples at B and C are incorrect. A **weld symbol** (**Figure 19-5**) substitutes for the drawing of the actual joint or weld. A weld symbol is a combination of graphics, text, and numerical data. The following is a list of the common weld symbols.

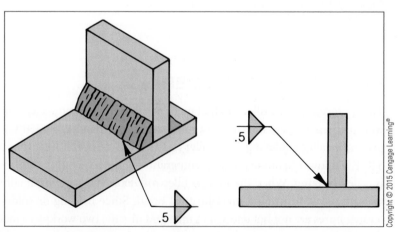

Figure 19-5 *The configuration of the weld is not shown on the working drawing.*

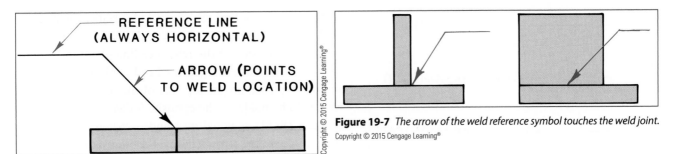

Figure 19-7 *The arrow of the weld reference symbol touches the weld joint.*
Copyright © 2015 Cengage Learning®

Figure 19-6 *Welding reference line and arrow.*

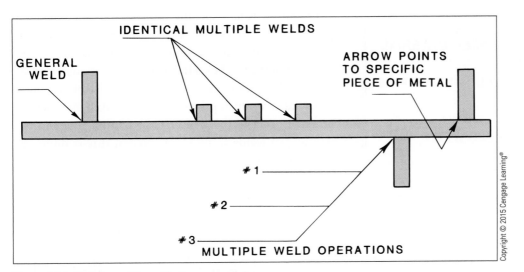

Figure 19-8 *Variations of the weld reference symbol.*

Reference Line and Arrow. A welding reference line is a horizontal line upon which specifications for each weld are placed. An **arrow line** always extends at an angle from the reference line to the workpiece (**Figure 19-6**). The arrowhead at the end of the reference line touches the location of the weld (**Figure 19-7**). Variations of how a weld reference symbol may be used are shown in **Figure 19-8**.

Weld Type Graphic Symbols. Weld type graphic symbols refer to the shape and configuration of the weld. A graphic symbol resembling the shape of the weld is used on, above, or under the reference line (**Figure 19-9**). **Figure 19-10** shows supplementary graphic symbols that are placed on, above, or below the reference line to convey additional information about the weld. These symbols will be covered in detail later in this chapter.

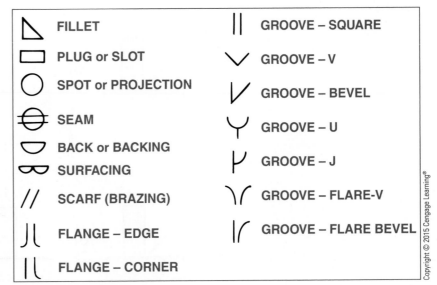

Figure 19-9 *Graphic symbols for drafting.*

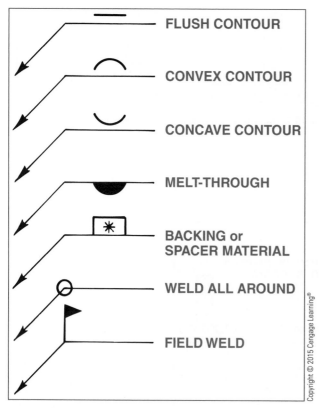

Figure 19-10 *Supplementary graphic weld symbols.*

The position of the graphic symbol specifies the location of the weld (**Figure 19-11**). When the symbol is placed on the bottom side of the reference line, the joint is to be welded exactly where the arrowhead is placed. When the symbol is placed on the top side of the reference line, the weld is to be made on the opposite (other) side of the joint (**Figure 19-11**) as noted in A and B. When the symbol is placed on both the arrow side and the other side of the reference line, both sides of the joint are to be welded, as noted in C. The placement of all possible data in the weld reference symbol is shown in **Figure 19-12**.

Tail Symbol. When specific information, such as detail specifications, procedures, welding processes, filler material, or preparation of the weld area, is required, it is placed in the tail of the symbol (**Figure 19-13**). The abbreviations for the welding processes in **Figure 19-2** are placed in the tail to indicate which process is to be used. If all welds use the same process, this can be covered in a general note, such as "all welds to be OAW." When this is done, the tail information may be omitted, unless there are other specific instructions.

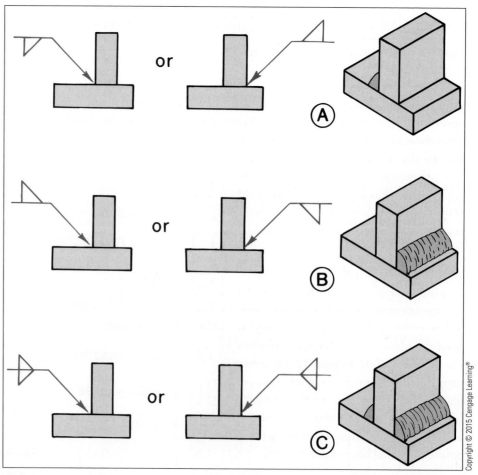

Figure 19-11 *The position of the weld symbol (fillet) specifies the location of the weld.*

Finished Method Symbol. The method of finish for each weld is shown by a letter placed at the F position (**Figure 19-12**). The following letters indicate the method of finish:

C – Chipping
G – Grinding
M – Machining
R – Rolling
H – Hammering

The exact degree of finish is not shown on a weld symbol since this is part of the finishing limits established for the entire surface on the working drawing. Surface finishes are explained in Chapter 9.

Contour Symbol. The shape of the finished surface of a weld is shown directly under the finishing method symbol (**Figure 19-12**). A flat surface is indicated by a straight line. A rounded (concave) surface is shown by a curved line, like a smile; a depressed (convex) surface is shown by a curved line like a frown (**Figure 19-14**).

Groove Angles. The letter directly below the contour symbol (**position A, Figure 19-12**) shows the groove angle if a groove joint is specified. Groove angles are expressed as included angles (**Figure 19-15**). This location on the reference weld symbol is also used to show the angle of any countersinks on plug welds.

Root Opening. The gap between two workpieces before welding is known as the **root opening** or **root gap** (**Figure 19-15**). The width of the root opening is located directly under the

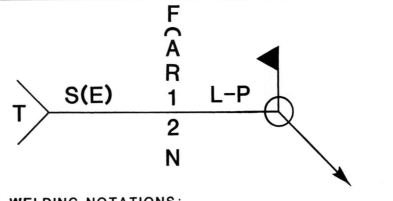

WELDING NOTATIONS:

F – FINISH SYMBOL
⌢ – CONTOUR SYMBOL
A – GROOVE ANGLE: INCLUDED ANGLE OF COUNTERSINK FOR PLUG WELDS
R – ROOT OPENING: DEPTH OF FILLING FOR PLUG AND SLOT WELDS
(E) – EFFECTIVE THROAT
T – SPECIFIC PROCESS OR REFERENCE
L – LENGTH OF WELD, LENGTH OF OVERLAP (BRAZED JOINTS)
S – DEPTH OF PREPARATION, SIZE OR STRENGTH FOR CERTAIN WELDS, HEIGHT OF WELD REINFORCEMENT, RADII OF FLARE-BEVEL GROOVES, RADII OF FLARE-V GROOVES, ANGLE OF JOINT (BRAZED WELDS)
P – PITCH OF WELDS (CENTER-TO-CENTER SPACING)
1 – WELD SYMBOL ON THIS SIDE MEANS THE WELD MUST BE ON THE OPPOSITE SIDE OF THE MATERIAL THE ARROW IS TOUCHING
2 – WELD SYMBOL ON THIS SIDE MEANS THE WELD MUST BE ON THE SAME SIDE OF THE MATERIAL THE ARROW IS TOUCHING
(N) – THE NUMBER OF SPOT OR PROJECTION WELDS
⌐ – WELD DONE IN THE FIELD
⌐○ – WELD ALL AROUND

Figure 19-12 *Weld reference symbol with its data symbol information.*

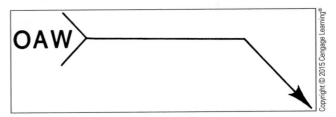

Figure 19-13 *When a specific welding notation is required, a tail is added (oxyacetylene welding) as shown.*

SURFACE SHAPE	WELD	CONTOUR SYMBOL	REMEMBER
CONVEX		⌢	🙁
CONCAVE		⌣	🙂
FLUSH		—	😐

Figure 19-14 *Contour symbols.*

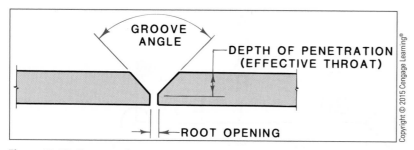

Figure 19-15 *Groove angles.*

groove angle on the welding symbol (**position R, Figure 19-12**). This location may also be used to show the filling depth for plug or slot welds.

Size and Strength. The size or strength of the weld is shown on or under the reference line directly to the left of the graphic weld symbol (**position S, Figure 19-12**). Weld size can mean the height of the bead, radii of a flared bevel joint, or depth of penetration. The depth of penetration is sometimes called the **effective throat** (**position E, Figure 19-12**). For some welds, the minimum allowable strength of the joint in pounds is substituted for weld size in position S.

Length of Weld. The length of the weld is located on or below the reference line directly to the right of the weld type graphic symbol (**position L, Figure 19-12**). If no length is indicated on the symbol, it is assumed that the weld extends the full length of the part. This location is also used to show the length of overlap on brazed joints.

Weld Pitch. When welds are intermittent over the length of a joint, the spacing of each weld and the length of unwelded spaces between welds must be shown. The center-to-center distance between intermittent welds is called the **pitch**. The length of an individual weld is called the **increment** (**Figure 19-16**). The length of the pitch and increments is placed on or below the reference line (**position P, Figure 19-12**). For example, a dimension of 6-10 on the symbol means 6" of weld spaced 10" apart from center to center.

Number of Spot or Projection Welds. When spot or projection welds are used, the spacing between welds is shown below the reference line (**position N, Figure 19-12**).

Weld All Around. When a joint is to be welded around all of its edges, the weld all around symbol is used. This symbol is a circle drawn at the intersection of the reference line and the arrow line (**Figure 19-12**). This is one of the supplementary symbols shown in **Figure 19-10**.

Field Weld. If a part is to be welded on the construction site, a vertical line with a flag is drawn beginning at the intersection of the reference line and the arrow line (**Figure 19-12**). This is also one of the supplementary symbols shown in **Figure 19-10**.

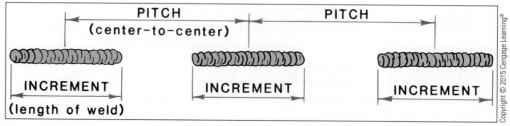

Figure 19-16 *Intermittent welds (increments).*

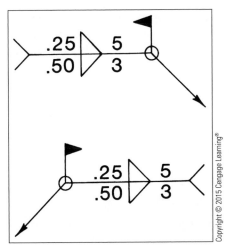

Figure 19-17a *The order of the weld data remains in the same position regardless of the arrow's direction.*

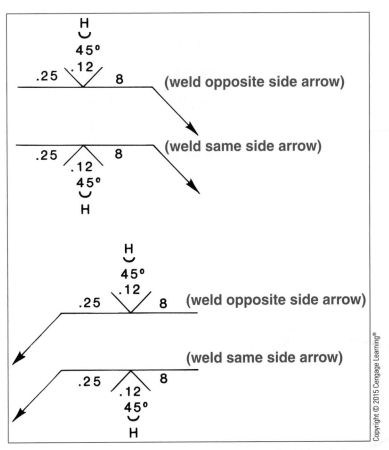

Figure 19-17b *Placement of weld data depicts the location side of the weld.*

Order of Information. Except for the tail symbol information, and the weld all around and field weld symbols, all data is located in the same horizontal order on the reference line, regardless of the position of the arrow. For example, **Figure 19-17a** includes a left arrow line symbol and a right arrow line symbol containing the same information, read in the same sequence.

Vertically aligned data, such as the root opening, groove angle, contour symbol, and finish method symbol, is always in the same order from the reference line. Therefore, the order is reversed when this data is below the line. **Figure 19-17b** shows the order of the information applied to an arrow side symbol and another side symbol.

Data Selection

Not all data shown in **Figure 19-12** is always used for each weld. Some information simply does not apply to some welds. For example, the groove angle degree does not apply to fillet welds, and the pitch distance does not apply to continuous welds.

When the data does apply to the weld, all information must be located on the symbol. If data is missing, the welder must determine the specifications of the weld without full knowledge of the design requirement. From the information supplied in **Figure 19-18**, the welder knows only the location and size of a V-groove. Not known is the welding process, finish method, surface shape, groove angle, root opening, or length of weld. For comparison, **Figure 19-19** shows a welding symbol containing complete data. From the information on this symbol, the welder knows exactly where a ¼", 60°, V-groove weld with a ⅛" root opening is to be made. The welder also knows that the gas-tungsten arc process is to be used in the field, and

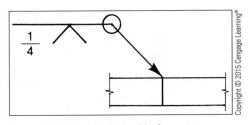

Figure 19-18 *Incomplete weld information.*

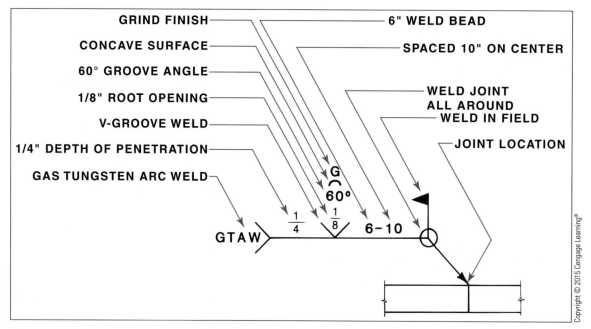

Figure 19-19 *Fully noted weld instructions for a V-groove weld.*

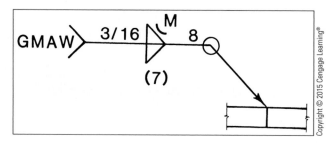

Figure 19-20 *Fully noted weld instructions for a fillet weld.*

that the weld is to be made all around and ground to a convex shape. The welder understands that this weld is intermittent, with 6" long welds spaced 10" apart center to center.

Figure 19-20 shows another example of a complete symbol. It specifies an 8" long, ³⁄₁₆" gas-metal arc fillet weld to be made all around the joint and machined to a concave surface. Although this weld symbol contains only seven items compared to 11 in **Figure 19-19**, it is complete, since the other items are not required for this type of weld.

Welding Type Symbol Applications

As shown in **Figure 19-9**, the type of weld is represented by a graphic symbol over or under the reference line. A thorough knowledge of the relationship of the actual weld types to symbols is essential for the preparation of welding drawings. This understanding is also important because the selection of the welding type is directly related to and usually precedes the establishment of other weld design specifications. **Figures 19-21** through **19-42** show pictorial interpretations of the welding type symbols used for each major type of weld.

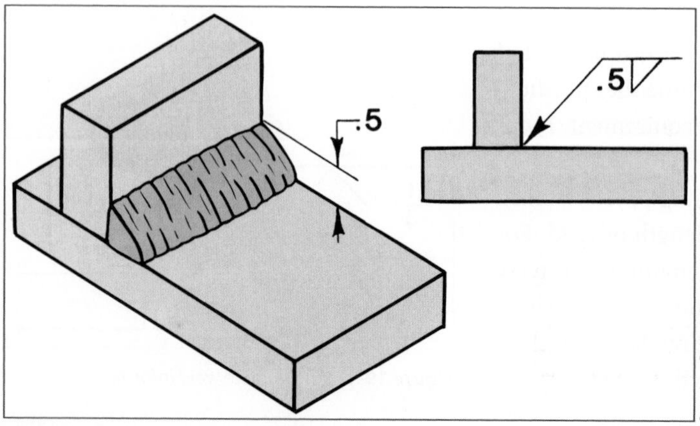

FILLET WELDS

Fillet welds are used to weld two objects that join at right angles (**Figure 19-21**). Notice how the fillet weld symbol closely resembles the actual appearance

Figure 19-21 *Fillet weld.*

of a fillet. The vertical line of the fillet weld symbol is always on the left and the slanted line on the right, regardless of the position of the weld. The size of a fillet weld is normally dimensioned as the height of the bead. If the height and width of the bead are different, both dimensions are given in parentheses, with the height entered first (**Figure 19-22**). Fillet weld contours are shown next to the slanted part of the fillet symbol (**Figure 19-23**).

PLUG WELDS

Plug welds substitute for rivets. A plug weld is a hole filled with a weld filler material. Five items are necessary on plug weld symbols (**Figure 19-24**). The graphic symbol for a plug weld is a rectangular box roughly resembling the cross section of a plug weld. The diameter of the hole to be filled is to the left of the rectangle, and the depth of the area to be filled is inside the symbol box. If there is more than one hole, the center-to-center spacing (pitch) is to the right of the box symbol. The symbol in **Figure 19-24** specifies ½" holes with 60° countersinks to be spaced 2" apart (center to center) and filled to a depth of ¼".

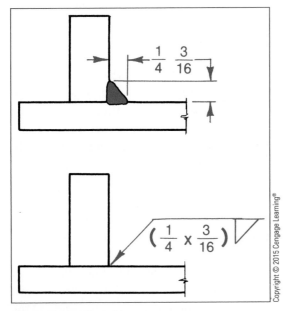

Figure 19-22 Fillet weld.

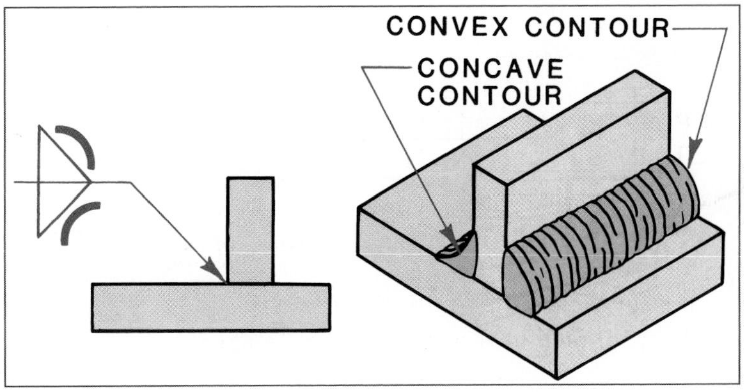

Figure 19-23 Contours for a fillet weld.

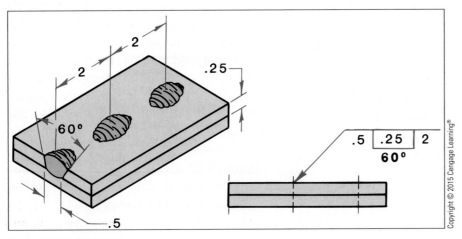

Figure 19-24 Plug welds.

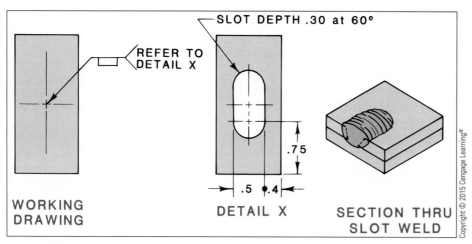

Figure 19-25 *Detail drawing of slot weld.*

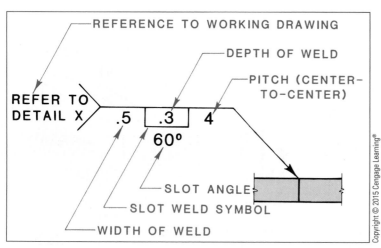

Figure 19-26 *Slot weld reference symbol.*

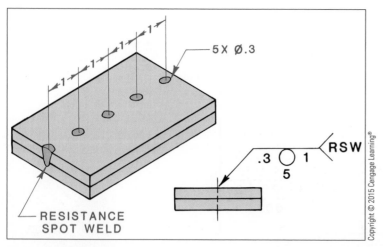

Figure 19-27 *Spot welds.*

SLOT WELDS

Since **slot welds** are basically elongated plug welds, the same rectangular box symbol is used for slot welds as for plug welds. The depth of slot, depth of fill, angle of countersink, size of hole, and length of slot must be specified. Because of the amount of data required, some or all of the dimensions for a slot weld are often shown on a detail drawing (**Figure 19-25**). The weld reference symbol for the slot weld can also be shown as in **Figure 19-26**. In this case, a separate detail is needed to show the exact position of the slot on the part. This is also often true for dimensioning the position of a plug weld.

SPOT WELDS

Electrical resistance processes are usually used for spot welding thin sheets of metal, although fusion processes are sometimes used. In spot welding, the top sheet is melted into the bottom sheet. The graphic symbol for a spot weld is a circle (**Figure 19-27**). The position of the circle above or below the reference line indicates the direction of the source of heat. If the heat side has no significance in the design, the circle is placed directly on the line. When this is done, the circle is placed in the center of the reference line to avoid confusion with the weld all around symbol, which is also a circle drawn at the intersection of the reference line and the arrow line.

The size of a **spot weld** is the diameter of the spot and is located to the left of the circle symbol. A spot weld may also be classified by the minimum acceptable shear strength per spot. If this method is preferred, the strength in pounds per spot is located to the left of the circle. The pitch, located to the right of the circle, is the center-to-center spacing between each spot. **Figure 19-27** specifies five .3" diameter resistance spot welds spaced 1" apart and located on the arrow side of the reference line.

PROJECTION WELDS

The only difference between a spot weld and a **projection weld** is the embossing of one of the metal sheets where the spot weld is to be made. The same graphic circle is used for both types of welds. The projection method is noted in the tail symbol data. The circle is located on the side of the reference line representing the side of the joint to be embossed. This means that the circle is placed below the reference line if the arrow side is to be embossed. Either the size or the strength of spot welds may be noted to the left of the circle. The symbol in **Figure 19-28** specifies four resistance projection welds with a joint strength of 200 lbs., spaced .8" apart on the arrow side of the joint.

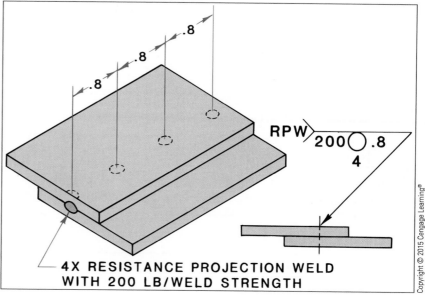

Figure 19-28 *Projection welds.*

SEAM WELDS

A continuous spot weld is called a **seam weld.** Seam welds are used instead of spot welds when a tight (storage of liquids or pressurized gas) or slightly stronger than usual joint is required. The graphic symbol for a seam weld is a circle with two horizontal lines (**Figure 19-29**). This symbol can be placed above, below, or on the reference line to show the location of the heat source, as in spot weld symbols. All other spot weld dimensioning rules apply to seam welds, except that the strength is expressed in pounds per linear inch. **Figure 19-29** specifies a continuous resistance seam weld on the same side of the arrow, with a strength of 100 lbs. per linear inch.

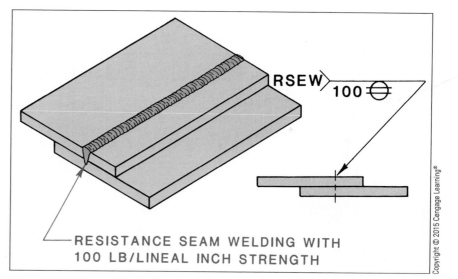

Figure 19-29 *Seam weld.*

FLANGE WELDS

Welding flat surfaces of two sheets of metal bent to provide a surface for welding is called **flange welding.** There are two types of flange welds—edge flange welds and corner flange welds. Edge flange welds join sheets of metal that lie in the same plane (**Figure 19-30**). Corner flange welds

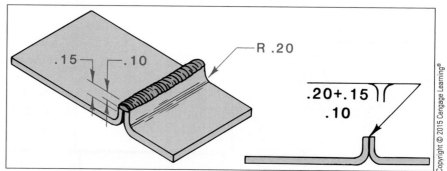

Figure 19-30 *Edge flange weld.*

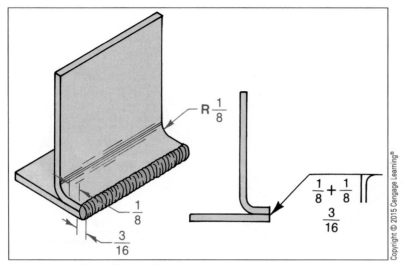

Figure 19-31 *Corner flange weld.*

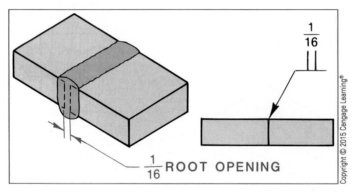

Figure 19-32 *Root opening for a square groove weld.*

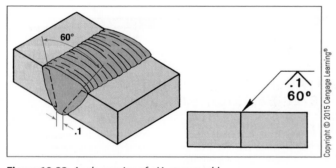

Figure 19-33 *Angle opening of a V-groove weld.*

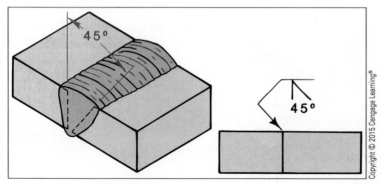

Figure 19-34 *Opening for a bevel groove weld.*

join sheets of metal that intersect at right angles (**Figure 19-31**). The graphic symbol for each is different, but the rules for dimensioning are the same. Dimensions are shown on the reference line by using the bend radius, the distance to the point of tangency, and the height of the bead. The bend radius is located to the left of the graphic symbol, followed by a plus (+) sign and the distance from the edge of the metal to the point of tangency of the arc. The height of the bead is shown below this dual dimension. The symbol in **Figure 19-30** specifies a .10" edge flange weld with a .20" radius, and the point of tangency of the arc located .15" from the edge of the metal. **Figure 19-31** shows a ³⁄₁₆" corner flange weld with a ⅛" radius and a point of tangency located ⅛" from the edge of the metal.

GROOVE WELDS

Often a groove or gap is made between two metal parts to be welded. During welding, this groove is filled with material from the welding rod. There are seven basic types of **groove welds**: the square groove, V-groove, bevel groove, U-groove, J-groove, flare V-groove, and flare bevel groove. The graphic symbol for each groove weld resembles the shape of the groove.

When two workpieces do not touch, the gap between the parts is called the root opening. The size of the root opening is located directly above or below the graphic symbol (**Figure 19-32**). The angle of the groove, if any, is placed directly above or below the root opening dimension, as in the V-groove symbol in **Figure 19-33**. The symbol for a V-groove weld is always drawn at right angles, regardless of the actual angle of the groove. The symbol for a bevel groove weld is always drawn with one perpendicular line. The angle of the 45° bevel is shown in **Figure 19-34**.

For all groove welds, the depth of penetration is always located to the left of the graphic symbol. The size of the U-groove weld. (**Figure 19-35**) is .5" with a .1" root opening. The size of groove welds is the depth of penetration (effective throat) of the filler material. The effective throat of the J-groove weld. (**Figure 19-36**) is .8" and the root opening is .2".

When two metal plates are bent to create flat adjacent surfaces for welding, the type of weld is called a flare groove weld. There are

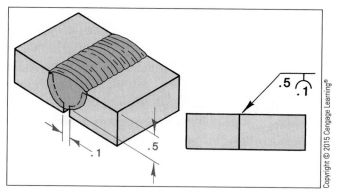

Figure 19-35 *Depth of penetration for a U-groove weld.*

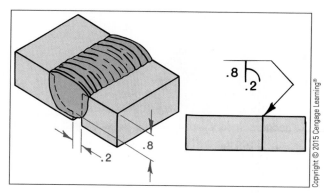

Figure 19-36 *Effective throat for a J-groove weld.*

two types of flare groove welds—the flare V-groove and the flare bevel groove weld. The flare V-groove weld joins metal plates that lie in the same plane (**Figure 19-37**). The flare bevel groove weld joins two plates that intersect at an angle (**Figure 19-38**). The depth of penetration is the radius of the bevel for both types.

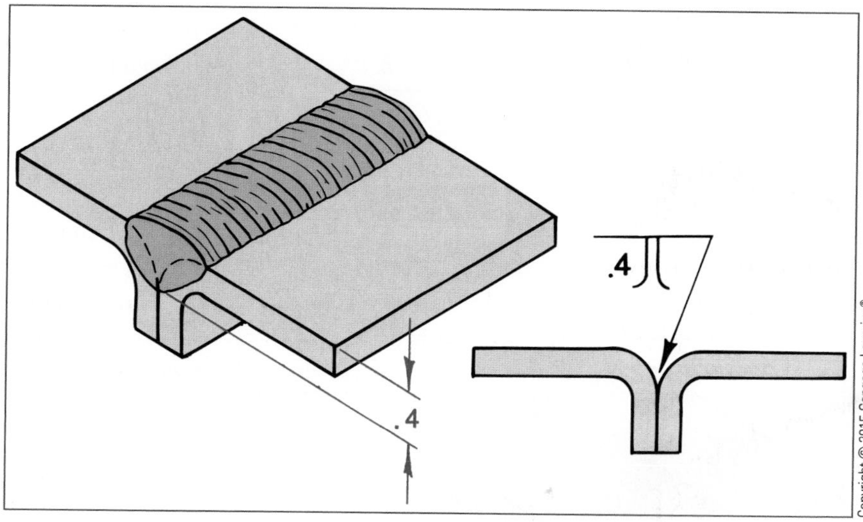

BACKING AND MELT-THROUGH WELDS

For maximum support, a weld bead is sometimes required on the back side of a joint in addition to the front. This is done in two separate operations. To specify a **backing weld,** a semicircle is drawn on the reference line directly opposite the graphic symbol. **Figure 19-39** shows this symbol applied to a V-groove weld. The depth of the backup bead is shown to the left of this semicircle.

Melt-through welds are made completely through the joint from one side with a backup bead. The semicircle symbol is made solid (**Figure 19-40**). It is similar to the backing weld but is made with one welding operation.

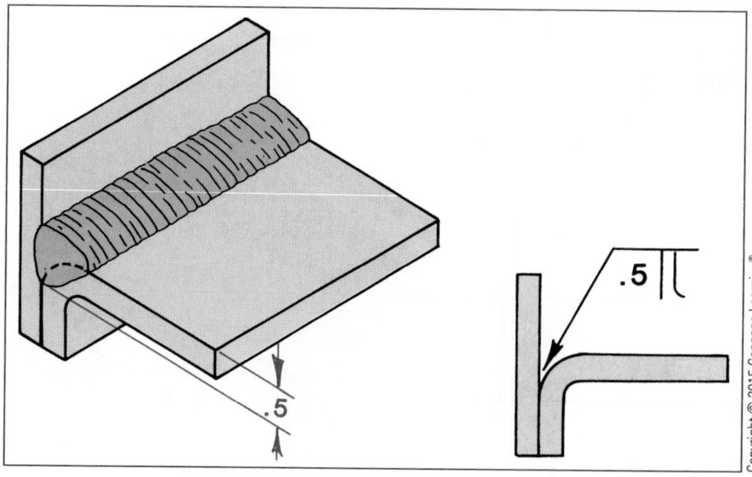

SURFACING WELDS

Surfaces are often made thicker with the application of filler material. This process is known as **surfacing** or **building up**. The graphic symbol for surfacing is two semicircles (**Figure 19-41**). Since no joint is involved in surfacing welds, the minimum height of the filler material is the only dimension needed. However, the welding process and method of finishing the filled surface are very important. If only part of the buildup surface is to be finished, a separate detail drawing must be

Figure 19-37 *Flare V-groove weld.*

Figure 19-38 *Flare bevel groove weld.*

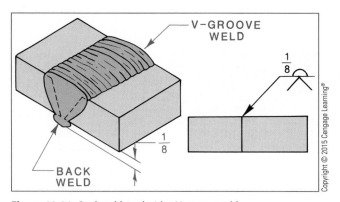

Figure 19-39 *Back weld used with a V-groove weld.*

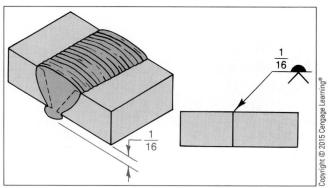

Figure 19-40 *Melt-through weld.*

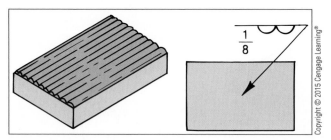

Figure 19-41 *Surfacing weld.*

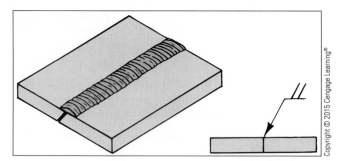

Figure 19-42 *Scarf braze.*

prepared showing the exact dimensions of the area to be finished. This information must be indexed in the welding tail symbol.

BRAZING SYMBOLS

Welding symbols are also used for brazing. Brazing occurs at lower temperatures; therefore, the materials being joined are not melted. Usually, softer or thinner metals are involved (**Figure 19-42**).

COMBINATION WELDS

For instructional purposes, the types of weld symbols shown thus far included only information about one type of weld. However, many welds contain combinations of different types of welds. **Figure 19-43** shows how a combination of different welding types and dimensions are incorporated into one symbol.

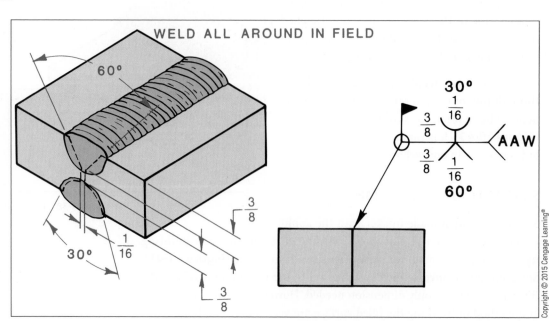

Figure 19-43 *Example of combination welds.*

Welded Joints

In addition to knowing which type of weld is needed, the designer must know which joint is appropriate to join metal parts by welding. The common joints used in welding are shown in **Figure 19-44**. These include the butt, corner, lap, T-, and edge joints. The type of weld that can be used for each joint is also shown in **Figure 19-44**. The final selection of the joint and weld type depends on the metal specified and the structural and aesthetic requirements of the design.

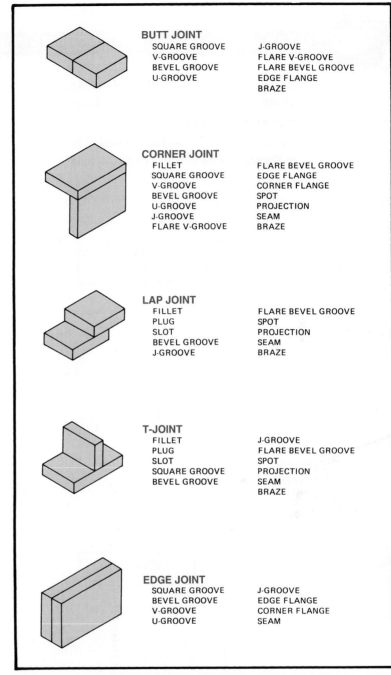

BUTT JOINT
SQUARE GROOVE	J-GROOVE
V-GROOVE	FLARE V-GROOVE
BEVEL GROOVE	FLARE BEVEL GROOVE
U-GROOVE	EDGE FLANGE
	BRAZE

CORNER JOINT
FILLET	FLARE BEVEL GROOVE
SQUARE GROOVE	EDGE FLANGE
V-GROOVE	CORNER FLANGE
BEVEL GROOVE	SPOT
U-GROOVE	PROJECTION
J-GROOVE	SEAM
FLARE V-GROOVE	BRAZE

LAP JOINT
FILLET	FLARE BEVEL GROOVE
PLUG	SPOT
SLOT	PROJECTION
BEVEL GROOVE	SEAM
J-GROOVE	BRAZE

T-JOINT
FILLET	J-GROOVE
PLUG	FLARE BEVEL GROOVE
SLOT	SPOT
SQUARE GROOVE	PROJECTION
BEVEL GROOVE	SEAM
	BRAZE

EDGE JOINT
SQUARE GROOVE	J-GROOVE
BEVEL GROOVE	EDGE FLANGE
V-GROOVE	CORNER FLANGE
U-GROOVE	SEAM

Figure 19-44 *Weld joints and their recommended welds.*

DRAFTING EXERCISES

1. Name each type of weld in **Figures 19-45**, **19-46**, and **19-47**. Name one example of how each weld might be used.

2. All the individual steel pieces of materials in **Figures 19-48** through **19-61** are to be welded. With drawing instruments or a CAD system, draw the working drawings and add the appropriate types of weld symbols and notations needed.

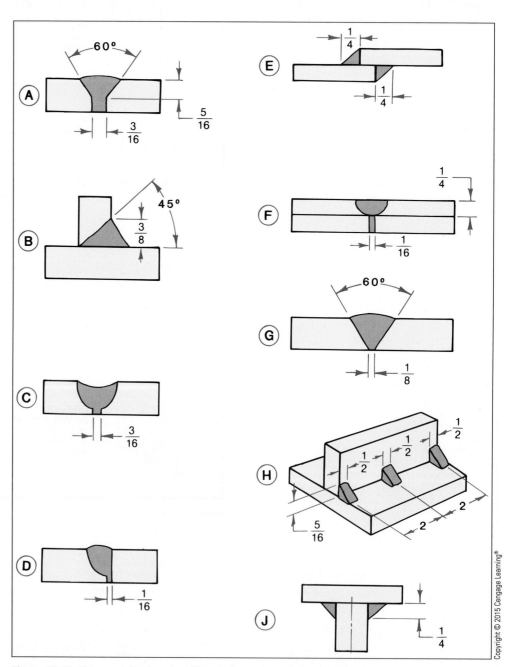

Figure 19-45 *Name and sketch each weld symbol.*

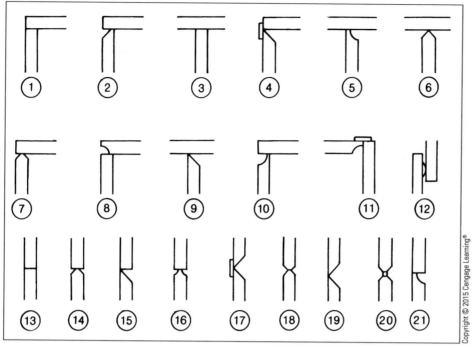

Figure 19-46 *Sketch each example to be welded. Add the proper weld reference symbol and data required with another sketch as it will appear on a working drawing.*

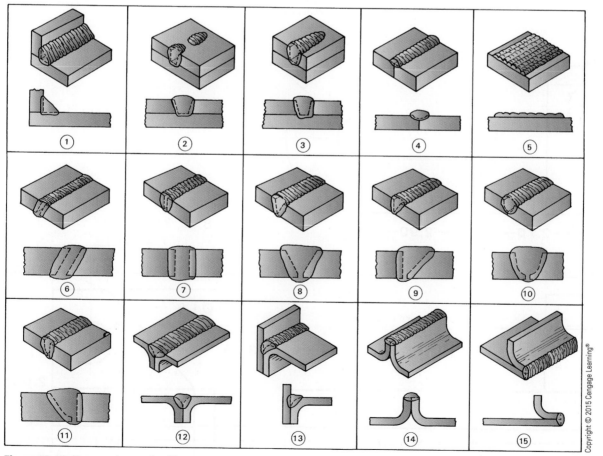

Figure 19-47 *Name each type of weld.*

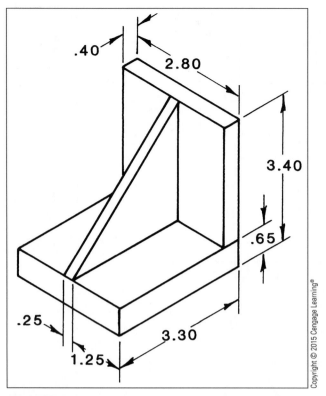

Figure 19-48 *Draw the working drawings and add weld notations as needed for the bracket with a CAD system.*

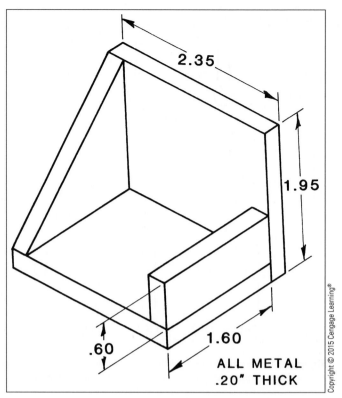

Figure 19-49 *Draw the working drawing and add fillet welds with seam welds as needed.*

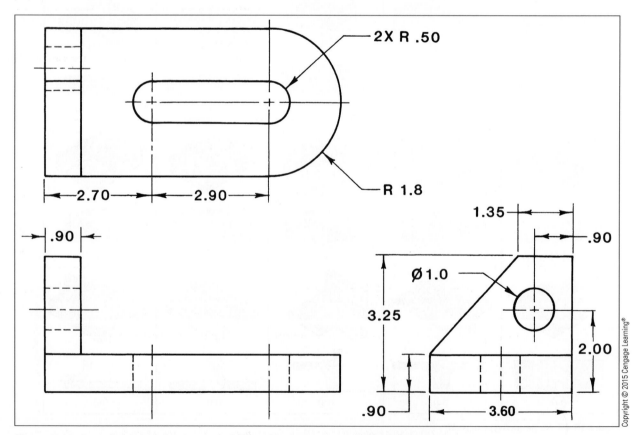

Figure 19-50 *Draw the working drawings and add the weld notations as needed for the bearing plate.*

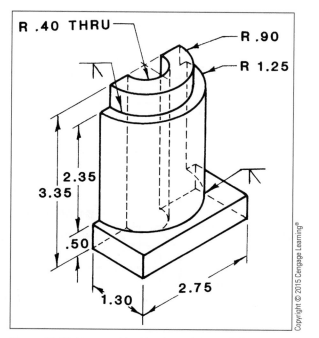

R .40 THRU

R .90

R 1.25

2.35

3.35

.50

1.30

2.75

Figure 19-51 *Draw working drawings of the semicolumn support, and complete the weld notations with a CAD system.*

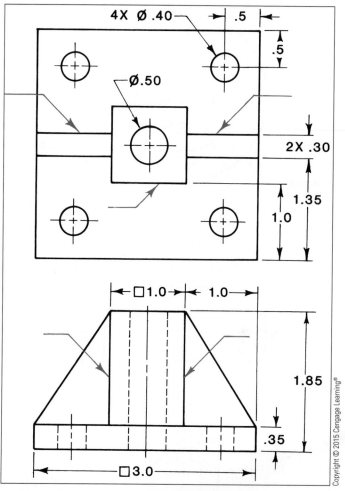

4X Ø .40

.5

.5

Ø.50

2X .30

1.35

1.0

□1.0

1.0

1.85

.35

□3.0

Figure 19-52 *Draw the multiview drawings. Locate and indicate the proper weld notations.*

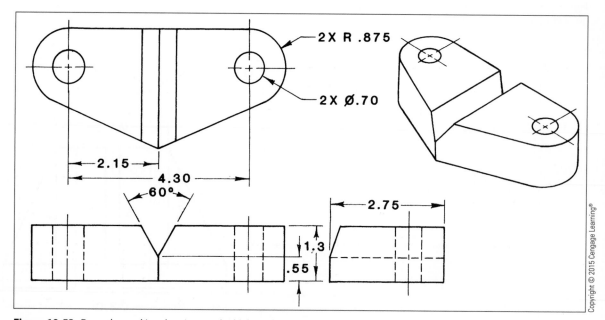

2X R .875

2X Ø.70

2.15

4.30

60°

2.75

1.3

.55

Figure 19-53 *Draw the working drawings and add the weld notations as needed for the V-guide.*

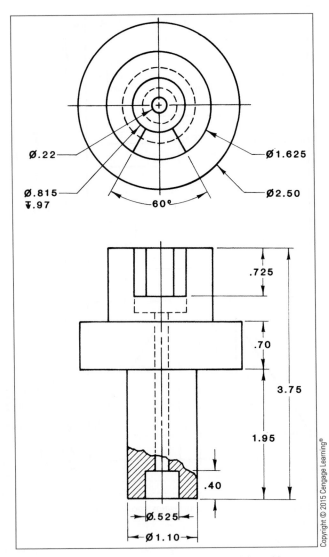

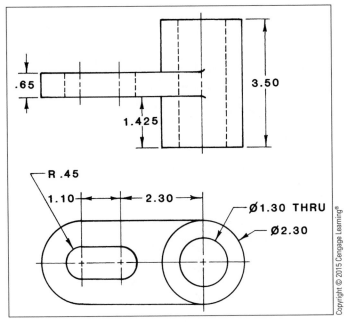

Figure 19-55 *Draw the working drawings and add the weld notations as needed for the adjustable bearing locator.*

Figure 19-54 *Draw the working drawings and add the weld notations as needed for the hub guide.*

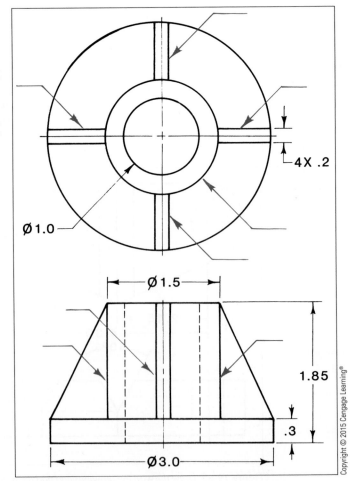

Figure 19-56 *Draw the multiview drawings. Locate and indicate the proper weld notations.*

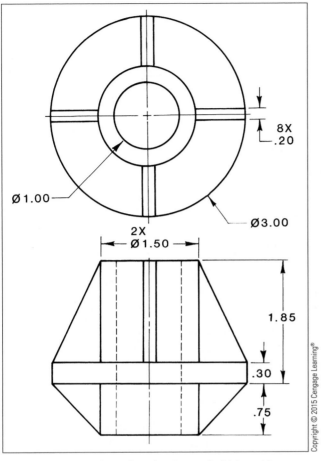

Figure 19-57 *Draw the working drawings and add the weld notations as needed for the cylinder and four webs with a CAD system.*

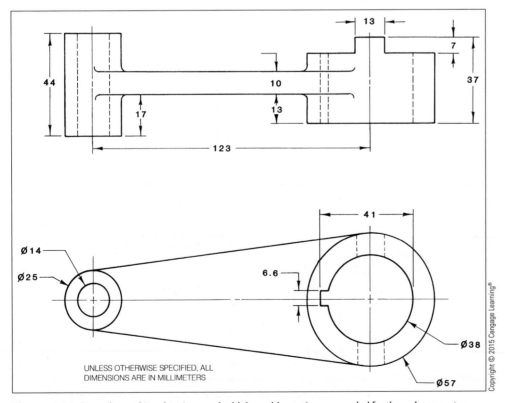

UNLESS OTHERWISE SPECIFIED, ALL
DIMENSIONS ARE IN MILLIMETERS

Figure 19-58 *Draw the working drawings and add the weld notations as needed for the web connector.*

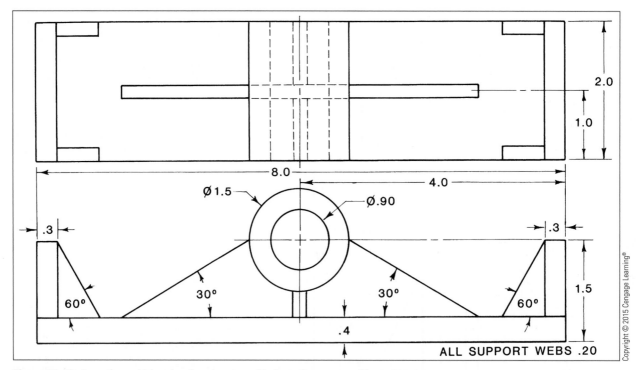

Figure 19-59 *Draw the multiview drawings. Locate and indicate the proper weld notations.*

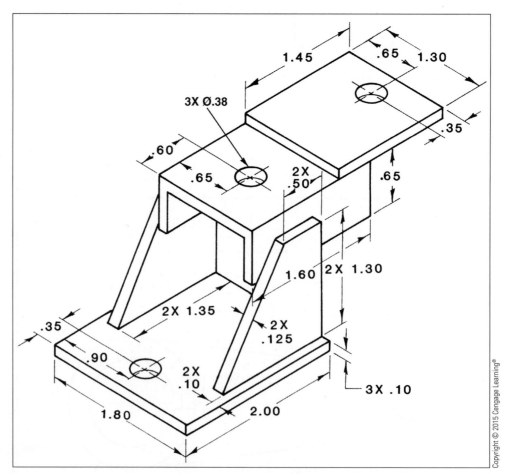

Figure 19-60 *Draw the working drawings and add the weld notations as needed for the tool jig stand.*

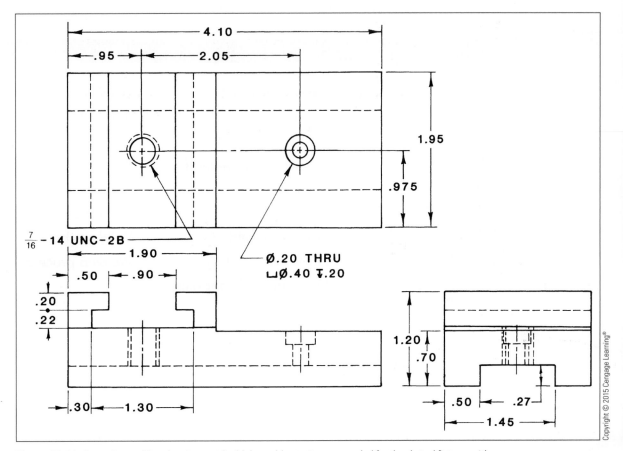

Figure 19-61 *Draw the working drawings and add the weld notations as needed for the slotted fixture guide.*

DESIGN EXERCISES

1. Design a metal container that will hold five compressed gas tanks. Each tank has an outside diameter of 12" and a height of 36". Use weldments for the assembly. Research the safety features required for the storage of compressed gas tanks.

2. Design a "hook setup" for a lifting crane that must pick up a 4' diameter spherical glass ball that weighs 500 pounds. Draw a working drawing showing all the components of the "hook" and attachments. Use weldments whenever possible for the assembly.

3. Design a welded metal frame to hold a minimum of 20 CD-ROM disks.

KEY TERMS

Arc welding

Arrow line

Backing weld

Brazing

Effective throat

Fillet weld

Flange weld

Fusion welding

Gas welding

Groove weld

Increment

Melt-through weld

Oxyacetylene welding

Pitch

Plug weld

Projection weld

Resistance welding

Root opening (root gap)

Seam weld

Slot weld

Soldering

Solid-state welding

Spot weld

Surfacing (building up)

Weld symbol

Welding process

Gears and Cams

20

OBJECTIVES

The student will be able to:

- Draw a working drawing of a spur gear
- Draw the teeth of a gear with a gear template
- Draw a working drawing of a rack and pinion gear
- draw a working drawing of a bevel gear
- Draw a working drawing of a worm gear
- Draw a plate cam from displacement diagram data
- Draw a drum cam
- Lay out a displacement diagram from data in a plate cam profile drawing
- Lay out a displacement diagram from data in a drum cam drawing
- Recognize the various types of gears and cams
- Prepare a gear drawing using a CAD system
- Complete a cam drawing using a CAD system

Introduction

Gears, cams, and drives are mechanical devices used to control power. These devices can transmit motion from one device to another or change the speed and direction of motion. To design and draw gears, cams, and drives as integral parts of machine assemblies, you must have a basic understanding of their function.

Gears and Drives

Cylinders or cones used to transmit power and motion, and to change the speed and direction of motion, are known as **gears**. Gears may be smooth (friction gears) or toothed. There are many types of gears designed to perform specific transmission, directional, and speed control functions. Gears are used to transmit rotary or reciprocating (back and forth) motion between machine parts.

Gears are divided into two general classifications based on the relative position of the gear shafts. Parallel shaft gears are connected with spur gears, helical gears, or herringbone gears. Perpendicular shaft gears are connected with bevel or miter gears. If intersecting gear shafts are neither parallel nor perpendicular, connections are made with crossed helical gears or worm gears, called **angular gears**.

When one gear is rotated and brought into contact (meshed) with another gear, motion is transmitted. The second gear turns in the opposite direction of the first. If two gears are equal in circumference, both will rotate one complete revolution in opposite directions. If the second gear is twice the circumference of the first, it will rotate only one-half revolution, while the first gear rotates one full revolution. This results in the second gear rotating at one-half the speed of the first gear. The difference between the rotation speeds of the two gears is known as the **gear ratio**. When gears are the same size and rotate at the same speed, the gear ratio is 1:1. When one gear rotates at twice the speed of the other, the ratio is 2:1. Three times rotation equals 3:1, and so forth.

The most common types of gears and drives are described in the following list, and are illustrated in **Figures 20-1** through **20-9**:

- **Spur gears** (**Figure 20-1**) are one of the most common types of gear. The teeth on the circumference are all parallel to the axis and perpendicular to the sides of the wheel.

- **Rack and pinion gears** transmit reciprocating motion through a flat-toothed surface (rack) intersecting with a small cylindrical gear (pinion; **Figure 20-2**). The pinion is the smaller gear in any gear system.

- **Bevel gears** transmit power between shafts that intersect at an angle (**Figure 20-3**). If the intersection of shafts is a right angle, they are often called **miter gears**.

- **Hypoid gears** are similar to bevel gears except for the angle of the teeth. They are very strong and operate smoothly (**Figure 20-4**).

- **Worm gears** connect shafts that are perpendicular with shaft centerlines that do not intersect. The small gear is the worm, and it drives the larger worm gear. The worm is shaped like a screw. The worm gear is round like a wheel. The worm includes at least one thread around the pitch surface. Worm gear teeth are cut on an angle to mesh with worm teeth (**Figure 20-5**).

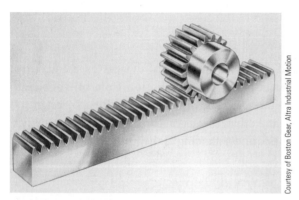

Figure 20-1 *Spur gears.*

Figure 20-2 *Rack and pinion gears.*

Figure 20-3 *Bevel gears.*

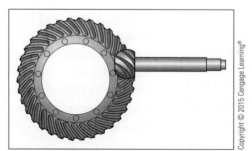

Figure 20-4 *Hypoid gears. (Courtesy Engineering Drawing & Design, Madsen and Madsen, Delmar, Cengage Learning.)*

Figure 20-5 *Worm and worm gear.*

- **Helical gears** operate on parallel shafts with gear teeth cut in a twisted, oblique angle to the shaft (**Figure 20-6**).

- **Herringbone gears** contain two sets of helical gear teeth placed side by side on an opposite slant on the same gear wheel (**Figure 20-7**).

- **Cross helical gears** provide for nonintersecting perpendicular shaft drives (**Figure 20-8**).

- **Ring gears** are used primarily in automobile differentials. A propeller shaft pinion transmits power to a large ring gear that has teeth cut on the inside surface of the ring (**Figure 20-9**).

- **Chain and sprocket drives** are flexible power transmission devices. They consist of a continuous chain with links that mesh with toothed sprocket wheels attached to a drive shaft (**Figure 20-10**). Chain drives are used where ease of assembly and disassembly and elasticity are required, or where shaft-to-shaft distances are extremely long as on a bicycle.

- **Belt drives** transmit power between shafts with a belt wrapped around the outside circumference of a pulley mounted on each shaft. Belts that connect pulleys are either flat (**Figure 20-11**), grooved with longitudinally ribbed undersides, or V-belts. To connect perpendicular shafts, belts can be twisted by one-quarter. Belts can also be crossed (twisted by one-half) between pulleys to reverse the direction of the second pulley.

- **Friction wheels** are used to transmit power between shafts without using gears, chains, or belts. Friction wheels rely on heavy pressure to maintain sufficient contact to transfer power (**Figure 20-12**). Friction wheels engage easily but will slip when a predetermined amount of resistance is applied to the drive wheel.

- **Figure 20-13** shows several different types of gear drives in a single unit.

Figure 20-6 *Helical gears. (Reprinted from Mechanical Drafting by Madsen, Schumaker, and Stewart, Delmar Learning.)*

Copyright © 2015 Cengage Learning®

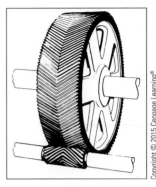

Figure 20-7 *Herringbone gear. (Reprinted from Mechanical Drafting by Madsen, Schumaker, and Stewart, Delmar Learning.)*

Copyright © 2015 Cengage Learning®

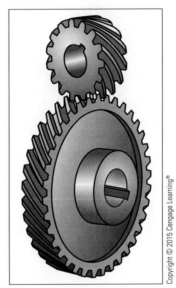

Figure 20-8 *Crossed helical gears. (Courtesy Engineering Drawing & Design, Madsen and Madsen, Delmar, Cengage Learning.)*

Copyright © 2015 Cengage Learning®

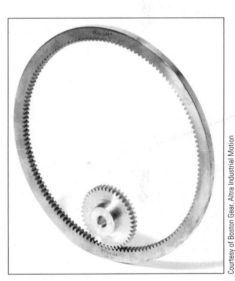

Figure 20-9 *Ring gears.*

Courtesy of Boston Gear, Altra Industrial Motion

Figure 20-10 *Chain and sprocket drive.*

Courtesy of Boston Gear, Altra Industrial Motion

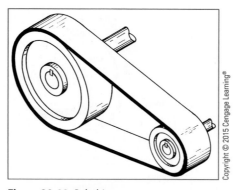

Figure 20-11 *Belt drive.*

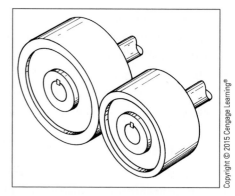

Figure 20-12 *Friction drive.*

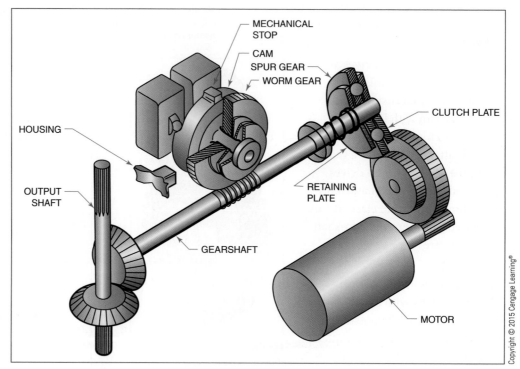

Figure 20-13 *Various gear drives.*

GEAR TEETH

Two basic curves are used to define the shape of gear teeth—involute and cycloidal. An **involute curve** (**Figure 20-14**) is produced by connecting the end points of a flexible and taut thread as it is unwound from a cylinder. A **cycloidal curve** (**Figure 20-15**) is formed by connecting the position of a point at evenly spaced intervals on the circumference of a circle as that circle is rolled on a flat surface.

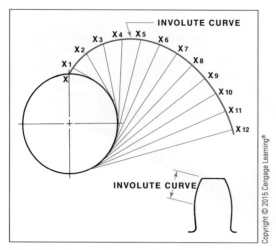

Figure 20-14 *An involute curve is formed by a taut string as it unwinds from a cylinder.*

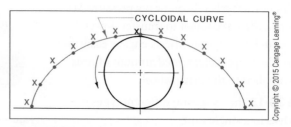

Figure 20-15 *Plotting a cycloidal curve.*

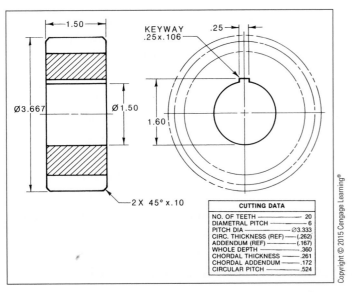

CUTTING DATA	
NO. OF TEETH	20
DIAMETRAL PITCH	6
PITCH DIA	Ø3.333
CIRC. THICKNESS (REF)	(.262)
ADDENDUM (REF)	(.167)
WHOLE DEPTH	.360
CHORDAL THICKNESS	.261
CHORDAL ADDENDUM	.172
CIRCULAR PITCH	.524

Figure 20-16 *A working drawing for a gear seldom shows the actual outline of each gear tooth.*

The actual shape (profile) of gear teeth is rarely used on working drawings. Simplified symbols are used to show the outline of gear teeth; however, complete cutting data notations are always necessary (**Figure 20-16**). When gear teeth must be shown on detail drawings, gear tooth templates of different types and sizes are used (**Figures 20-17** and **20-18**).

GEAR NOMENCLATURE

Before preparing working drawings involving gears, you must understand the following parts of a gear that define its size and shape (**Figures 20-19** and **20-20**):

• **Gear axis** is the central axis that the gear revolves around.

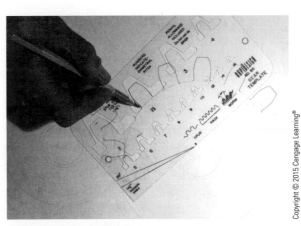

Figure 20-17 *Involute rack and spur gear teeth drawing template.*

DIAMETRAL PITCH	TOOTH SIZE	DIAMETRAL PITCH	TOOTH SIZE
4		8	
5		10	
6		12	

Figure 20-18 *The diametral pitch determines the number and size of gear teeth on a gear. (Reprinted from Mechanical Drafting by Madsen, Schumaker, and Stewart, Delmar Learning.)*

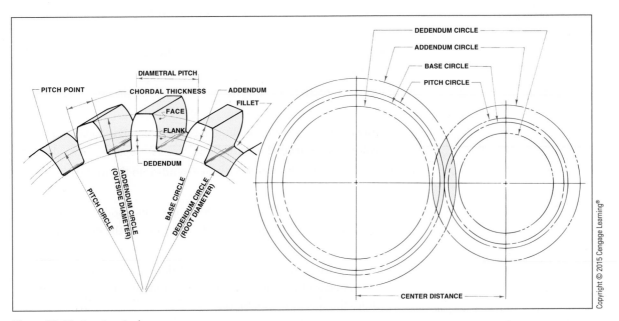

Figure 20-19 *Gear terminology.*

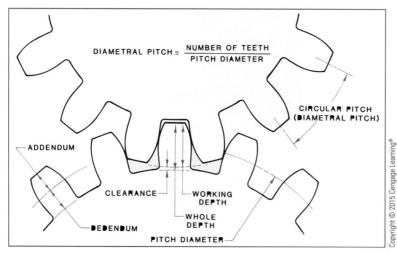

Figure 20-20 *Gear terminology.*

- **Pitch circle** is an imaginary circle concentric with the gear axis and passing through the pitch points.
- **Pitch point** is the point of intersection between the tooth face and pitch circle.
- **Tooth face** is the surface of the gear tooth that applies pressure on the adjacent gear's tooth.
- **Addendum** is the radial distance from the top of the tooth to the pitch circle.
- **Addendum circle (outside diameter)** is a circle connecting the outer edge of the gear teeth.
- **Chordal thickness** is the thickness of a tooth measured on the pitch circle.
- **Pitch diameter** is the diameter of the pitch circle.
- **Diametral pitch** is the ratio equal to the number of teeth on the gear per inch of pitch diameter.
- **Dedendum** is the radial distance from the pitch circle to the root diameter.
- **Dedendum circle** is the diameter of the pitch diameter minus the dedendum.
- **Root diameter** is the same as the dedendum circle. It is the smallest diameter for the base of a gear tooth.
- **Base circle** is the circle from which the involute tooth profile begins. When drawing gear teeth with a template, you do not need to lay out and draw the involute curve.
- **Fillet** is the concave intersection between the root and the side of a gear tooth.
- **Flank** is the lower side of a gear tooth between the base circle and root.
- **Center distance** is the distance between the centerlines of two parallel gear shafts.
- **Number of teeth** is the total number of teeth on the circumference of a gear.
- **Whole depth** is the total height of a gear tooth and is equal to the addendum plus the dedendum.
- **Working depth** is the depth a tooth extends into the mating space of the opposite tooth.
- **Clearance** is the distance between the top of one tooth and the bottom of the opposite mating tooth. It is equal to the dedendum minus the addendum.
- **Circular pitch** is the distance measured along the pitch circle from a point on one tooth to the corresponding point on the adjacent tooth. It includes the space and the tooth.

GEAR MACHINING FORMULAS

Before gears can be drawn, accurate and complete specifications must be established. Once the basic data is gathered, formulas are used to determine the numerical values for the dimensions needed for gear machining. **Figure 20-21** shows the formulas used for tooth cutting of spur and pinion gears. **Figure 20-22** shows how the final data is used on a spur gear detail drawing.

Cutting data and formulas for rack and pinion gears are basically the same as for spur gears. Since the rack is a straight surface, all circular dimensions become linear, with the addendum, dedendum, whole depth, and tooth thickness identical to the mating pinion (**Figure 20-23**). **Figure 20-24** shows a working drawing of a rack accompanied by the data notations needed for machining.

Information for cutting bevel gear teeth is based on the same involute curve as spur gear teeth; however, bevel gear teeth are tapered in a cone shape. Bevel gear teeth are designed in pairs and are not interchangeable. Dimensions are determined by the same formulas used for spur gears, with some modifications. The pitch diameter of bevel gears is the diameter of the base of the cone. The circular pitch is measured on this base circle (**Figures 20-25** and **20-26**). Working drawings of bevel gears include only the outline of the gear blank. This is normally drawn as a single sectional view (**Figure 20-27**) with all related gear tooth data provided in tabular notations.

Worm gears consist of a worm (screw form) and a worm gear (wheel form). The worm is basically a screw with single or multiple threads. The teeth on the worm gear are curved to conform to the shape of the worm teeth. Large reductions and high mechanical advantages result from the high ratios possible between worm and worm gear (up to 50:1). One revolution of the worm may result in a lead of only one tooth width on the worm gear. The lead of the worm is the distance the worm gear will travel in one revolution of the worm. Normally, only a one-view drawing of a worm gear assembly is necessary when cutting data is supplied. **Figure 20-28** shows cutting data for a worm. **Figure 20-29** shows cutting data for a worm gear.

	REQUIRED CUTTING DATA		
ITEM	TO FIND:	HAVING	FORMULA
1	Number of teeth (N)	D & P	$D \times P$
		DO & P	$(DO \times P) - 2$
2	Diametral pitch (P)	p	$\frac{3.1416}{p}$
		D & N	$\frac{N}{D}$
		DO & N	$\frac{N+2}{DO}$
3	Pressure angle (∅)	—	20° STANDARD / 14°-30' OLD STANDARD
4	Pitch diameter (D)	N & P	$\frac{N}{P}$
		N & DO	$\frac{N \times DO}{N+2}$
		DO & P	$DO - \frac{2}{P}$
5	Whole depth (ht)	a & b	$a + b$
		P	$\frac{2.157}{P}$
6	Outside diameter (DO)	N & P	$\frac{N+2}{P}$
		D & P	$D + \frac{2}{P}$
		D & N	$\frac{(N+2) \times D}{N}$
7	Addendum (a)	P	$\frac{1}{P}$
8	Working depth (hk)	a	$2 \times a$
		P	$\frac{2}{P}$
9	Circular thickness (t)	P	$\frac{3.1416}{2 \times P}$
10	Chordal thickness	N,D & a	$\sin\left(\frac{90°}{N}\right) \times D$
11	Chordal addendum	N,D & a	$\left[1 - \cos\left(\frac{90°}{N}\right)\right] \times \frac{D}{2} + a$
12	Dedendum (b)	P	$\frac{1.157}{P}$

Figure 20-21 *Required cutting data and formulas for spur gears. (Reprinted from Technical Drawing and Engineering Communication by Goetsch, Chalk, Nelson, and Rickman, Delmar/Cengage Learning.)*

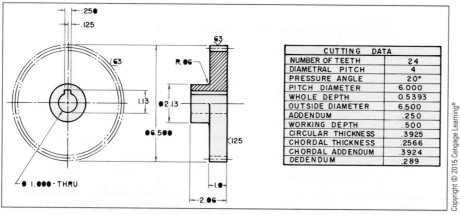

CUTTING DATA	
NUMBER OF TEETH	24
DIAMETRAL PITCH	4
PRESSURE ANGLE	20°
PITCH DIAMETER	6.000
WHOLE DEPTH	0.5393
OUTSIDE DIAMETER	6.500
ADDENDUM	.250
WORKING DEPTH	.500
CIRCULAR THICKNESS	.3925
CHORDAL THICKNESS	.2566
CHORDAL ADDENDUM	.3924
DEDENDUM	.289

Figure 20-22 *Example of a spur gear detail working drawing. (Reprinted from Technical Drawing and Engineering Communication by Goetsch, Chalk, Nelson, and Rickman, Delmar/Cengage Learning.)*

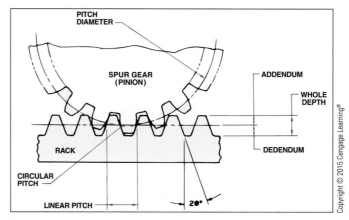

Figure 20-23 *Example of rack gear terminology.*

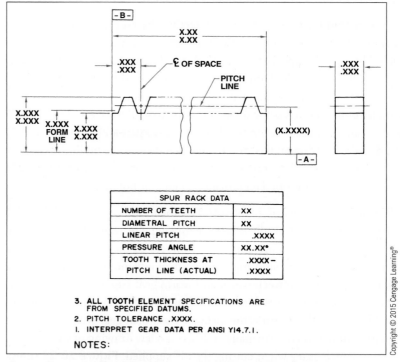

Figure 20-24 *Partial example of a rack gear working drawing. (Reprinted from Mechanical Drafting by Madsen, Schumaker, and Stewart, Delmar Learning.)*

When a second view is required, the relationship of gear teeth is shown with radial centerlines on the side view, and interlocking profiles sectioned on the front view (**Figure 20-30**).

PICTORIAL REPRESENTATION

When multiple gears are connected with shafts that lie on many different planes, the positioning and relationship of gears is often difficult to visualize. This is especially true when the position of gears change as they do in an automobile transmission. Pictorial drawings (usually isometric) are used in those cases as an aid to interpretation (**Figure 20-31**). Pictorial drawings, unless prepared to a very large scale, do not include teeth details, since these drawings are not used for manufacturing.

ITEM	TO FIND:	HAVING	FORMULA	
			SPUR	PINION
1	Number of teeth (N)	–	AS REQ'D.	
2	Diametral pitch (P)	p	$\dfrac{3.1416}{p}$	
3	Pressure angle (∅)	–	20° STANDARD 14°-30' OLD STANDARD	
4	Cone distance (A)	D & d	$\sin d \overline{)\dfrac{D}{2}}$	
5	Pitch distance (D)	p	$\dfrac{N}{p}$	
6	Circular thickness (t)	p	$\dfrac{1.5708}{p}$	
7	Pitch angle (d)	N & d (of pinion)	90°-d(pinion)	$\tan d \dfrac{N \text{ pinion}}{N \text{ gear}}$
8	Root angle (ɣR)	d & ƃ	d – ƃ	
9	Addendum (a)	p	$\dfrac{1}{p}$	
10	Whole depth (ht)	p	$\dfrac{2.188}{p} + .002$	
11	Chordal thickness (C)	D & d	$\dfrac{1}{2}\left(\dfrac{D}{\cos d}\right)$	$1-\cos\dfrac{\frac{90°}{N}}{\cos d}+a$
12	Chordal addendum (aC)	d	$\sin\left(\dfrac{\frac{90°}{N}}{\cos d}\right)$	
13	Dedendum (bC)	P	$\dfrac{2.188}{P}-a(\text{pinion})$	$\dfrac{2.188}{P}-a\,(\text{gear})$
14	Outside diameter (DO)	D, a & d	D + (2 x a) x cos. d	
15	Face	A	1/3 A (max.)	
16	Circular pitch (p)	p & N	$\dfrac{3.1416 \times p}{N}$	
17	Ratio	N gear & N pinion	$\dfrac{N \text{ gear}}{N \text{ pinion}}$	
18	Back angle (ɣO)	–	SAME AS PITCH ANGLE	
19	Angle of shafts	–	90°	
20	Part number of mating gear	–	AS REQ'D.	
21	Dedendum angle ƃ	A & b	$\dfrac{b}{A} = \tan ƃ$	

REQUIRED CUTTING DATA

Figure 20-25 *Required cutting data and formulas for bevel gears. (Reprinted from Technical Drawing and Engineering Communication by Goetsch, Chalk, Nelson, and Rickman, Delmar/Cengage Learning.)*

Figure 20-32 is a pictorial assembly drawing designed to show the relationship of gears and other mechanical components. **Figure 20-33** is a pictorial cutaway view of a gear assembly revealing the internal configuration and relationship of gears and gear assemblies.

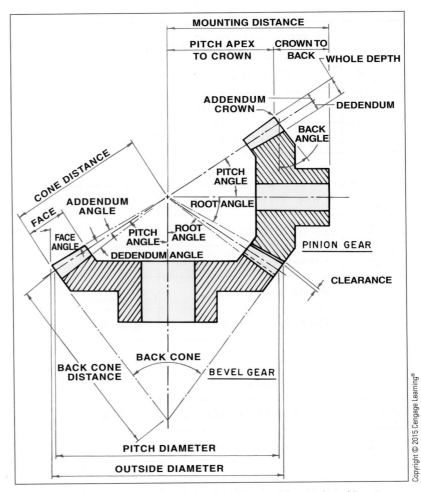

Figure 20-26 *Example of bevel gear terminology. (Reprinted from Technical Drawing and Engineering Communication by Goetsch, Chalk, Nelson, and Rickman, Delmar/ Cengage Learning.)*

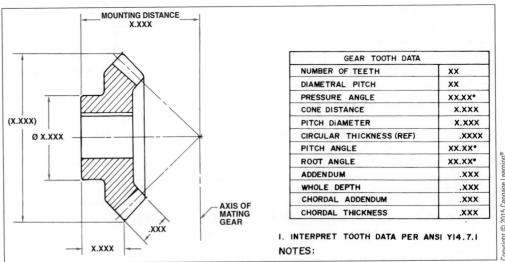

Figure 20-27 *Example of a bevel gear working drawing. (Reprinted from Mechanical Drafting by Madsen, Schumaker, and Stewart, Delmar Learning.)*

REQUIRED CUTTING DATA			
ITEM	TO FIND	HAVING:	FORMULA
I	Number of teeth (N)	P	$\dfrac{3.1416}{P}$
2	Pitch diameter (D)	Pa	(2.4 x Pa)+ 1.1
		DO & a	DO - (2 x a)
3	Axial pitch (Pa)	–	Distance from a point on one tooth to same point on next tooth
4	Lead (L) Right or Left	p & N	p x N
5	Lead angle (La)	L & D	$\dfrac{L}{3.1416 \times D} = \tan La$
6	Pressure angle (∅)	–	20° STANDARD 14°-30' OLD STANDARD
7	Addendum (a)	p	p x .3183
		P	$\dfrac{1}{P}$
8	Whole depth (ht)	Pa	.686 x Pa
9	Chordal thickness	N,D & a	$\left[1 - \cos\left(\dfrac{90°}{N}\right)\right] \times \dfrac{D}{2} + a$
10	Chordal addendum	N,D & a	$\sin\left(\dfrac{90°}{N}\right) \times D$
11	Outside diameter (DO)	D & a	D + (2 x a)
12	Worm gear part no.	–	AS REQ'D.

Axial pitch (Pa) must be same as worm gear circular pitch (p)

Figure 20-28 *Required cutting data and formula for a worm. (Reprinted from Technical Drawing and Engineering Communication by Goetsch, Chalk, Nelson, and Rickman, Delmar/Cengage Learning.)*

REQUIRED CUTTING DATA			
ITEM	TO FIND:	HAVING	FORMULA
I	Number of teeth (N)	–	AS REQ'D.
2	Pitch diameter (D)	N & p	$\dfrac{N \times p}{3.1416}$
3	Addendum (a)	p	p x .3181
		P	$\dfrac{1}{P}$
4	Whole depth (ht)	p	p x .6866
		P	$\dfrac{2.157}{P}$
5	Lead (L) Right-Left	p & N	p x N
6	Worm part no.	–	AS REQ'D.
7	Pressure angle ∅	–	20° STANDARD 14°-30' OLD STANDARD
8	Outside diameter (DO)	Dt & Pa	Dt + .4775 x Pa
9	✱ Circular pitch (p)	P	$\dfrac{3.1416}{P}$
		L & N	$\dfrac{L}{N}$
10	Diametral pitch (P)	p	$\dfrac{3.1416}{p}$
11	Throat diameter (Dt)	D & Pa	D + .636 x Pa
12	Ratio of worm/worm gear	N worm & N worm gear	$\dfrac{N \text{ worm gear}}{N \text{ gear}}$
13	Center to center distance between worm & worm gear.	D worm & D worm gear	$\dfrac{D \text{ worm} + D \text{ worm gear}}{2}$

Circular pitch (p) must be same as worm Axial pitch (Pa)

Figure 20-29 *Cutting data and formulas for a worm gear. (Reprinted from Technical Drawing and Engineering Communication by Goetsch, Chalk, Nelson, and Rickman, Delmar/Cengage Learning.)*

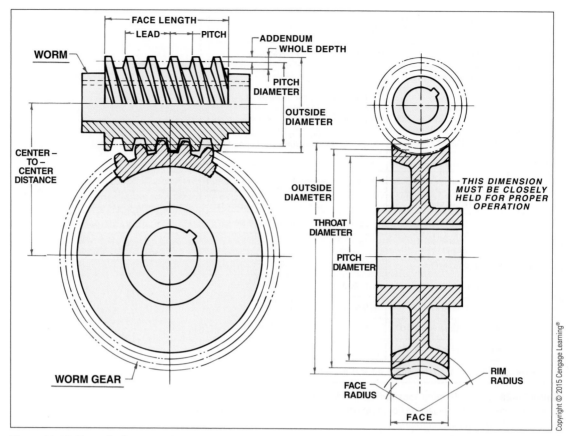

Figure 20-30 *Example of worm and worm gear terminology. (Reprinted from Technical Drawing and Engineering Communication by Goetsch, Chalk, Nelson, and Rickman, Delmar/Cengage Learning.)*

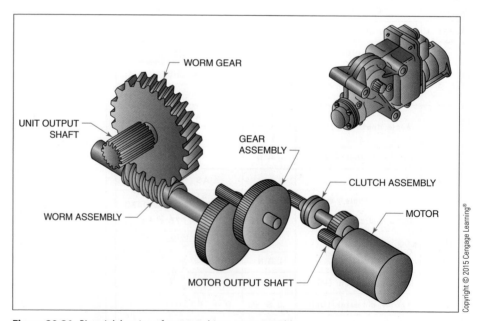

Figure 20-31 *Pictorial drawing of a motor-driven gear assembly.*

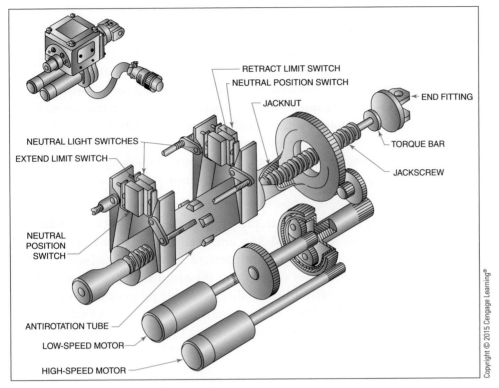

Figure 20-32 *Pictorial drawing of a motor-driven gear assembly.*

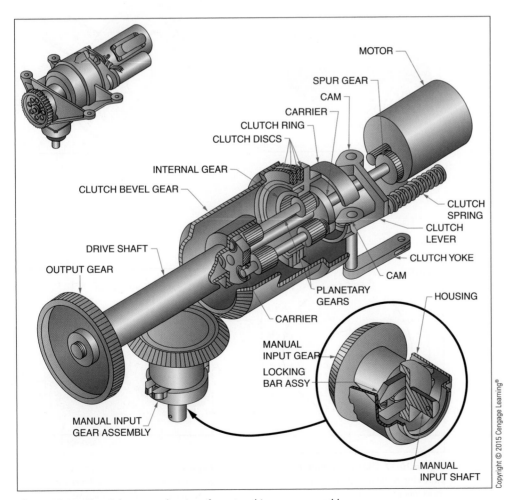

Figure 20-33 *Pictorial cutaway drawing of a motor-driven gear assembly.*

Cam Mechanisms

It is often necessary to convert the rotary motion of a motor into linear motion for many mechanical operations. It is also often necessary to convert this motion back again to rotary motion for successive manufacturing procedures. Gears and screws can only change or maintain motions that are evenly spaced at regular intervals. **Cam** mechanisms can convert regular motion to irregular or unusual motion. They also can convert one type of motion into another, for example, rotary motion to linear oscillating motion. Cam mechanisms consist of a cam, follower, and camshaft.

Cams are divided into three types—**plate cams** (**Figure 20-34**), **drum cams** (**Figure 20-35**), and **face cams** (**Figure 20-36**). Cam shapes are determined by the desired motion of a follower. A **plate cam follower** moves vertically as the cam rotates, converting rotary motion into vertical linear (displacement) reciprocating movement (**Figure 20-37**). Drum cams convert circular motion to linear horizontal reciprocating movement by guiding the follower through a groove (**Figure 20-38**).

Figure 20-34 *Plate cam.*

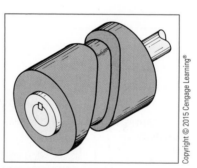

Figure 20-35 *Drum cam.*

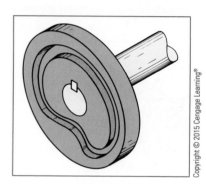

Figure 20-36 *Face cam.*

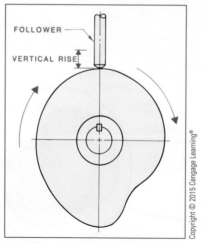

Figure 20-37 *The circular motion of the plate cam is converted to a vertical linear movement by the follower.*

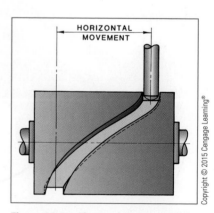

Figure 20-38 *The circular motion of the drum cam is converted to a horizontal linear movement by the follower.*

CAM FOLLOWERS

There are four types of cam followers—**roller**, **pointed**, **round**, and **flat-faced** (**Figure 20-39**). Roller followers are used the most since they operate better at higher speeds with less friction and wear. The exact shape of a cam determines the exact timing and motion limits of the follower during one cam revolution. Think of a cam as a simple inclined plane with uneven surface contours. Although **Figure 20-40** shows the follower position at only four points, it moves smoothly from one position to another. **Figure 20-41** summarizes the nomenclature used in the design and drawing of a plate cam.

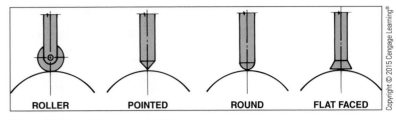

Figure 20-39 *Basic types of cam followers.*

DISPLACEMENT DIAGRAM

The first step in designing and drawing a cam shape is the development of a **displacement diagram**. A displacement diagram is a curve representing the desired position (displacement) of the follower at successive intervals during one revolution of the **camshaft** (**Figure 20-42**). Geometric line shapes used in the development of displacement diagrams include:

1. Straight angular lines, which produce uniform motion

2. Parabolic curves, which produce constant acceleration or deceleration

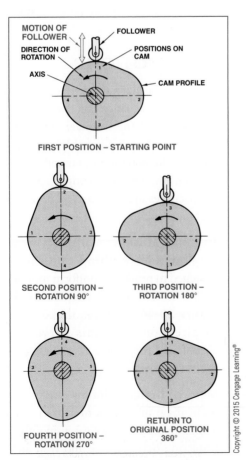

Figure 20-40 *Movement of the cam's follower.*

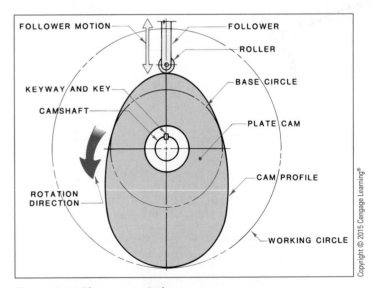

Figure 20-41 *Plate cam terminology.*

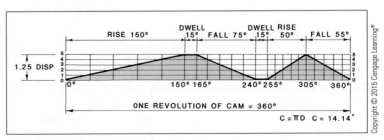

Figure 20-42 *Displacement diagram for a cam.*

3. Harmonic curves, which produce a harmonic motion
4. Straight horizontal lines, which hold the follower at a fixed (**dwell** or rest) position

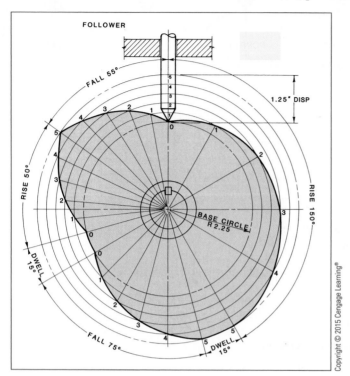

FOLLOWER

FALL 55°

1.25" DISP

RISE 50°

RISE 150°

BASE CIRCLE R 2.25

DWELL 15°

DWELL 15°

FALL 75°

Figure 20-43 *Layout of a cam drawing made from the displacement diagram in Figure 20-42.*

All nonhorizontal lines result in an upward movement (**rise**) of the follower or in a downward movement (**fall**) of the follower. The displacement diagram in **Figure 20-42** illustrates two uniform rise and fall curves with two dwell (rest) periods. In **Figure 20-42**, the horizontal axis represents 360° or one revolution of the cam. The vertical grid represents the maximum distances the follower will move during one revolution of the cam. The profile line identifies the exact position of the follower required at specific degrees of rotation.

Displacement diagrams are flat, while cams are basically round. The flat profile intersections or the displacement diagram must be converted to radial distances to create a cam shape (**Figure 20-43**). In **Figure 20-43**, all horizontal distances are converted to degrees on concentric circles. These are spaced evenly to represent the movement range of the follower. Intersecting points are then plotted to coincide with the intersections on the displacement diagram. Connecting these points creates the outline shape of the cam that will produce the follower action plotted in the displacement diagram.

6
5
4
3
2
1
0

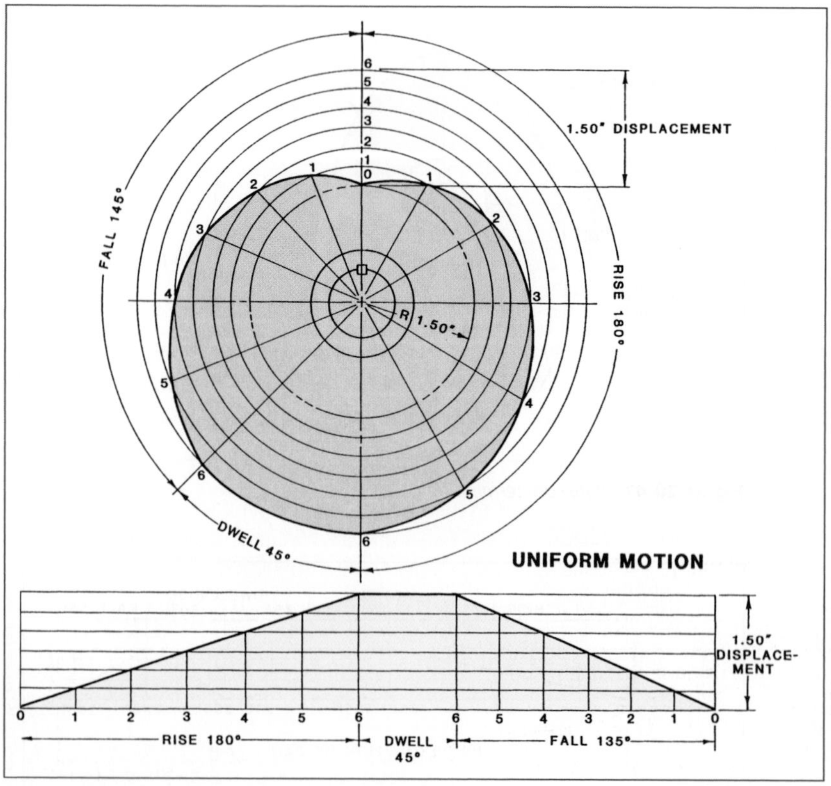

1.50" DISPLACEMENT

FALL 145°

RISE 180°

R 1.50"

DWELL 45°

UNIFORM MOTION

1.50" DISPLACE-MENT

0   1   2   3   4   5   6   6   5   4   3   2   1   0

RISE 180°   DWELL 45°   FALL 135°

Figure 20-44 *The uniform cam motion gives a jerk to the follower at the start and end of each phase.*

CAM MOTIONS

Care must be taken to provide the correct cam shape, since that controls the movement of the follower. The follower must move smoothly, without jerks, at the maximum planned speed of rotation. The following displacement curves are designed to produce specific results.

Uniform Motion. A displacement diagram and resulting cam shape designed to produce a uniform motion at a constant speed are shown in **Figure 20-44**. However, the changes in direction from this curve are abrupt, and they cause a jerky motion in the follower.

Modified Uniform Motion. A displacement diagram with a modified uniform motion that produces a smoother transition between rise and rest, and rest and fall, is shown in **Figure 20-45**.

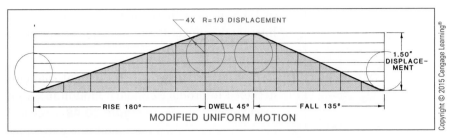

Figure 20-45 *The modified uniform cam motion will smooth out the follower's jerky motion.*

Here, straight line intersections are converted to small arc intersections. The radius of the arcs is equal to one-third the displacement rise or fall.

Harmonic Motion. When even greater smoothness is required, usually for high speeds, harmonic curves are plotted (**Figure 20-46**). Harmonic curves provide a smooth transition among rise, fall, and dwell. They also allow smoother start and stop movements of the follower. A harmonic curve is plotted by projecting segments of a circle onto the displacement diagram in place of evenly spaced distances.

Uniform Accelerated and Decelerated Motion. When the desired motion of a follower is to increase or decrease steadily and smoothly, a parabolic curve is plotted on the displacement diagram (**Figure 20-47**). In plotting a parabola, lay out the 360° line or time line horizontally as before. Divide the vertical follower distance line into displacement ratios

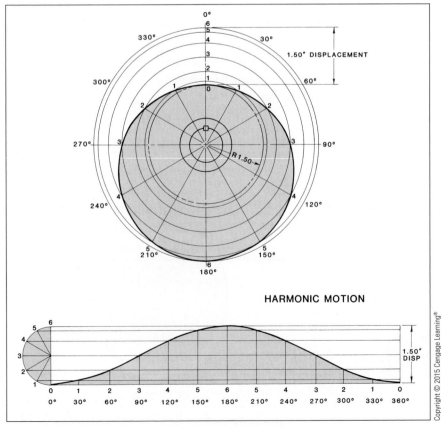

Figure 20-46 *Harmonic motion provides the smoothest motion.*

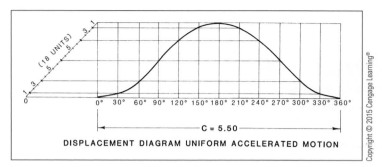

Figure 20-47 *The uniform accelerated and decelerated motion produces a very smooth motion with a parabolic curve.*

of 1:3:5. Project these lines to intersect degree lines, and connect the intersections to produce a parabola.

Cycloidal Motion. A displacement diagram designed to produce cycloidal motion in a cam mechanism is shown in **Figure 20-48**. Cycloidal curves produce extremely smooth curves by eliminating the dwell contour. Cycloidal curves are used in cams that operate at optimum efficiency at high speeds.

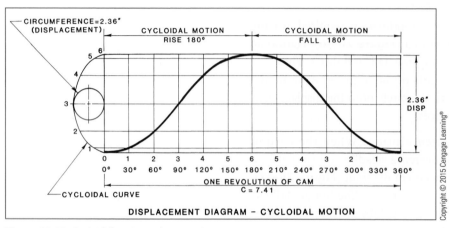

Figure 20-48 *Cycloidal motion is the smoothest of the cam's motions.*

DESIGN AND DRAWING PROCEDURES

Figures 20-49 through **20-52** illustrate the specific steps used to design, draw, and dimension cams in detail.

A uniform motion displacement diagram (**Figure 20-49**) is used to complete the cam profile shown in **Figure 20-50**.

The procedures used to develop and draw a uniform motion plate cam are shown in **Figure 20-51**. This involves laying out the base circle, followed by segmenting the circumference. The rise displacement circles are then added to represent the range of follower motion. Points are plotted to coincide with intersections on the displacement diagram. These points are connected to produce the final shape of the cam.

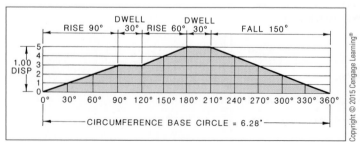

Figure 20-49 *Displacement diagram designed with a uniform motion.*

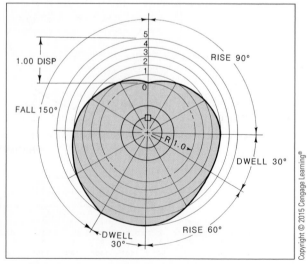

Figure 20-50 *Uniform motion plate cam drawn from the displacement diagram in Figure 20-49.*

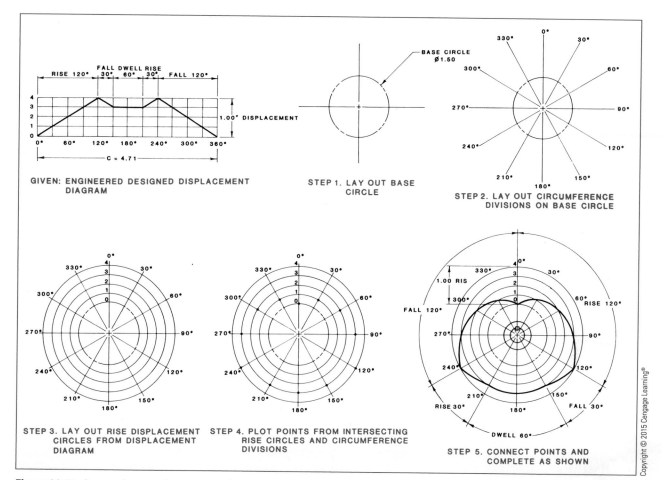

Figure 20-51 *Steps to draw a uniform motion plate cam.*

As in pattern drafting, the smoothness of the shape is directly related to the number of circle divisions used to develop the profile on the displacement diagram. If intersecting points are too far apart to produce a smooth curve, an irregular (French) curve template may be used to smooth out the contour.

Similar sequences used to develop a cam to produce a constant acceleration motion are shown in **Figure 20-52**. The only difference in this procedure is the use of an angular profile line to produce the fall-rise segments on the displacement diagram.

Since cam shapes are irregular, grid layouts may be used to establish the exact shape of the circumference. The shape of the cam also may be defined by locating the position of the follower at various degree increments during a 360° rotation. Again, more degree divisions produce a smoother contour on the cam circumference.

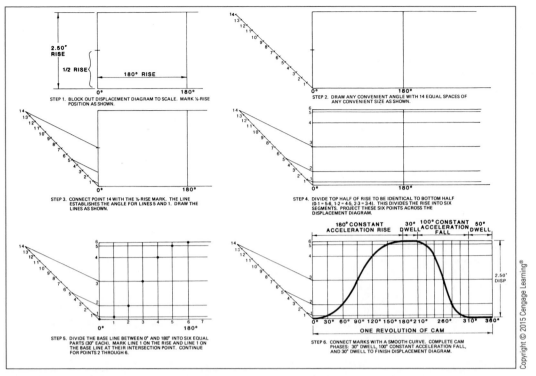

Figure 20-52 *Steps to draw a constant acceleration cam with a rise of 180°, dwell 30°, fall 100°, and dwell of 50°.*

 DRAFTING EXERCISES

1. Redraw the pinion and spur gear in **Figure 20-53**.

2. With drawing instruments or a CAD system, draw the spur gears in **Figures 20-54**, and **20-55**, and the helical gear in **Figure 20-56**.

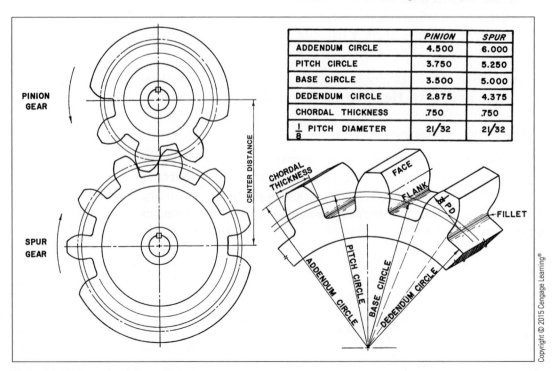

	PINION	*SPUR*
ADDENDUM CIRCLE	4.500	6.000
PITCH CIRCLE	3.750	5.250
BASE CIRCLE	3.500	5.000
DEDENDUM CIRCLE	2.875	4.375
CHORDAL THICKNESS	.750	.750
$\frac{1}{8}$ PITCH DIAMETER	21/32	21/32

Figure 20-53 *Draw the pinion and spur gear.*

Copyright © 2015 Cengage Learning®

3. With drawing instruments or a CAD system, draw the section of the two miter gears in **Figure 20-57**.

4. With drawing instruments or a CAD system, draw the worm gear in **Figure 20-58**.

5. Draw the cams and their displacement diagrams for **Figures 20-59** through **20-64**.

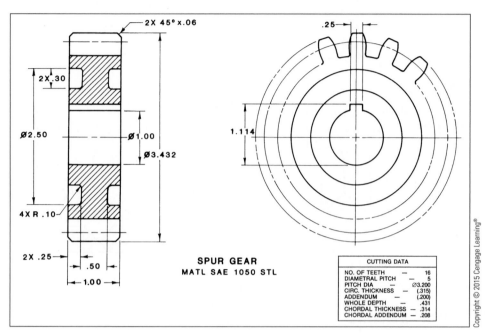

Figure 20-54 *Draw the spur gear.*

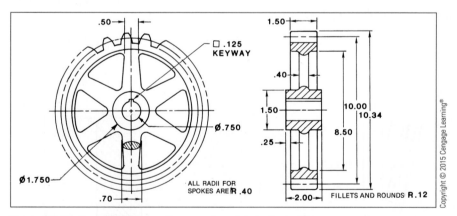

Figure 20-55 *Draw the spur gear. Use a gear template for the teeth.*

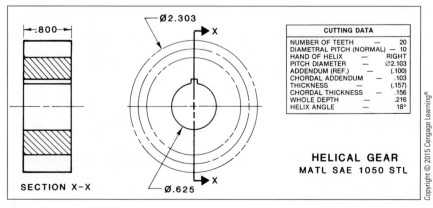

Figure 20-56 *Draw the helical gear.*

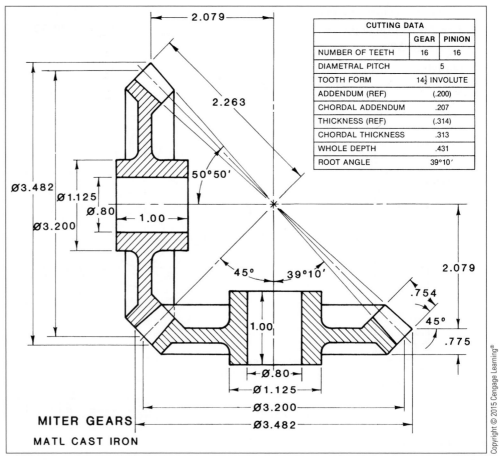

CUTTING DATA		
	GEAR	PINION
NUMBER OF TEETH	16	16
DIAMETRAL PITCH	5	
TOOTH FORM	$14\frac{1}{2}$ INVOLUTE	
ADDENDUM (REF)	(.200)	
CHORDAL ADDENDUM	.207	
THICKNESS (REF)	(.314)	
CHORDAL THICKNESS	.313	
WHOLE DEPTH	.431	
ROOT ANGLE	39°10′	

MITER GEARS
MATL CAST IRON

Figure 20-57 *Redraw the two miter gears.*

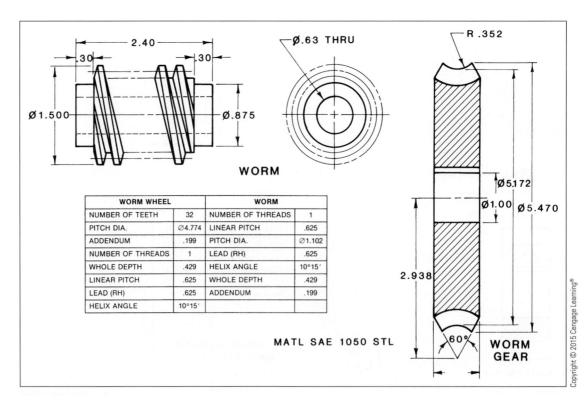

WORM

WORM WHEEL		WORM	
NUMBER OF TEETH	32	NUMBER OF THREADS	1
PITCH DIA.	Ø4.774	LINEAR PITCH	.625
ADDENDUM	.199	PITCH DIA.	Ø1.102
NUMBER OF THREADS	1	LEAD (RH)	.625
WHOLE DEPTH	.429	HELIX ANGLE	10°15′
LINEAR PITCH	.625	WHOLE DEPTH	.429
LEAD (RH)	.625	ADDENDUM	.199
HELIX ANGLE	10°15′		

MATL SAE 1050 STL

WORM GEAR

Figure 20-58 *Redraw the worm gear.*

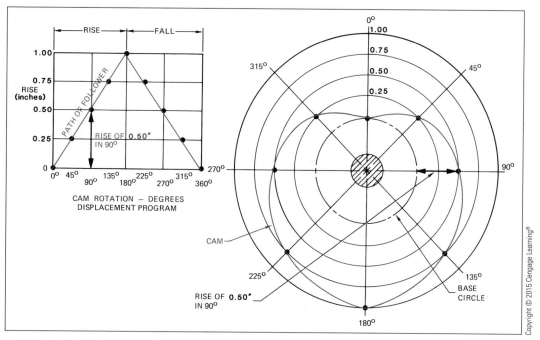

Figure 20-59 *Draw the cam and its displacement diagram.*

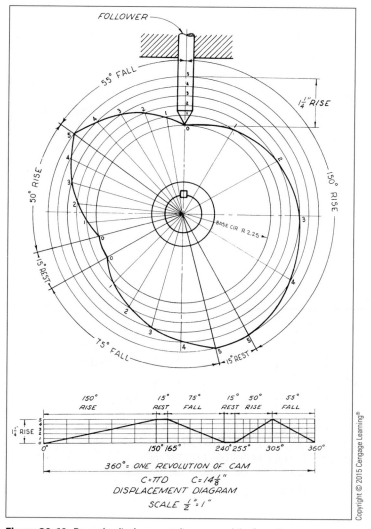

Figure 20-60 *Draw the displacement diagram and the face plate.*

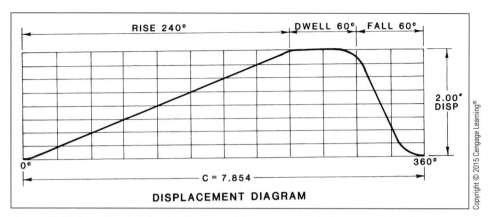

Figure 20-61 *Draw the displacement diagram and the plate cam. To find the base circle's diameter, divide pi (π) into the circumference.*

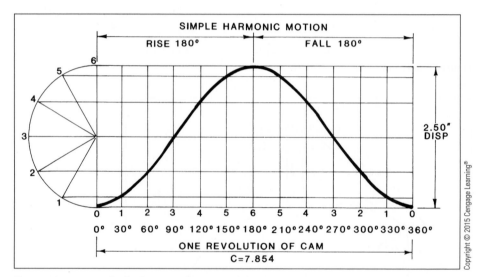

Figure 20-62 *Draw the displacement diagram and the plate cam.*

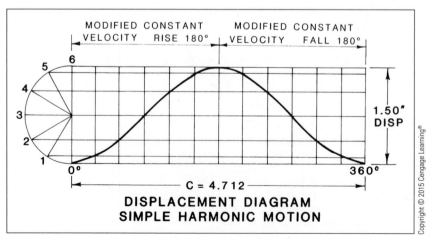

Figure 20-63 *Draw the displacement diagram and the plate cam.*

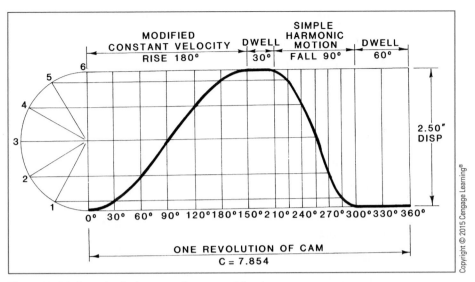

Figure 20-64 *Draw the displacement diagram and the plate cam.*

DESIGN EXERCISES

1. Draw a cam displacement diagram for a cam follower that rises with a harmonic motion 2.00" in 180°, then has a dwell for 30°, then falls 2.00" in 120°, and completes the full 360° of rotation with a 30° dwell. The circumference of the base circle is 12.00".

2. Design a plate cam using the displacement diagram created with the previous data. The basic dimensions for the cam are to be:

   Base circle diameter = 3.82"
   Plate thickness = .50"

Total thickness, including plate and hub thicknesses = 1.50"
Hub diameter = 2.50"
Bore diameter = 1.000" + .005, −.000
Use a square key to keep the cam from rotating on the shaft. Recommended square key for a 1.00" dia shaft is ³⁄₁₆" square.
Material = 6061 T6 Aluminum

KEY TERMS

Addendum	Center distance	Dedendum
Addendum circle	Chain and sprocket drives	Dedendum circle
Angular gears	Chordal thickness	Diametral pitch
Base circle	Circular pitch	Displacement diagram
Belt drive	Clearance	Drum cam
Bevel gear	Cross helical gears	Dwell (rest)
Cam	Cycloidal curve	Face cam
Camshaft	Cycloidal motion	Fall

Fillet

Flank

Flat-faced follower

Friction wheel

Gear

Gear axis

Gear ratio

Harmonic motion

Helical gear

Herringbone gear

Hypoid gear

Involute curve

Miter gear

Modified uniform motion

Number of teeth

Pitch circle

Pitch diameter

Pitch point

Plate cam

Pointed follower

Rack and pinion gear

Ring gears

Rise

Roller follower

Root diameter

Round follower

Spur gears

Tooth face

Uniform accelerated motion

Uniform decelerated motion

Uniform motion

Whole depth

Working depth

Worm gears

Piping Drawings

21

OBJECTIVES

The student will be able to:

- Recognize pipe drafting symbols
- Complete an orthographic single-line piping drawing
- Complete an orthographic double-line piping drawing
- Complete an isometric single-line piping drawing
- Complete an isometric double-line piping drawing
- Prepare a piping drawing on a CAD system

Introduction

Pipes are used in a wide range of applications, from small residences to large industrial plants (**Figure 21-1**). Pipes carry all types of liquids, gases, and semisolids including water, oil, acid, liquid metal, nitrogen, oxygen, steam, granular materials, and exhausts. Pipes are also used structurally as support columns and for a wide variety of architectural applications, such as handrails and fence components.

Figure 21-1 *Example of industrial piping. (Courtesy Engineering Drawing & Design, Madsen and Madsen, Delmar, Cengage Learning.)*

Pipes are the arteries and veins of plumbing, heating, cooling, drainage, and manufacturing systems. Pipes are made of steel, wrought iron, cast iron, copper, brass, aluminum, plastic, clay, concrete, brass, glass, lead, rubber, wood, and other synthetic material combinations. Drawings that describe the pipe material, size, and location in a piping system are assembly drawings. These working drawings use orthographic or pictorial (isometric) projection (**Figures 21-2** and **21-3**).

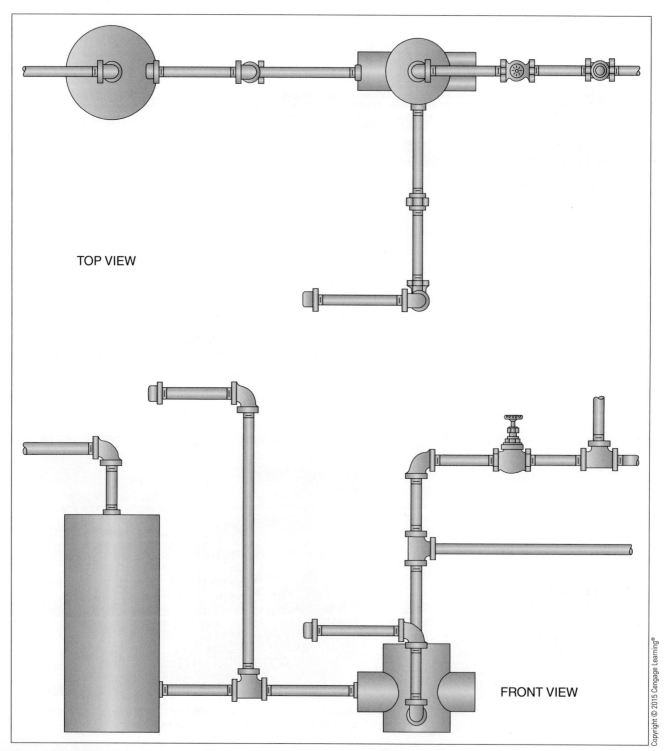

TOP VIEW

FRONT VIEW

Figure 21-2 *Multiview drawing of a piping system.*

Pipe Connections

Fittings are usually required to join pipes and attach them to other devices, such as pumps and tanks. Pipes are temporarily joined with threaded, flanged, or bell and spigot connections. Others are permanently attached by brazing, soldering, or welding. Some pipes can be bent to change direction.

THREADED CONNECTIONS

Threaded connections are used primarily on pipes with 2" diameter or less and are available with straight or tapered threads. Threaded connectors, sometimes called **screwed fittings**, are available in many shapes—elbows (**Figure 21-4**), tees, Y bends, reducers, return bends, caps, and crosses. Couplings with inside threads are also used to connect straight pipes or nipples. Nipples are short pipe lengths with external threads on both ends.

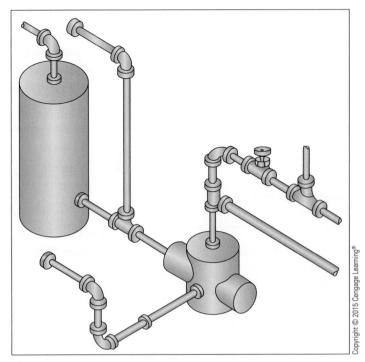

Figure 21-3 *A pictorial drawing of Figure 21-2.*

FLANGE CONNECTIONS

Flange connections are designed to enable a pipe joint to be disassembled by removing a series of bolts on the face of the flange. Flanges are manufactured in the form of elbows (**Figure 21-5**), return bends, Y bends, tees, taper reducers, and crosses.

WELDED CONNECTIONS

Since **welded connections** are virtually leakproof, welding is used to join pipes containing hot or volatile materials. Welded joints are also more economical to construct. Gas and arc welding methods are most commonly used to join pipe with butt, lap, tee, edge, and corner joints. **Figure 21-6** shows a welded tee shape. See Chapter 19 for welding joint details.

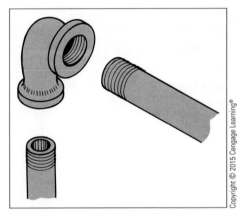

Figure 21-4 *Threaded elbow pipe connection.*

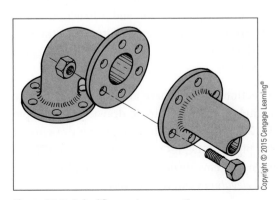

Figure 21-5 *Bolted flange pipe connection.*

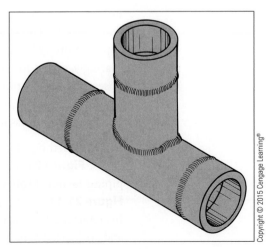

Figure 21-6 *Welded pipe connections.*

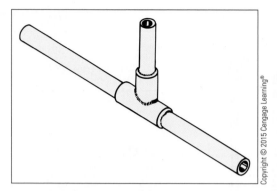

Figure 21-7 *Soldered copper pipe connections.*

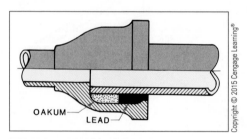

Figure 21-8 *Bell and spigot connection.*

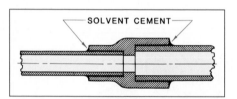

Figure 21-9 *Cemented connection for plastic pipes.* Copyright © 2015 Cengage Learning®

In the processes of brazing and soldering, the base metal does not melt. Consequently, brazing and soldering are restricted to joining small copper or brass pipe and tubing. Pipe joints are soldered with the use of connectors (**Figure 21-7**). Connectors must be specified to fit tightly over the pipe to aid the sweat soldering process. Remember, solder has a low boiling point, so extremely hot materials, especially under high pressure, must be avoided.

BELL AND SPIGOT CONNECTIONS

Bell and spigot connections are so named because one pipe end is shaped like a bell and inserted into a pipe with a spigot-like shape. These joints are fitted tightly with lead and oakum to make them waterproof (**Figure 21-8**).

CEMENTING

Many materials such as plastic, clay, concrete, and glass cannot be welded, soldered, or threaded effectively. Joining pipes made of these materials requires **cementing** (**Figure 21-9**).

Pipe Standards

Pipes are specified by the type of material, sizes, and wall thickness. Pipe sizes relate to either inside or outside diameters. The **nominal pipe size** for steel pipe up to 12" is the inside diameter (ID). For pipes over 12", the nominal size is the outside diameter (OD). Pipe wall thicknesses are standardized by schedules of pipe weights. Pipe schedules range from 10 to 160 with wall thickness increasing as the schedule number increases (**Figure 21-10**).

Piping Drawings

There are two general types of piping drawings—orthographic and pictorial. Either type may be prepared using a single line to represent each pipe, or a double line to represent the outline of a pipe. **Double-line piping drawings** appear more realistic but require extensive drafting time. They are used sparingly, except for large pipes or for presentation and instructional purposes. **Single-line piping drawings** are used the majority of the time. In single-line drawings, a single thick line is drawn at the location of each pipe centerline, regardless of the pipe size.

When orthographic projection is used, there are usually three basic views. **Figure 21-11** is a single-line orthographic drawing of a simple piping layout. **Figure 21-12** depicts a double-line piping drawing of **Figure 21-11**. Orthographic piping drawings may involve only two views. In architectural work, plumbing systems are depicted in top and front views and are labeled *plan view* and *elevation view*, respectively.

Orthographic drawings effectively show the layout of piping systems when pipes align on one or two planes. When pipes lie on many different planes and angles, orthographic drawings become cluttered and complex. For this reason, pictorial drawings (isometric or oblique) are used to eliminate the confusion created by overlapping pipes, fittings, and fixtures. **Figure 21-13** shows a single-line isometric piping drawing, and **Figure 21-14** shows a double-line isometric drawing of the same layout. Isometric piping drawings are especially effective for describing complex systems that involve a variety of pipe configurations, fittings, valves, fixtures, and devices (**Figure 21-15**).

If a single projection plane is involved, an orthographic drawing (**Figure 21-16**) can be drawn to a more accurate scale. Piping systems are often broken into small segments that fall on one plane. Dimensioned orthographic drawings are prepared for these segments. Isometric drawings are used only to show the general assembly of the entire system or subsystem. Note that the orthographic drawing shown in **Figure 21-16** represents the back of the hydraulic system shown in **Figure 21-15**.

STANDARDIZED PIPE SIZES—SCHEDULE 20		
Nominal Size (inches)	Outside Diameter (inches)	Inside Diameter (inches)
$^1/_8$	.405	.269
$^1/_4$	.540	.364
$^3/_8$	.675	.493
$^1/_2$	.840	.622
$^3/_4$	1.050	.824
1	1.315	1.049
$1^1/_4$	1.660	1.380
$1^1/_2$	1.900	1.610
2	2.375	2.067
$2^1/_2$	2.875	2.469
3	3.500	3.068
$3^1/_2$	4.000	3.548
4	4.500	4.026
5	5.563	5.047
6	6.625	6.065
8	8.625	7.981
10	10.750	10.020
12	12.750	11.938
14	14.000	13.126
16	16.000	15.000
18	18.000	16.876
20	20.000	18.814
24	24.000	22.626

Figure 21-10 *Standardized pipe diameters for Schedule 20 pipes.*

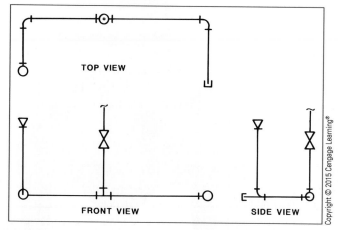

Figure 21-11 *Single-line orthographic piping drawing.*

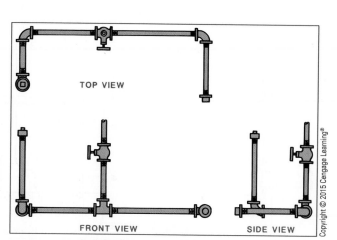

Figure 21-12 *Double-line orthographic piping drawing.*

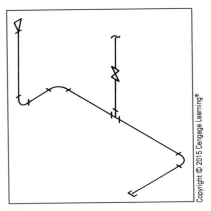

Figure 21-13 *Single-line isometric piping drawing.*

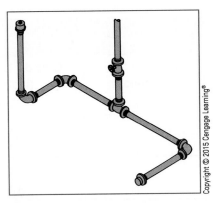

Figure 21-14 *Double-line isometric piping drawing.*

PIPING SYMBOLS

The symbol for a pipe is either a single or a double solid line. Pipes represent only one component in a complete piping system. Other components include valves, fittings, instruments, flanges, and other devices. Inserting an orthographic view of these components at hundreds of locations on a piping drawing is time-consuming and impractical.

Valves are complex devices that control the flow of pressurized liquids and gases. There are dozens of valve types, manufactured in hundreds of sizes and types of materials (**Figures 21-17** through **21-20**). This makes the preparation of orthographic drawings of valves extremely involved. Templates simply outline the drawing of valves and other components (**Figure 21-21**). Nevertheless, most valves, fittings, flanges, instruments, fixtures, special lines, and connection methods are represented by simplified symbols on piping drawings. Piping symbols are shown in **Figure 21-22** in orthographic single- and double-line views,

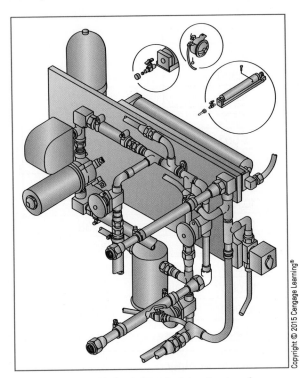

Figure 21-15 *Double-line isometric piping drawing of a segment of a hydraulic system.*

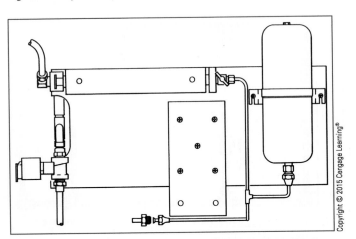

Figure 21-16 *An orthographic view of the backside of a segment of the hydraulic system in Figure 21-15.*

Figure 21-17 *Gate valve.*

Figure 21-19 *Ball valve.*

Figure 21-20 *Ball valve used as a stop valve.*

Figure 21-18 *A gate valve used as a shut off valve.*

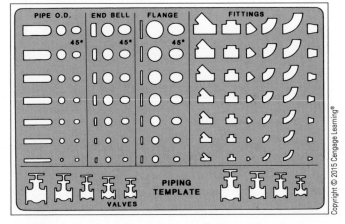

Figure 21-21 *Piping templates help to speed piping working drawings.*

and in the isometric single-line view. The various types of pipe connection symbols must be apparent on the piping drawings to ensure correct assembly. **Figure 21-23** shows the application of pipe connection symbols.

Figures 21-24 and **21-25** are partial examples of single-line and double-line piping systems. **Figures 21-26** through **21-28** are examples of piping drawings created with a CAD system.

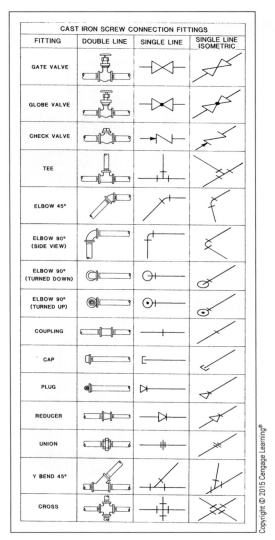

Figure 21-22 *Screw pipe fitting symbols.*

DIMENSIONING

Piping dimensions are similar to normal orthographic or isometric dimensioning standards with the following special considerations (**Figure 21-29**):

1. Dimensions perpendicular to a pipe are located to the centerline of the pipe. Dimension lines are connected to the centerline in double-line drawings and to the pipe line in one-line drawings.

2. Dimensions parallel with pipe lengths are placed to the center of the pipe or flange.

3. Pipe sizes and pipe fitting sizes are provided in a general note if all pipes are equal. Dimensions and notes are connected to a pipe with a leader if pipe sizes vary. Notations include information on the type of material, size, and wall thickness.

4. When more than one length of pipe is required for a total run, only the total length of the run is dimensioned, not the length of the various pieces.

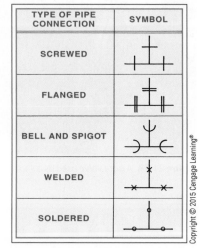

Figure 21-23 *Pipe connection symbols.*

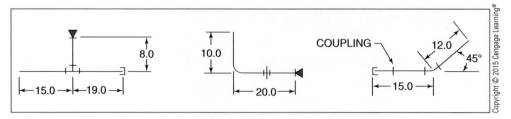

Figure 21-24 *Typical single-line drawing examples of dimensioned piping segments.*

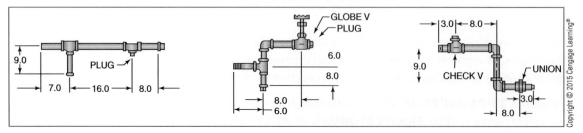

Figure 21-25 *Typical double-line drawing examples of dimensioned piping segments.*

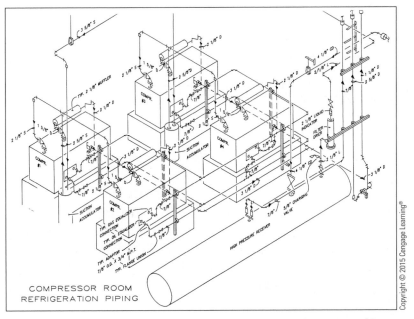

Figure 21-26 *CAD-generated, single-line isometric piping drawing for a segment of a refrigeration system.*

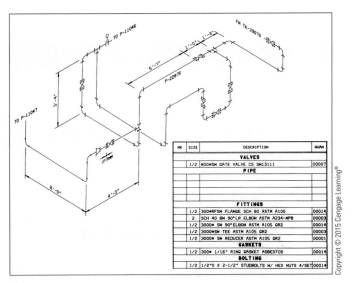

Figure 21-27 *CAD-generated, single-line isometric piping drawing with materials list.*

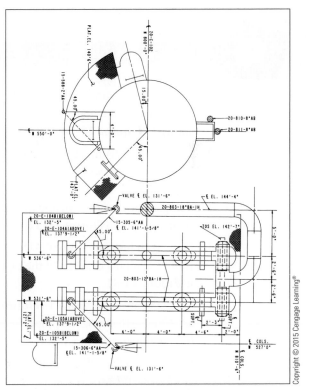

Figure 21-28 *CAD-generated, double-line orthographic piping drawing for a segment of a steam system.*

5. Keep dimensions and extension lines thinner than pipe object and symbol lines.

6. Do not chain-dimension. Dimension to a base datum to avoid accumulating errors.

7. Use arrows to show the direction of content flow.

8. Label the contents of each pipe subsystem.

9. Keep all dimensions in feet and inches if using the U.S. customary system, or in millimeters if using the metric system.

10. Use unilateral numerals for isometric dimensioning.

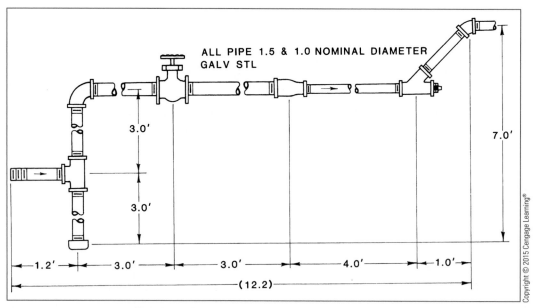

ALL PIPE 1.5 & 1.0 NOMINAL DIAMETER
GALV STL

3.0'

3.0'

7.0'

1.2'   3.0'   3.0'   4.0'   1.0'

(12.2)

Copyright © 2015 Cengage Learning®

Figure 21-29 *Example of a dimensioned pipe drawing.*

DRAFTING EXERCISES

1. Draw with instruments or CAD the single-line orthographic drawing of the piping system shown in **Figures 21-30** and **21-31**.

2. Draw with instruments or CAD a double-line orthographic drawing of the piping system in **Figure 21-31**. Draw with instruments or CAD a single-line isometric piping drawing of the piping system in **Figure 21-31**.

3. Draw with instruments or CAD a double-line isometric piping drawing of the piping system in **Figure 21-31**.

4. Make the following ECOs to **Figure 21-32**:
   a. Change the horizontal pipe diameter to .75".
   b. Lengthen pipe 4 to 5" and cap.
   c. Correct pipe 7 to 26" in the materials list and drawing.
   d. Correct pipe 6 (2) to 38" in the materials list.
   e. Shorten pipe 8 (top only) to 28".
   f. Update all changes on the parts list.
   g. List the ECOs on the drawing.

5. With instruments or CAD, draw an ortho-graphic single- and double-line piping drawing of the sketched piping layout in **Figure 21-33**.

6. With a CAD system, draw the 90° flanged pipe elbow in **Figure 21-34**.

7. Design a single-line isometric pipe drawing plan for a water heater and dryer hookup with a .5" gas pipeline for **Figure 21-35**.

8. Draw with instruments or CAD a single-line isometric drawing of the hot tub filter system in **Figure 21-36** and/or the hot water tank system

9. Draw with instruments or CAD a double-line orthographic drawing of the hot tub filter sys-tem in **Figure 21-36** and/or the hot water tank system in **Figure 21-37**.

10. Use the symbols in **Figure 21-38** to make a CAD symbol library or individual blocks.

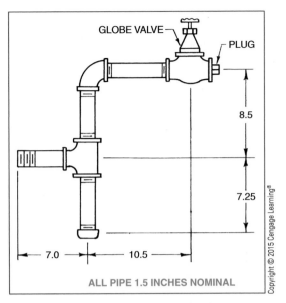

Figure 21-30 *Draw the single orthographic drawing.*

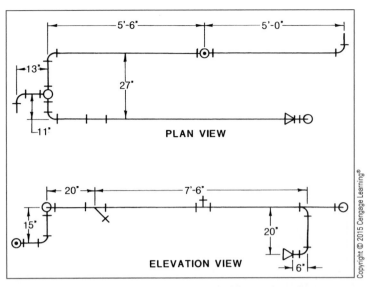

Figure 21-31 *Draw the following, using 5" pipe and fittings throughout.*

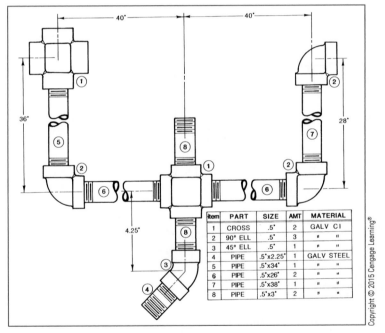

Figure 21-32 *Draw the double-line orthographic piping drawing. Make the ECOs listed in Exercise 5.*

item	PART	SIZE	AMT	MATERIAL
1	CROSS	.5"	2	GALV CI
2	90° ELL	.5"	3	" "
3	45° ELL	.5"	1	" "
4	PIPE	.5"x2.25"	1	GALV STEEL
5	PIPE	.5"x34"	1	" "
6	PIPE	.5"x26"	2	" "
7	PIPE	.5"x38"	1	" "
8	PIPE	.5"x3"	2	" "

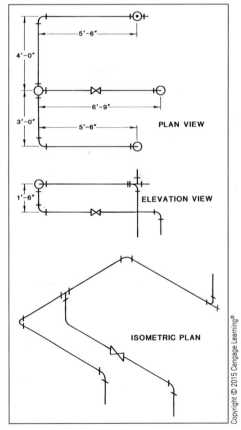

Figure 21-33 *Draw the single-line piping drawings and their double-line counterparts with instruments or a CAD system.*

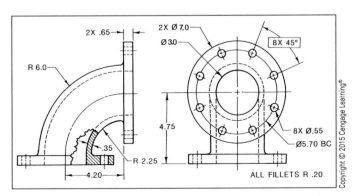

Figure 21-34 *Draw the 90° flanged elbow with instruments or a CAD system.*

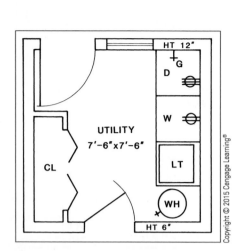

Figure 21-35 *Design a single-line isometric piping drawing for the water heater hookup with a .5" gas pipe and dimension.*

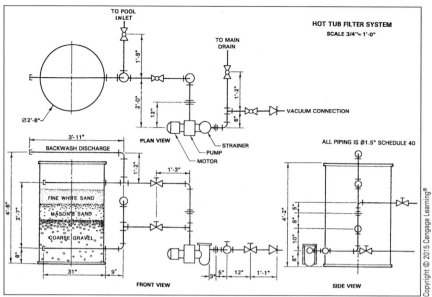

Figure 21-36 *Draw single-line and double-line multiview drawings.*

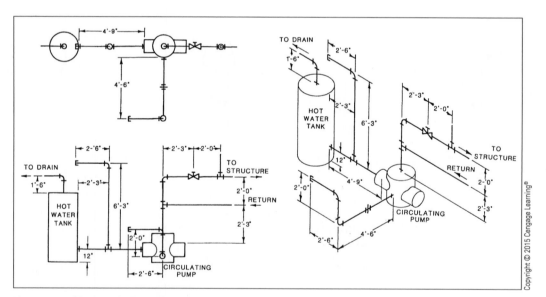

Figure 21-37 *Draw single-line and double-line multiview drawings.*

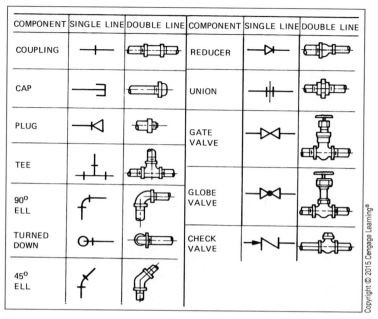

COMPONENT	SINGLE LINE	DOUBLE LINE	COMPONENT	SINGLE LINE	DOUBLE LINE
COUPLING			REDUCER		
CAP			UNION		
PLUG			GATE VALVE		
TEE					
90° ELL			GLOBE VALVE		
TURNED DOWN			CHECK VALVE		
45° ELL					

Figure 21-38 *Make up a series of single-line and double-line symbols for your CAD system's piping library.*

Copyright © 2015 Cengage Learning®

DESIGN EXERCISES

1. Go to a hot tub or spa showroom and ask to see the motor, pump, heater, and filtering system of their "top-of-the-line" spa. Measure the components and accurately sketch the system, including all pipes and fittings. Draw a single- or double-line ortho-graphic (suggest front and top views) piping diagram of the system. Label all components. Draw a single-line isometric diagram of the same system.

   Return to the hot tub store and show the drawings to installers. Ask if there should be any design changes made to the system to simplify their job. Make the changes to your drawings.

2. Go through the same procedure as in Design Exercise 1 for a swimming pool, family bathroom, and school shower room.

KEY TERMS

Bell and spigot connection

Cementing

Double-line piping drawing

Flange connections

Nominal pipe size

Screwed fitting

Single-line piping drawing

Valves

Welded connections

Electronics Drafting

22

OBJECTIVES

The student will be able to:

- Identify the block diagrams, schematic diagrams, connection diagrams, logic diagrams, and cableform diagram
- Differentiate between schematic and simple outline electronics symbols
- with a CAD system or manually, draw a block diagram, schematic diagram, connection diagram, logic diagram, cableform diagram, and chassis pattern
- Name the associations that set electronics standards
- Understand how sophisticated CAD systems are used to create electronics drawings

Introduction

The electronics industry is in a virtual explosion of expansion and advancement in research, product development, and manufacturing. Extremely rapid growth has occurred in such high-tech areas as computers, telecommunication, aerospace, automotive technology, consumer products, and industrial instrumentation. The electronics industry touches almost every facet of our daily lives.

This expanding electronics industry is creating an increasing demand for qualified electronics drafters to produce drawings manually and with computer-aided drafting systems. Before electronics products are produced, drawings are needed to direct the manufacture, assembly, and packaging of all component parts. In addition, drawings provide necessary information for the installation, control, repair, and maintenance of every product. Electronics drafters are a vital communications link in this process. In industry, electronic drawings may be drawn as simple diagrams or very complex working drawings. Electronic drawings, depending on production needs, will be one of four types:

1. Block diagram (**Figure 22-1**)
2. Pictorial drawing (**Figure 22-2**)
3. Multiview drawing (**Figure 22-3**)
4. Schematic (**Figure 22-4**)

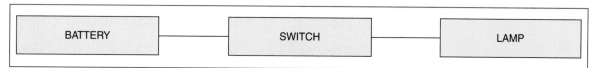

Figure 22-1 *Block Diagram.* Copyright © 2015 Cengage Learning®

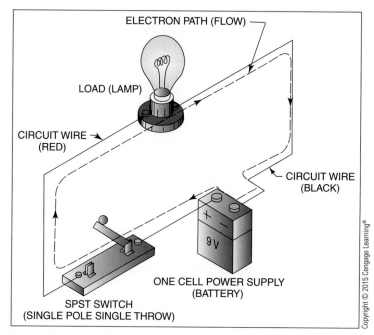

Figure 22-2 *Pictorial diagram.*

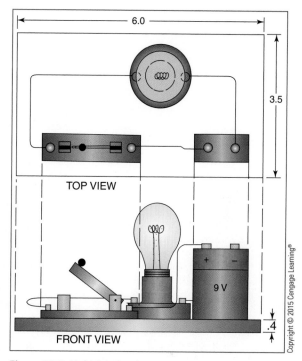

Figure 22-3 *Multiview drawing.*

Electronics Industry Standards

Electronics drafting uses a broad combination of governmental and professional association standards in the development of drawings and diagrams. This standardization improves the efficiency and minimizes the potential errors in the interpretation of drawings. Standards cover the uniform selection, formation, symbolism, and reference notations used for all electronics and electrical circuits, equipment, and devices. Standardization of electronics and electrical drawings is overseen by the following organizations:

- **ASME**—American Society of Mechanical Engineers
- **ASA**—American Standards Association
- **IEC**—International Electrotechnical Commission
- **EIA**—Electronics Industries Association
- **IEEE**—Institute of Electrical and Electronics Engineers
- **MIL STD**—Military Standards

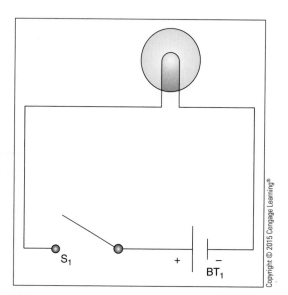

Figure 22-4 *Schematic.*

Electronics Drawing Symbols

Graphic symbols (electronics shorthand) play a major role in the preparation of **electronics drawings**. Using symbols is much faster than using orthographic views, pictorials, or notations. Because of the complexity of electronic circuits and devices, the use of graphic symbols is also the most accurate method of preparing electronics and electrical drawings. The most universally accepted set of electronics symbols is the ASME Y32.2 standard symbology, as shown in **Figures 22-19** and **22-20**, page 421. In using electronic symbols to represent electronic circuit devices and components, use the following guidelines:

1. Lay out circuits should read from left to right. Complex circuits should be read from the upper left to the lower right.

2. Symbols can be drawn at any size but should be approximately the same size and proportional to the size of the drawing.

3. Symbol directions have no significance, and component symbols may be rotated to any direction on the drawing.

4. Avoid line (circuit path) crossovers wherever possible.

5. Keep the length of circuit lines and distance between components to a minimum.

6. Different and discrete parts of a diagram need not adhere to the same scale.

7. Although angular lines have no significance, avoid them wherever possible.

8. If additional data is needed to describe a symbol, standard abbreviations should be used next to the symbol.

9. Lay out circuit paths with symbols placed in logical order as the circuit functions, from power source to loads.

10. Use vertical or horizontal lines wherever possible. Avoid using angular lines.

11. If lines are interrupted or progress to another circuit or drawing, identify the lines and indicate where they are connected on the original and receiving drawing.

Most electronic drawings are prepared with pencil lines. However, since many drawings are used in technical manuals, ink drawings are used to produce high-quality lines for this type of reproduction.

Types of Electronics Drawings

There are many types of electronics drawings. Most start as sketches or written instructions from an electronics engineer or designer. The electronics drafter interprets these instructions and graphically transforms them into working drawings. Each type of electronics drawing serves a specific purpose in describing the design, construction, fabrication, maintenance, or repair of a product. It is the responsibility of the electronics drafter to select the best type or combination of drawings to represent each design.

BLOCK DIAGRAMS

A **block diagram** is a general presentation drawing used to simplify and show the interrelationship of the various parts of an entire electronics system. Block diagrams are often used in the first planning and design stages because they are a fast and uncomplicated method of recording general design ideas. Circuit details are omitted from block diagrams.

Blocks in the shape of rectangles, squares, or triangles are joined by a single line with arrows. Arrows show the direction of current flow. The direction of block diagram design progresses from left to right, and the function of each block is lettered within the block. **Figure 22-5** shows a simple block diagram for a crystal radio. Examples of a block diagram for a CAD system and a radio are shown in **Figures 22-6** and **22-7**.

SCHEMATIC DIAGRAMS

Schematic diagrams are the most frequently used drawings in the electronics industry. They show the connection and function of each circuit in a simplified form (**Figure 22-8**). Nevertheless, **schematic diagrams** show more detail than block diagrams and serve as the master reference for the preparation of production drawings, parts lists, and component specifications. They contain graphic symbols, connections, and noted functions to describe each specific circuit arrangement. Schematic diagrams also show the sequence in which components are connected to complete a circuit. Each component is represented by its own special schematic symbol and is drawn to show its relationship to other components (**Figure 22-9**). Schematic diagrams are sometimes called **elementary diagrams**. The following are general rules for drawing schematic diagrams:

1. Current flows from left to right, or upper left to lower right.

2. Allow space between symbols and connectors to avoid crowding.

3. Reference callouts are positioned in the same location, next to all symbols.

4. Keep the crossing of connector lines to a minimum.

5. If scale is a factor, use a grid overlay to avoid excessive dimensioning.

CONNECTION DIAGRAMS

Connection or **wiring diagrams** are electronic assembly drawings that contain details needed to make or trace all electronic connections.

This type of diagram shows precisely where components are located in a circuit and how wires are connected. The components are

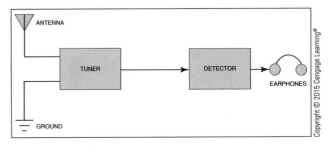

Figure 22-5 *Block diagram of a crystal set.*

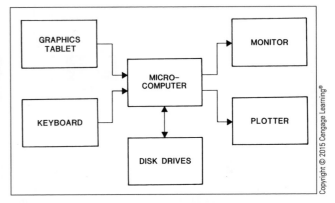

Figure 22-6 *Block diagram of a microcomputer CAD system.*

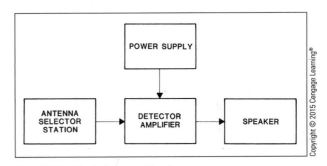

Figure 22-7 *Block diagram of a radio.*

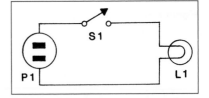

Figure 22-8 *Schematic diagram for a lamp and its power supply.*
Copyright © 2015 Cengage Learning®

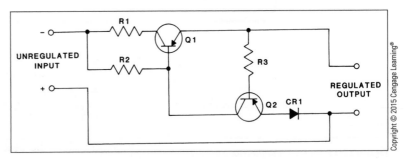

Figure 22-9 *Schematic for a regulated power supply.*

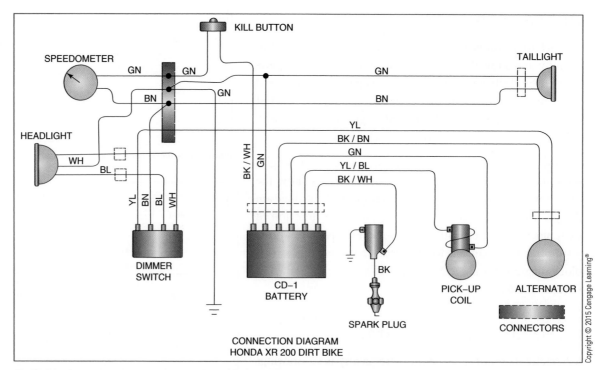

Figure 22-10 *Point-to-point connection diagram for a dirt bike.*

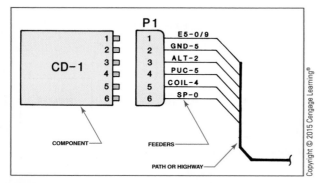

Figure 22-11 *Highway diagram's feeders and paths.*

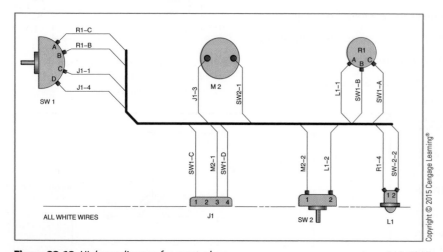

Figure 22-12 *Highway diagram for a meter box with wire destinations.*

drawn pictorially or as single orthographic views. Connection diagrams are supplementary drawings used with a schematic diagram for installing and repairing electronic equipment. Types of connection diagrams include point-to-point, highway or trunkline, baseline or airline, and cableform diagrams.

Point-to-Point Diagrams. Point-to-point diagrams show components, drawn as simple outlines, with connecting lines and terminals. These drawings are used by technicians for the assembly of electrical systems. An example of a point-to-point diagram is the motorcycle wiring diagram shown in **Figure 22-10**. Automotive manuals frequently contain point-to-point diagrams for troubleshooting purposes.

Highway Diagrams. Highway diagrams are preferred when the addition of a large number of point-to-point interconnecting lines would be excessive. **Highway or trunkline diagrams** are similar to point-to-point diagrams, except all wires are joined to a thick line called a *highway* or *trunkline*. Wires are separated where they reach the component to which the wire is connected. The long, thick line is called a *path* or *highway*, and shorter lines are called feeders (**Figure 22-11**). To show where connections are made, wire data is labeled on feeder lines. **Figure 22-12** shows a highway diagram for a meter box with the appropriate wire data. This wire data includes the wire destination (hookup), wire color, and wire size. EIA color code standards are as follows:

Wire Number	Color
0	Black
1	Brown
2	Red
3	Orange
4	Yellow
5	Green
6	Blue
7	Violet
8	Grey
9	White

Baseline Diagrams. Baseline or airline diagrams are similar to highway or trunkline diagrams. The connection lines are theoretical and do not depict the true size and form of the connectors. The baseline diagram is a drawing shortcut procedure to allow the repositioning of components so the connectors can be shortened and drawn conveniently. All the feed lines meet at a right angle to the baseline, which is always vertical or horizontal (**Figure 22-13**). The use of a wire data chart will list the connection points for each wire connector and their sizes and colors.

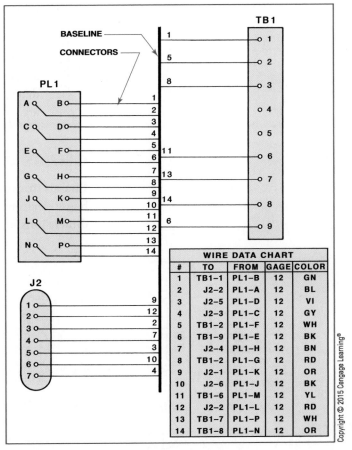

Figure 22-13 *A baseline diagram with a wire data chart.*

Cableform Diagrams. A cable is a group of wires bound together (**Figure 22-14**). A **cableform diagram** shows where each wire leaves the cable harness to make a connection (**Figure 22-15**). Line corners are drawn rounded to simulate cables and wires. Information such as color, gage, length, and types of connectors are also included on cableform diagrams.

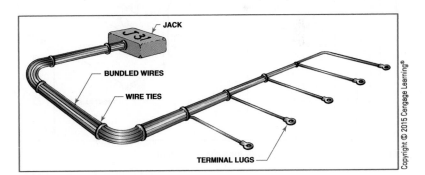

Figure 22-14 *Cable (bundled wires) with jack, ties, and terminal lugs.*

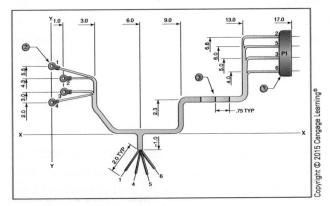

Figure 22-15 *Cableform diagram.*

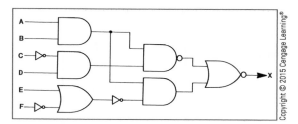

Figure 22-16 *Simple logic diagram.*

LOGIC DIAGRAMS

Logic diagrams are used to design the microcircuitry used in many contemporary devices such as digital watches, calculators, and computers. A logic diagram is a specialized type of block diagram showing circuitry flow details. Blocks of specific shapes are used to represent different logic functions (**Figure 22-16**).

CHASSIS DRAWINGS

Chassis drawings are pattern or development drawings of cabinets that house electronic circuits and equipment. Chassis drawings include overall dimensions and stretchout views of a chassis. They also include the location and size of all holes and other openings needed to mount components and provide openings for wires (**Figures 22-17a** and **22-17b**).

PRINTED CIRCUITS

Printed circuits are contemporary replacements for wired circuits. One component is electrically connected to another through printed copper conductors on an insulated board, called a **printed circuit board (PCB)**. Printed circuits eliminate wiring faults and errors, reduce production costs, and enable complex circuits to be miniaturized.

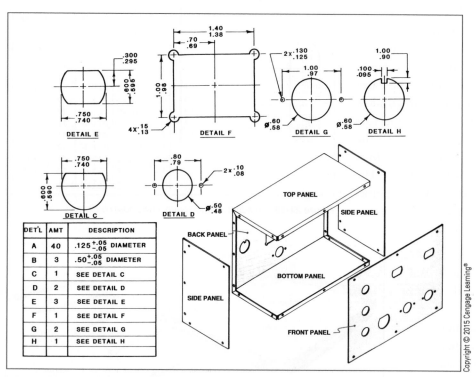

Figure 22-17a *Cutout dimensions for a chassis pattern.*

Printed circuit drawings are prepared at full scale or larger and represent the exact layout of a circuit. After the drawings are photographically reproduced and reduced in size, they are used in the etching process to create printed circuit boards. Many drawings are needed to design and create the documentation necessary to produce a printed circuit board (**Figures 22-18a** through **22-18e**).

Standard Symbols Charts

There are two types of standard symbols for electronics components on connecting diagrams—**schematic symbols** and **simple outline symbols** (**pictorial**; **Figure 22-19**). Different symbols are used to represent logic diagrams (**Figure 22-20**).

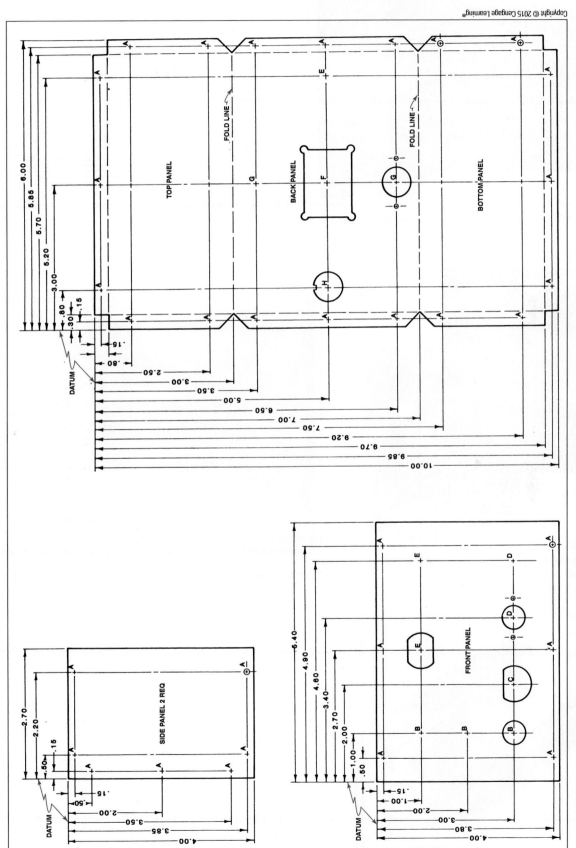

Figure 22-17b *A chassis pattern's working drawing using rectangular coordinate dimensioning.*

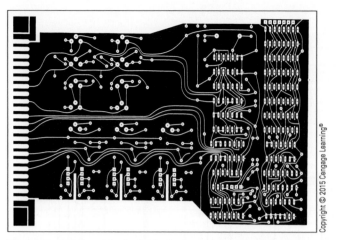

Figure 22-18a *Example of a finished printed circuit board layout.*

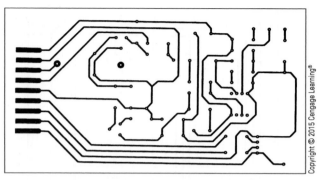

Figure 22-18b *Example of a tape-up layout for a printed circuit board.*

Figure 22-18c *Both sides of a printed circuit board with components installed.*

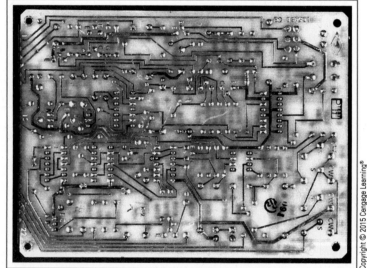

Figure 22-18d *Solder side of a printed circuit board.*

Figure 22-18e *Printed circuit board with components for a digital display.*

Drawing Connection Diagrams

In a connection diagram, the placement of components (simple outlines) is determined by their relative location on the chassis, rather than by electron flow. Lines that connect components represent wires. It is common to draw the wires with corner radii and color labels. Procedures for drawing a point-to-point connection diagram for a charging system tester are shown in **Figures 22-23a** through **22-23d** and include the following steps:

Step 1. Draw the outline of the chassis (**Figure 22-23a**).

Step 2. Place the components. Arrange them as they should be located in the chassis (**Figure 22-23b**).

Step 3. Draw the connecting wires (**Figure 22-23c**).

Step 4. Label the reference designations on the components and label wire colors (**Figure 22-23d**).

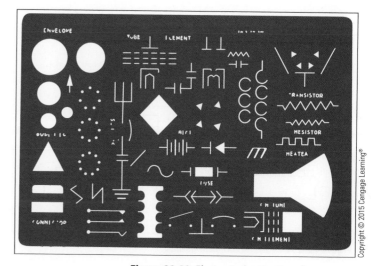

Figure 22-22 *Electronic drawing template.*

Figure 22-23a *Draw a chassis outline.*

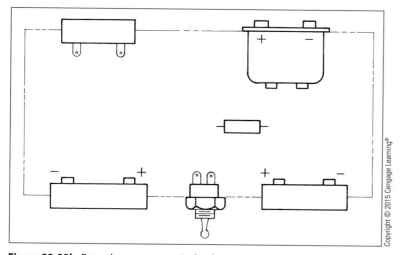

Figure 22-23b *Draw the components in the chassis as per location.*

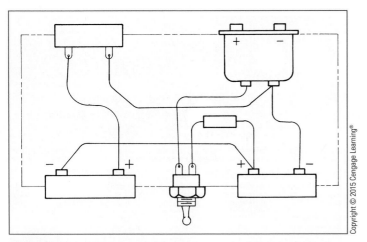

Figure 22-23c *Draw the connecting wires.*

Drawing Logic Diagrams

A logic diagram is a special type of block diagram. It consists of logic symbols, connecting lines, and input/output letters. Procedures are shown in **Figures 22-24a** through **22-24c** and include the following steps:

Step 1. Arrange logic symbols so they are evenly spaced and the system flows smoothly from left to right (**Figure 22-24a**).

Step 2. Place the connecting lines, with significant line contrast between component symbols and connecting lines. Connecting lines are thinner than the symbol lines (**Figure 22-24b**).

Step 3. Place input/output letters on the diagram (**Figure 22-24c**).

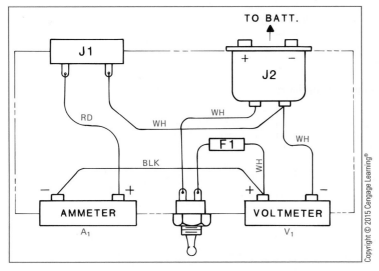

Figure 22-23d *Complete the connection diagram with labels and designations.*

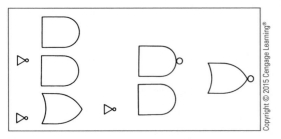

Figure 22-24a *Lay out and draw the logic symbols.*

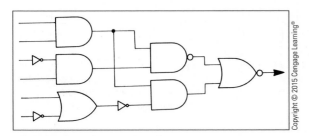

Figure 22-24b *Draw the connecting lines.*

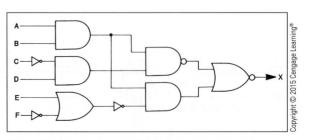

Figure 22-24c *Place input and output letters on the drawing.*

DRAFTING EXERCISES

1. Draw the block diagram for the amplifier system in **Figure 22-25** with drafting instruments or a CAD system.

2. Draw the electronics symbols in **Figure 22-26** with drafting instruments or a CAD system.

3. Draw the electronic schematics in **Figures 22-27** through **22-38** with drafting instruments or a CAD system.

4. Locate as many symbols in **Figures 22** through **39a–e** with a CAD system.

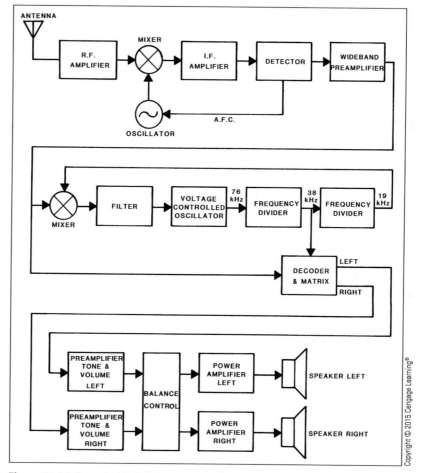

Figure 22-25 *Draw the block diagram with drafting instruments or a CAD system.*

Copyright © 2015 Cengage Learning®

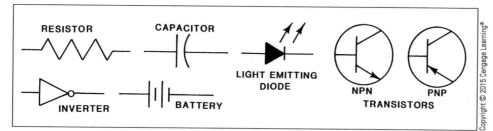

Figure 22-26 *Practice drawing the electronics symbols.*

Copyright © 2015 Cengage Learning®

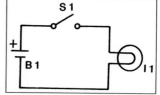

Figure 22-27 *Draw the schematic diagram for the flashlight.*

Copyright © 2014 Cengage Learning®

Figure 22-28 *Draw the schematic for the battery-operated bell system.*

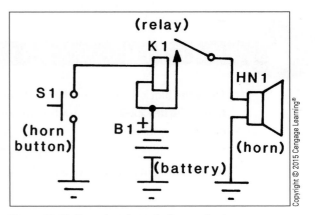

Figure 22-29 *Draw the schematic diagram for the automobile horn.*

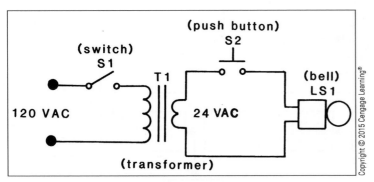

Figure 22-30 *Draw the schematic diagram for the residential doorbell.*

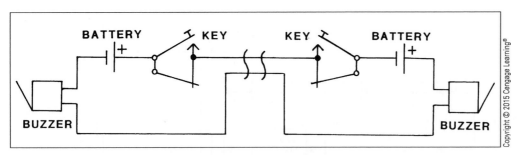

Figure 22-31 *Draw the schematic diagram for the two-station telegraph.*

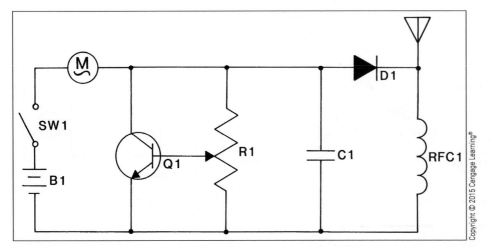

Figure 22-32 *Draw the schematic diagram for the sensitive field-strength meter.*

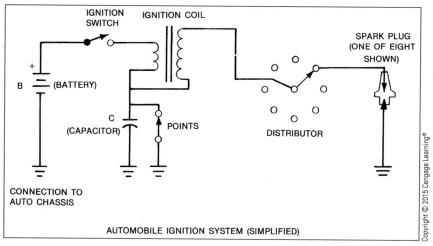

Figure 22-33 *Draw the schematic diagram for the automobile ignition system.*

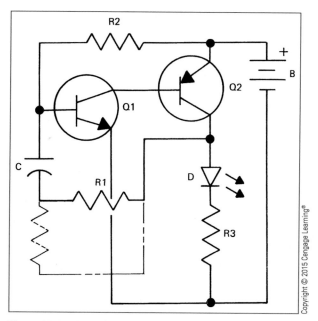

Figure 22-34 *Draw the schematic diagram for the flasher unit. Reposition resistor one as shown.*

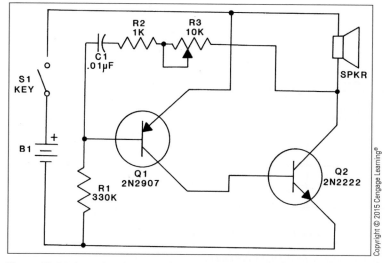

Figure 22-35 *Draw the schematic diagram for the code practice oscillator.*

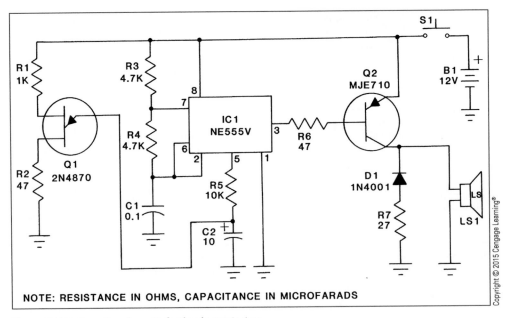

Figure 22-36 *Draw the schematic for the electronic siren.*

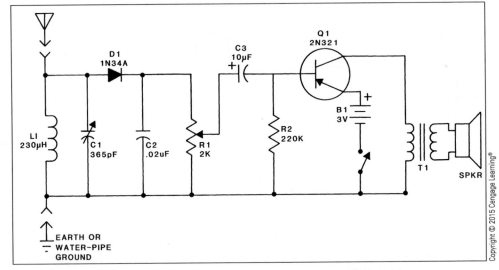

Figure 22-37 *Draw the schematic diagram for the crystal set with audio amplifier and speaker.*

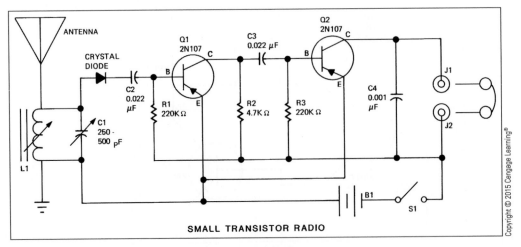

Figure 22-38 *Draw the schematic diagram for the transistor radio.*

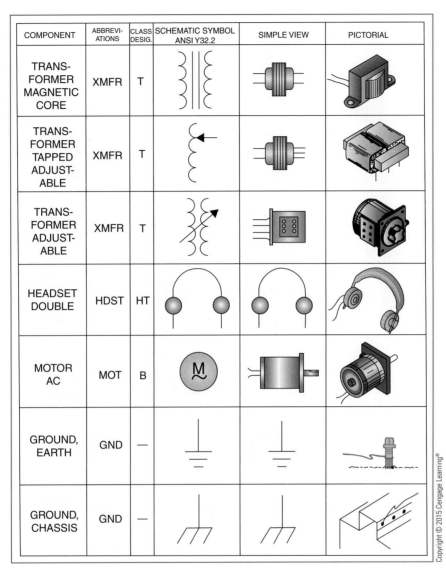

COMPONENT	ABBREVI-ATIONS	CLASS DESIG.	SCHEMATIC SYMBOL ANSI Y32.2	SIMPLE VIEW	PICTORIAL
TRANS-FORMER MAGNETIC CORE	XMFR	T			
TRANS-FORMER TAPPED ADJUST-ABLE	XMFR	T			
TRANS-FORMER ADJUST-ABLE	XMFR	T			
HEADSET DOUBLE	HDST	HT			
MOTOR AC	MOT	B			
GROUND, EARTH	GND	—			
GROUND, CHASSIS	GND	—			

Figure 22-39a *Electronic symbols.*

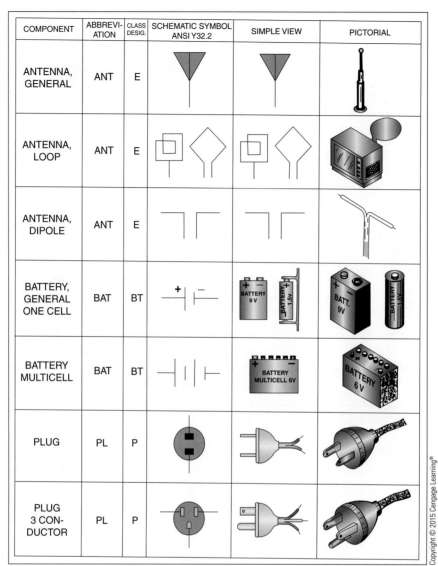

COMPONENT	ABBREVI-ATION	CLASS DESIG.	SCHEMATIC SYMBOL ANSI Y32.2	SIMPLE VIEW	PICTORIAL
ANTENNA, GENERAL	ANT	E			
ANTENNA, LOOP	ANT	E			
ANTENNA, DIPOLE	ANT	E			
BATTERY, GENERAL ONE CELL	BAT	BT			
BATTERY MULTICELL	BAT	BT			
PLUG	PL	P			
PLUG 3 CON-DUCTOR	PL	P			

Figure 22-39b *Electronic symbols.*

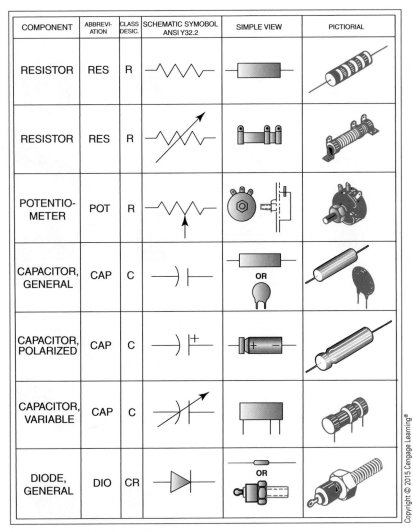

COMPONENT	ABBREVI-ATION	CLASS DESIC.	SCHEMATIC SYMOBOL ANSI Y32.2	SIMPLE VIEW	PICTIORIAL
RESISTOR	RES	R			
RESISTOR	RES	R			
POTENTIO-METER	POT	R			
CAPACITOR, GENERAL	CAP	C		OR	
CAPACITOR, POLARIZED	CAP	C			
CAPACITOR, VARIABLE	CAP	C			
DIODE, GENERAL	DIO	CR		OR	

Figure 22-39c *Electronic symbols.*

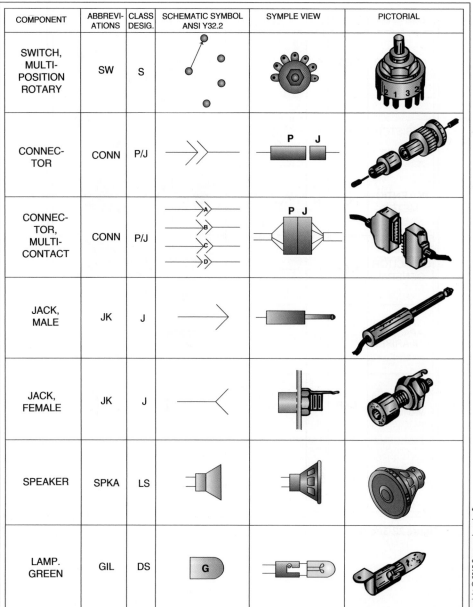

COMPONENT	ABBREVI-ATIONS	CLASS DESIG.	SCHEMATIC SYMBOL ANSI Y32.2	SYMPLE VIEW	PICTORIAL
SWITCH, MULTI-POSITION ROTARY	SW	S			
CONNEC-TOR	CONN	P/J			
CONNEC-TOR, MULTI-CONTACT	CONN	P/J			
JACK, MALE	JK	J			
JACK, FEMALE	JK	J			
SPEAKER	SPKA	LS			
LAMP. GREEN	GIL	DS			

Figure 22-39d *Electronic symbols.*

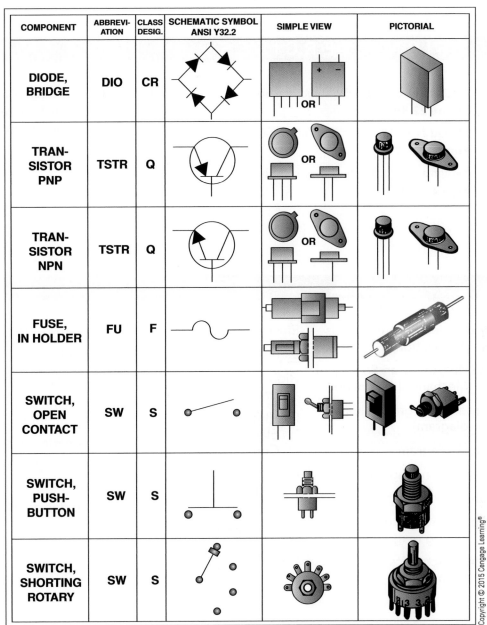

COMPONENT	ABBREVI-ATION	CLASS DESIG.	SCHEMATIC SYMBOL ANSI Y32.2	SIMPLE VIEW	PICTORIAL
DIODE, BRIDGE	DIO	CR		OR	
TRAN-SISTOR PNP	TSTR	Q		OR	
TRAN-SISTOR NPN	TSTR	Q		OR	
FUSE, IN HOLDER	FU	F			
SWITCH, OPEN CONTACT	SW	S			
SWITCH, PUSH-BUTTON	SW	S			
SWITCH, SHORTING ROTARY	SW	S			

Figure 22-39e *Electronic symbols.*

DESIGN EXERCISES

1. Design a case for the electronic siren in **Figure 22-36**.

2. Design a case for the crystal radio set in **Figure 22-37**.

3. Design a case for the transistor radio in **Figure 22-38**.

4. Design a flashlight that can be attached to your body or clothes so as to leave both hands free. Draw a working drawing for the flashlight and a schematic diagram.

KEY TERMS

ASA (American Standards Association)

ASME (American Society of Mechanical Engineers)

Baseline diagram (airline diagram)

Block diagram

Cableform diagram

Chassis drawings

Connection diagrams (wiring diagrams)

EIA (Electronics Industries Association)

Electronics drawings

Elementary diagrams

Highway diagrams (trunkline diagrams)

IEC (International Electrotechnical Commission)

IEEE (Institute of Electrical and Electronics Engineers)

Logic diagrams

MIL STD (Military standards)

Point-to-point diagrams

Printed circuit board (PCB)

Reference designation

Schematic diagrams (elementary diagram)

Schematic symbol

Simple outline symbol (pictorial)

Value designation

Tool Design Drafting

Introduction

To compete successfully in today's marketplace, industrial products must be accurately and economically produced. This requires effective tool engineering and design. **Tool engineering** is the branch of mechanical engineering responsible for the design and development of tools and processes needed in the mass production of products. This includes the planning, designing, and detail drawing of all tools and devices used in the manufacturing process.

Tool designers must possess a good working knowledge of all machine tool operations. Since they must communicate ideas and describe tools in great detail, a thorough understanding of working drawings and tolerancing is also mandatory. Tool engineers, tool designers, and tool design drafters are all involved at different levels of the machine tool operations and have the following responsibilities:

1. Control production costs by designing tools for rapid production.
2. Design special tools for the manufacture of complete parts.
3. Design tools that aid in the manufacture of the highest-quality product within the established limits.
4. Select the proper materials for the manufacture of products.
5. Design tools that require minimum maintenance.
6. Simplify manufacturing and assembly processes.
7. Design tools that are safe under all manufacturing conditions.

8. Design tools that are fault-free and reduce the potential for human error.

9. Design tools that produce accurate interchangeable parts within acceptable tolerances.

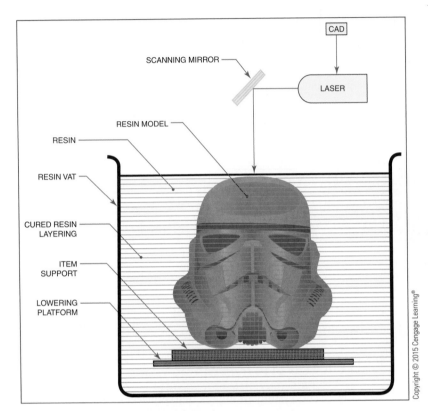

Figure 23-1 *Example of rapid programming.*

Figure 23-2 *The 3-D printer uses a print head with hundreds of jets to build models by dispensing a thermoplastic material in layers.*

3D Printing

Designing and setting up the tooling for the manufacturing of an item or a prototype of an item is an expensive process. An error or change in the design will require a makeover of the tools that were required for its production. This will be an expensive retooling situation.

Now available is an economical process to produce almost any object and eliminate the need to produce the tooling for a prototype. It is **rapid prototyping (RP)**, or commonly called **3D printing**. It uses the graphics of a 3D CAD computer-controlled laser beam to lay down and harden viscous or powdered materials in thin layers (0.002" to 0.006"–0.05 mm to 0.15 mm) until the item is finished (**Figure 23-1**). This process was introduced in 1988 and is called **stereolithography (SLA)**. New software may now convert digital photos into a 3D prototype and evaluate the design for structural flaws.

New models of 3D printers may print with a variety of materials such as: plastic, metal, glass, stone, and concrete. An example of some items that may be reproduced are: electronic components, building components, prosthetic limbs, medical supplies, and auto and aircraft parts. (**Figures 23-2**).

A new feature for 3D printing is bioengineering, or bioprinting, which can produce replacement organs and bones for the human body by printing layers of human cells. It can also produce meat for consumption and items to grow in your backyard.

3D printing will change the ways of manufacturing, the economics of manufacturing, food, and human reproduction. Just imagine when a 3D printer will reproduce itself.

Manufacturing Systems

Computer numerical Control (CNC), **computer-aided manufacturing (CAM)**, and **computer-aided engineering (CAE)** use a direct link from the machine tool to a CADD program that defines the machining operation. There is no need for a human machinist. **Computer integrated manufacturing (CIM)** integrates CNC, CAM, CAE, and CADD into a single management system.

Machining tools are power-driven tools such as mills, drills, and lathes. An experienced machinist will shape the metal to the required form. **Laser machining** uses a laser beam of amplified focused light waves and concentrate them into a high-frequency cutting beam.

Castings are the result of a process in which molten metal is poured into a hollow form, called a mold, in the shape of the item to be cast. The material used to make the forms may be sand or plaster. **Forging** is the process of pressing metal between dies or hammering malleable metals. **Stamping** is the process of pressing thin metal between two dies to create the desired from. **Hydroforming** shapes metal by placing the metal over a form. A very high-pressure fluid is applied to shape the metal to the desired form. **Powder metallurgy** is a process that uses a metal-alloyed powder and places the powder into a die, where it is compressed under very high pressure. The item is then heated to harden it.

Water-jet cutting is a process in which a very concentrated high-velocity stream of water is used with an abrasive substance to cut through metals. **Chemical machining** uses a strong chemical to accurately dissolve metal, leaving the desired item. **Ultrasonic machining** uses a high-frequency vibration to shape the metal. The machine and the metal are suspended in an abrasive fluid. **Electronic beam cutting** uses an electron beam to cut or machine metals.

With today's manufactured items, a large amount of plastic is used. Plastics can use many of the metal machining processes plus the following processes:

- **Injection molding** is a process in which molten plastic is injected into a mold.
- **Extrusion** is a process in which plastic is heated to a soft form and forced through a die, forming a long desired shape from the die.
- **Blow molding** is a process in which a hot liquid plastic is injected into a hollow mold (the shape of a container). Air is then forced into the liquid plastic, forcing it to the sides of the mold and creating a plastic container.
- **Calendaring** is a process in which sheets of plastic are passed through a series of heated rollers to create flat plastic products such as flooring and gaskets.
- **Rotational molding** is a process in which plastic pellets are placed into a hollow cylindrical mold. The mold is heated as it rotates and forces the molten plastic pellets against the sides, forming a plastic container. This process is used to form large containers.
- **Solid-phase forming** is similar to metal forging. Plastic is heated until it is soft and then it is pressed into a die for the desired form.
- **Robotics** is a reprogrammable CADD machine designed to assemble segments and move materials, parts, and tools (**Figure 23-3**).

Figure 23-3 *An industrial robotic arm.*

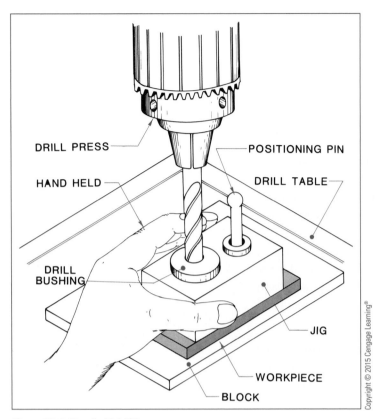

DRILL PRESS

HAND HELD

POSITIONING PIN

DRILL TABLE

DRILL BUSHING

JIG

WORKPIECE

BLOCK

Figure 23-4 *Handheld drill jig.*

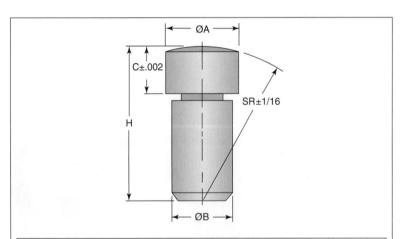

STANDARD SIZES FOR SPHERICAL LOCATORS				
ØA	ØB	C	H	SR
$5/16$	.1890 .1885	$1/8$	$3/8$	$1/2$
$3/8$	.2515 .2510	$3/16$	$1/2$	$5/8$
$1/2$	.3765 .3760	$3/16$	$9/16$	$3/4$
$1/2$	.3765 .3760	$1/4$	$5/8$	$3/4$
$1/2$	.3765 .3760	$3/8$	$3/4$	$3/4$

Figure 23-5 *Standard spherical locators.*

Jigs and Fixtures

Jigs are used in the machining and assembly of products. A **jig** is a device designed to position, hold, and support a workpiece in a fixed position during a machining operation. In addition, a jig holds or guides cutting tools through prescribed paths. A jig may be fastened to a worktable or can be handheld. **Figure 23-4** shows a jig used to hold a workpiece during a drill press operation.

Jigs used in product assembly are designed to hold separate components in fixed positions while final connections are completed. This may involve the use of all types of fasteners, weldments, or adhesives.

Fixtures, like jigs, hold and support workpieces. However, **fixtures** are fixed to a worktable or a machine surface to hold a workpiece rigidly. Fixtures do not hold or guide cutting tools. They are generally larger than jigs and are used in milling, grinding, and honing operations.

Standard Parts

Most jigs and fixtures use standard components. Only when a tool design problem cannot be solved with a standard component will a tool designer use a unique type, size, and configuration of materials. Standard parts in the design and fabrication of jigs and fixtures include locators, feet and rest buttons, clamps, drill bushings, pins, screws, V-blocks, plates, and aluminum or steel shapes. When cast shapes are used, all burrs and sharp edges are removed to protect machine operators.

LOCATORS

Locators are devices, such as pins, dowels, buttons, and pads, used to position a workpiece on a machine surface or worktable. **Figure 23-5** shows the design and size options for standard spherical radius locator buttons. This type of locator is normally positioned with the workpiece touching the tip of the button (**Figure 23-6**). It also can be used with the workpiece touching the flattened side of the locator (**Figure 23-7**). Note the location and direction of forces in relation to the locators in **Figure 23-7**.

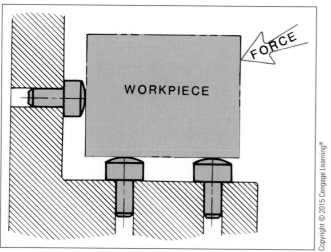

Figure 23-6 *Positioning workpiece against ends of locators.*

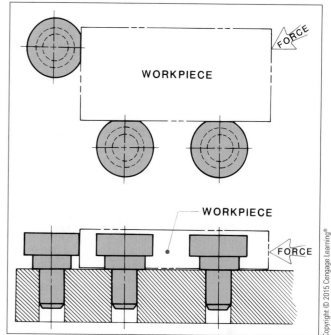

Figure 23-7 *Positioning workpiece against flattened sides of locators.*

If a workpiece contains holes, locators should be positioned to align with hole centers. **Figure 23-8** shows the alignment of bullet nose locators with the holes in a workpiece. **Figure 23-8** also contains standard dimensions for bullet nose locators. **Figure 23-9** shows standard diamond pin locator dimensions.

FEET AND REST BUTTONS

Feet are used to elevate jigs or fixtures from a machine surface. **Feet and rest buttons** are designed to provide flexibility in leveling a jig or fixture to precise surface angles. Standard dimensions for jig feet and rest buttons are shown in **Figure 23-10**. **Figure 23-11** shows standard dimensions for hexhead screw-type jig feet and rest buttons. Socket head cap screws are used for the same purpose. Socket head dimensions are shown in **Figure 23-12**.

CLAMPS

Clamps are used to hold workpieces firmly against locators and rest buttons without distorting the workpiece. There are many types of clamps for jig and fixture design. The three major categories of clamps are **toggle-head clamps**, **cam clamps**, and **screw clamps**.

Toggle clamps, such as the push-pull toggle clamp shown in **Figure 23-13**, are used where maximum surface contact

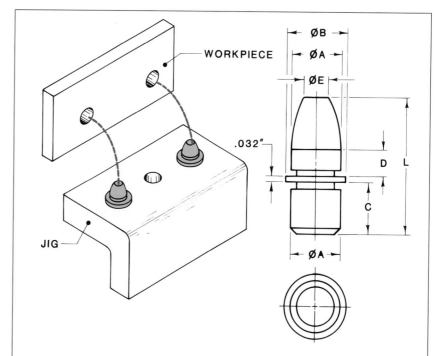

φA	φB	C	D	L	φE	HOLE'Sφ
.2495	.280	.50	.125	1.000	.125	.2500
.3120	.350	.50	.125	1.000	.156	.3125
.3745	.400	.50	.125	1.000	.188	.3750
.4995	.532	.50	.125	1.000	.250	.5000

Figure 23-8 *Bullet nose locators.*

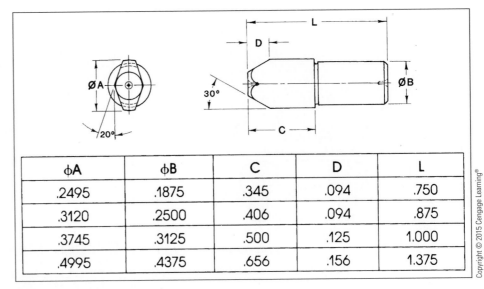

φA	φB	C	D	L
.2495	.1875	.345	.094	.750
.3120	.2500	.406	.094	.875
.3745	.3125	.500	.125	1.000
.4995	.4375	.656	.156	1.375

Figure 23-9 *Diamond pin locators.*

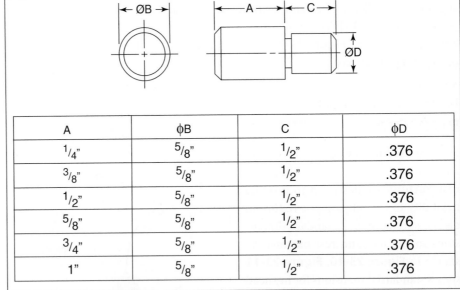

A	φB	C	φD
1/4"	5/8"	1/2"	.376
3/8"	5/8"	1/2"	.376
1/2"	5/8"	1/2"	.376
5/8"	5/8"	1/2"	.376
3/4"	5/8"	1/2"	.376
1"	5/8"	1/2"	.376

Figure 23-10 *Jig feet (rest buttons).*

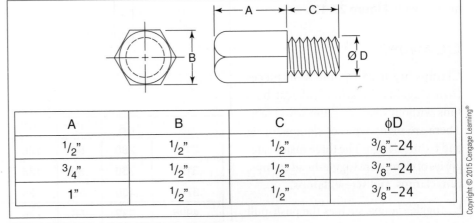

A	B	C	φD
1/2"	1/2"	1/2"	3/8"–24
3/4"	1/2"	1/2"	3/8"–24
1"	1/2"	1/2"	3/8"–24

Figure 23-11 *Hexhead screw-type jig feet (rest buttons).*

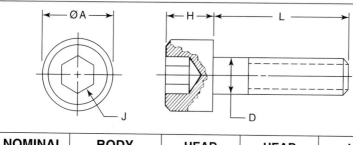

NOMINAL SIZE	BODY DIAMETER		HEAD DIAMETER		HEAD HEIGHT		HEX SOCKET SIZE	
	MAX	MIN	MAX	MIN	MAX	MIN	NOMINAL	
	D		A		H		J	
1/4"	0.2500	0.2435	0.375	0.365	0.250	0.244	3/16"	0.188
5/16"	0.3125	0.3053	0.469	0.457	0.312	0.306	1/4"	0.250
3/8"	0.3750	0.3678	0.562	0.550	0.375	0.368	5/16"	0.312
1/2"	0.5000	0.4919	0.750	0.735	0.500	0.492	3/8"	0.375
5/8"	0.6250	0.6163	0.938	0.921	0.625	0.616	1/2"	0.500

Copyright © 2015 Cengage Learning®

Figure 23-12 *Socket head cap screws.*

with the workpiece is required. Cam clamps (**Figure 23-14**) are used where ease of changing the workpieces is needed to speed up production. Screw clamps (**Figure 23-15**) reduce the effect of vibration on jigs and workpieces. Using screw clamps often slows down production unless swing-out features are used for loading and unloading workpieces.

The appropriate application of clamps allows clamping from multiple directions (**Figure 23-16**). Generally, cutting actions should be directed not toward a clamp, but toward a rigid locator. **Figure 23-17** also shows an example of a cutting motion directed toward rigid locators and away from a screw clamp.

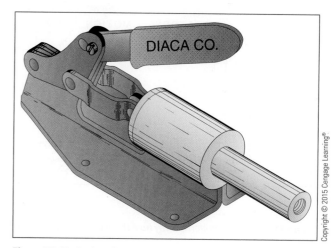

Copyright © 2015 Cengage Learning®

Figure 23-13 *Push-pull toggle clamp.*

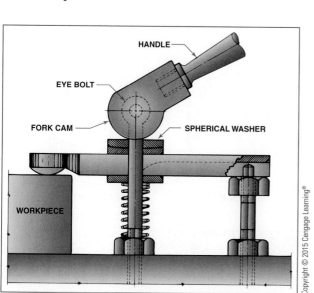

Copyright © 2015 Cengage Learning®

Figure 23-14 *Cam clamp assembly.*

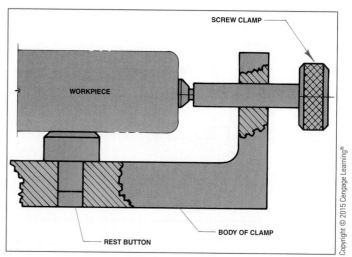

Copyright © 2015 Cengage Learning®

Figure 23-15 *Knurled head swivel pad screw clamp.*

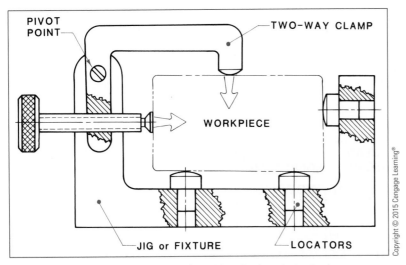

Figure 23-16 *Multiple swing-out screw clamp. Note the clamping action is toward the locators.*

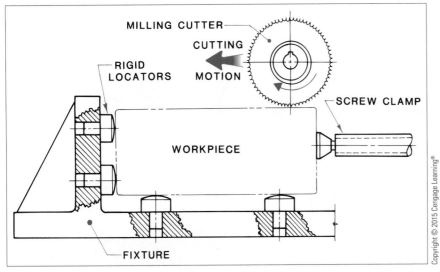

Figure 23-17 *Cutting motion is away from the clamp (toward the rigid locators).*

DRILL BUSHINGS

Drill bushings align, hold, and guide drill bits to an exact location, and at a specific angle for drilling (**Figure 23-18**). Drill bushings are made of a harder steel than the drill bit. This prevents the drill from damaging the bushings. Drill bushing sizes correspond to standard drill sizes.

Bushings may be either fixed or replaceable. **Fixed drill bushings** are force-fitted into jig bodies. Fixed drill bushings are relatively inexpensive, but they are difficult or impossible to replace. They are used for short production runs that do not require a bushing change. There are two kinds—headed and headless. Headed fixed drill bushings can withstand greater pressures than headless bushings because they have shoulders to support their position. Headless fixed drill bushings can be placed closer together and allow jig surfaces to remain flat (**Figure 23-19**).

Replaceable drill bushings are inserted into a liner that has been force-fitted into a jig body (**Figure 23-20**). Once the replaceable bushing is inserted, it is held in place with a lock screw. These bushings can be easily replaced by removing the lock screw.

Figure 23-18 *Drill bushing guides drills to exact location on workpiece.*

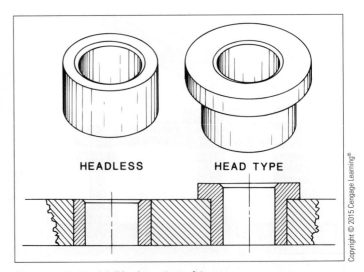

Figure 23-19 *Fixed drill bushings (press fit).*

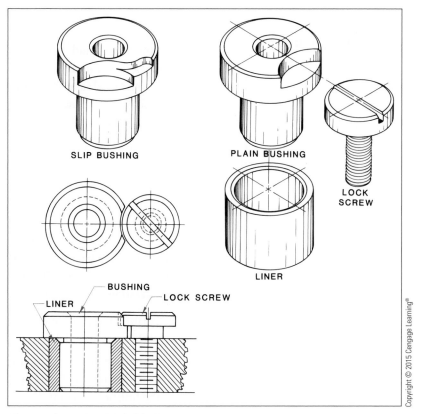

Figure 23-20 *Renewable drill bushings.*

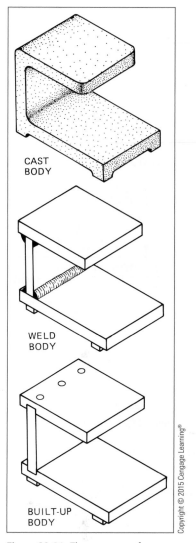

Figure 23-21 *Three common forms used to produce bodies for jigs and fixtures.*

BODIES

Jig and fixture bodies are made in three general forms—**weld bodies**, **cast bodies**, and **built-up bodies**. **Figure 23-21** shows the same jig body produced by these three methods.

The most commonly used form is the built-up body. When attaching plates using the built-up body method, dowel pins and socket head cap screws are used (**Figure 23-22**). Dowel pins are used to align the plates in an exact

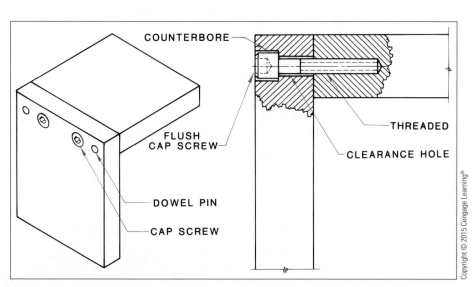

Figure 23-22 *Built-up body connections.*

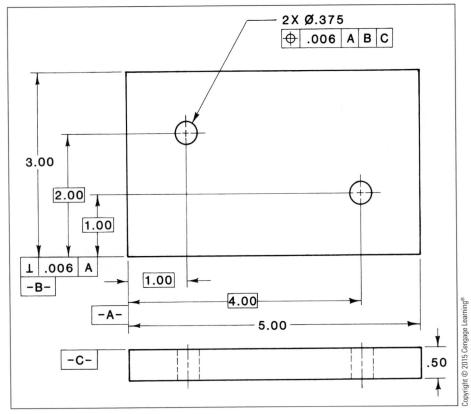

Figure 23-23 *Drawing of workpiece with two holes to be drilled with a drill jig.*

position without slippage, and the cap screws apply pressure to hold the plates rigid. When this standard fastening technique is used, both dowels and cap screws are flush with the plates.

Jig and Fixture Design

Designing the most appropriate jig for a specific operation depends on the following considerations:

- Existing form of the workpiece's raw material

- Dimensional and geometric tolerance requirements in the finished product

- Type of machine operator needed

- Number of repetitive operations needed

- Number of duplicate parts to be made

- Safety considerations, such as chip clearance, lubricant flow, operator distance, visibility, loading and unloading clearances, and weights

With these considerations in mind, established practices and design sequences can be used to arrive at a sound jig or fixture design. The proper sequence for jig or fixture design is shown in **Figures 23-23** through **23-30**, and they include the following steps:

1. To draw the workpiece shown in **Figure 23-23**, first draw the outline with red phantom lines (**Figure 23-24**).

2. Since the placement of the locators is critical, study the workpiece drawing carefully before drawing the locators on the workpiece datums (**Figure 23-25**).

3. Draw the rest buttons so they support the workpiece while the machining operation (drilling) is performed (**Figure 23-26**).

4. Draw the bushings so they are placed above the holes to be drilled (fixed-head type here). The distance from the bottom of the bushing to the top of the workpiece should equal one-half the drill's diameter, in this case, .188" (**Figure 23-27**).

5. Select and draw all clamps so they force the workpiece against the locators and hold it firmly. In this example, a double-acting clamp (**Figure 23-28**) forces the workpiece against the locators to the left and bottom.

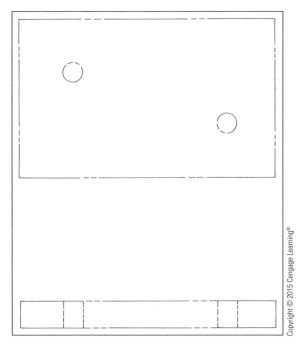

Figure 23-24 *Draw an outline of the workpiece with a colored pencil (red preferred).*

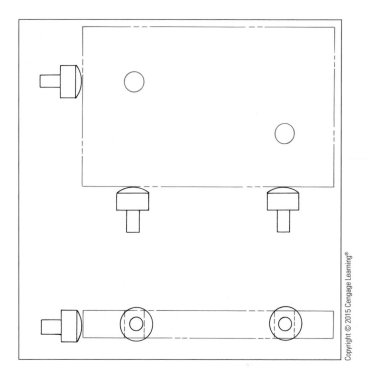

Figure 23-25 *Add the locators.*

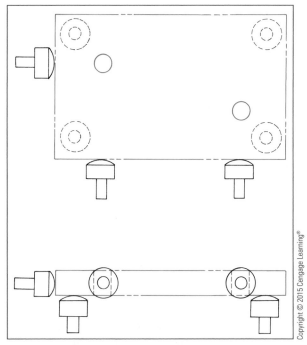

Figure 23-26 *Add the rest buttons.*

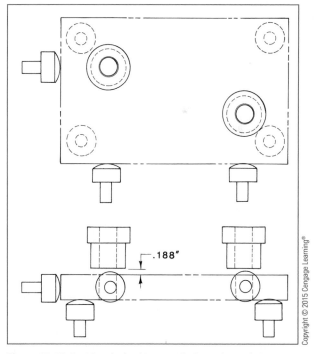

Figure 23-27 *Position the bushings .188" above the workpiece.*

6. Design the body of the jig (**Figure 23-29**). The jig body holds all jig components together and simultaneously allows the workpiece to be inserted and removed easily.

7. Draw the jig feet in position. In this case, socket head cap screws are used. Finish the design by adding fasteners to hold the body plates together and with clamps in place (**Figure 23-30**).

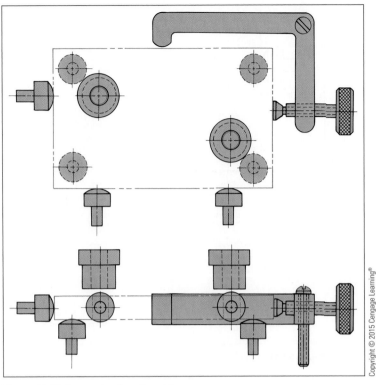

Figure 23-28 *Locate the clamp's position.*

8. Dimension the jig assembly. Jig and fixture dimensioning is completed using the dimensioning and tolerancing practices outlined in Chapters 9 and 10. Since most jigs and fixtures are made of many standard parts, few dimensions are needed because it is not necessary to specify standard part sizes. Some size dimensions are necessary for the body, but the majority of dimensions on a jig or fixture drawing are location dimensions.

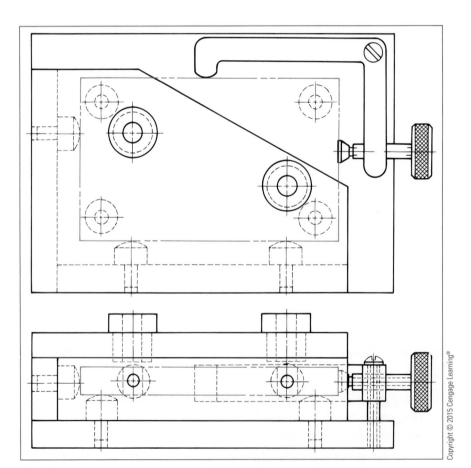

Figure 23-29 *Design the body to support all holding components.*

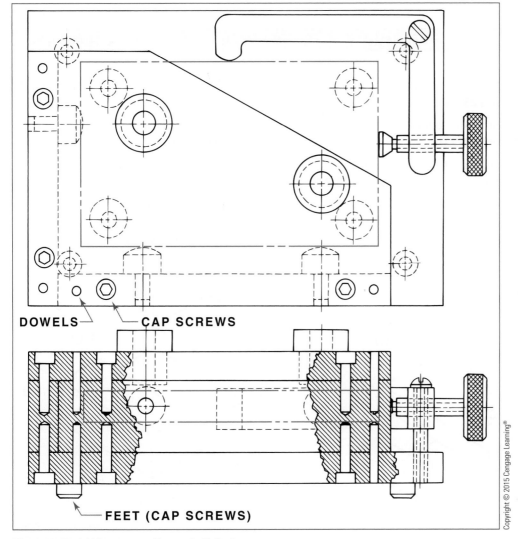

DOWELS **CAP SCREWS**

FEET (CAP SCREWS)

Figure 23-30 *Add fasteners and feet to the jig body.*

DRAFTING EXERCISES

1. Draw the multiview and isometric working drawings for **Figures 23-31** through **23-38**. The multiview drawings need to be created before designing the tooling. The isometric drawing of the part is commonly placed on the tooling drawing to aid in visualization.

DESIGN EXERCISES

1. Design a jig or fixture to hold a plate for drilling (**Figure 23-31**).

2. Design a jig or fixture to hold a plate for milling the slot (**Figure 23-32**).

3. Design a jig or fixture to hold a U-plate for surface grinding of the top only (**Figure 23-33**).

4. Design a jig or fixture to hold a triangular plate for drilling and top surface grinding operations (**Figure 23-34**).

5. Design a jig or fixture to hold a channel block for milling and drilling operations (**Figure 23-35**).

6. Design a jig or fixture to hold a pipe for 4" cut-offs (**Figure 23-36**).

7. Design a jig or fixture to hold a guide block for milling and drilling operations (**Figure 23-37**).

8. Design a jig or fixture to hold a milling guide block for milling, drilling, and grinding operations (**Figure 23-38**).

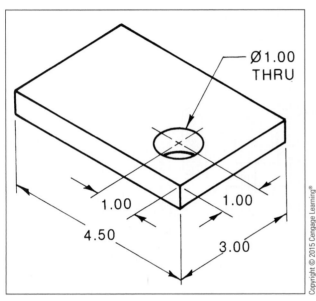

Figure 23-31 *Design a jig or fixture to hold the plate for drilling.*

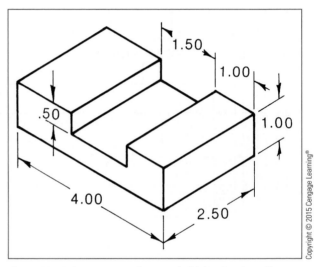

Figure 23-32 *Design a jig or fixture to hold the plate for milling the slot.*

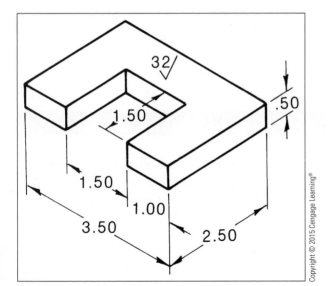

Figure 23-33 *Design a jig or fixture to hold the U-plate for surface grinding (top surface only).*

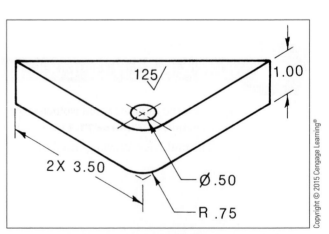

Figure 23-34 *Design a jig or fixture to hold the triangular plate for a drilling and grinding operation (top surface only).*

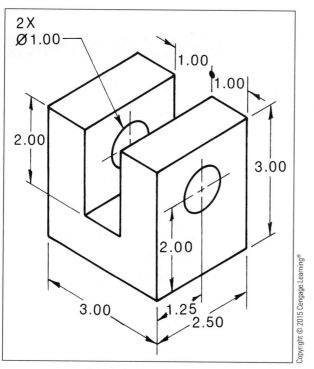

Figure 23-35 *Design a jig or fixture to hold the channel block for a milling (U shape) and drilling operation.*

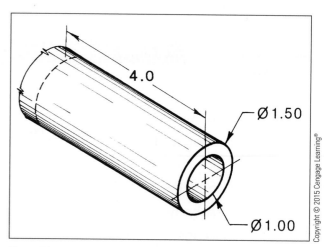

Figure 23-36 *Design a jig or fixture to hold the pipe for 4" cutoffs.*

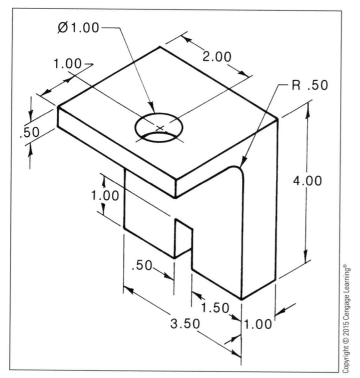

Figure 23-37 *Design a jig or fixture to hold the guide block for a slot milling and drilling operation.*

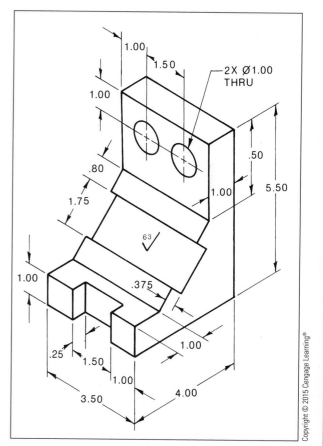

Figure 23-38 *Design a jig or fixture to hold the milling guide block for milling two slots (1.50" and 1.75"), a drilling operation for two holes, and a grinding operation for the 1.75" slot.*

KEY TERMS

3D printing

Blow molding

Built-up body

Calendaring

Cam clamp

Cast body

Castings

Chemical machining

Clamp

Computer-aided engineering (CAE)

Computer-aided manufacturing (CAM)

Computer integrated manufacturing (CIM)

Computer numerical control (CNC)

Drill bushing

Electronic beam cutting

Extrusion

Feet and rest buttons

Fixed drill bushing

Fixtures

forging

Hydroforming

Injection molding

Jig

Laser machining

Locator

Machining tool

Powder metallurgy

Rapid prototyping

Robotics

Rotational molding

Screw clamp

Solid-phase forming

Stamping

stereolithography

Toggle-head clamps

Tool engineering

Ultrasonic machining

Water-jet cutting

weld body

Architectural Drafting

OBJECTIVES

The student will able to:

- Gather data for an architectural design
- Read architectural prints
- Draw a plot plan
- Draw a floor plan
- Draw exterior elevations
- Draw a roof plan
- Draw interior elevations
- Draw foundations
- Draw construction details
- Draw framing plans
- Understand the local building codes and zoning ordinances

Introduction

Architecture is a science and an art. It is a science because the architect or designer must work with the strength of building materials and the concepts of design safety. It is also an art because he or she must work with people to create new designs and ideas for people to live and work within the buildings (**Figure 24-1**). The design of any architectural structure follows the same principles of design and drawing, but it is critical that the architect/designer have the architectural knowledge and drafting skills before starting an architectural design. For the further depth of study in architectural residential design, advanced architecture classes and books will be required, as this chapter is only an overview.

Throughout history the techniques of designing and building have not changed much. The only changes are modern tools and the types of building materials. The study of the history of architecture is nearly identical to the history of the world. You will find that the history of architecture up to modern times is an interesting subject (**Figure 24-2**).

Figure 24-1 *An example of a residential design.*

Figure 24-2 *A sample of the history of architecture.*

Gathering Information

Buildings are designed and built for people. Detailed information concerning their goals, tastes, and habits of all the building's inhabitants must be matched with the conditions of site, building and zoning code limitations, and the available funds and financing. The first step to a design project is to gather a multitude of personal information from the occupants who will be living or working in the building concerning:

- Occupants' living and working habits
- Number of occupants and their gender
- Building site conditions
- Preferred size of the home
- Preferred size on home
- Age and any physical impairments of inhabitants
- Preference of architectural style
- Number of occupants present and future
- Price range and monthly mortgage limits
- Special needs: amount of space, work areas, interests

The second step is to analyze the physical properties of the building site and the local zoning ordinances (**Figure 24-3**):

- The dimensions of the property lines
- Minimum lot frontage, depth, and square footage

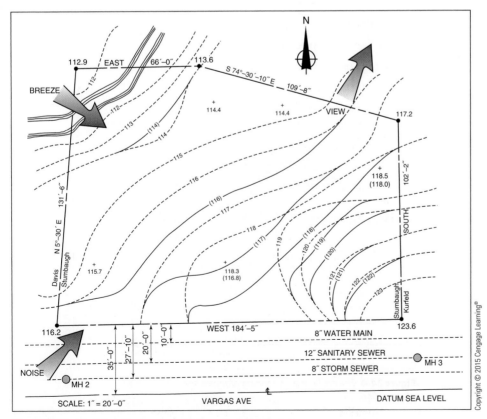

Figure 24-3 *A building site drawing with the orientation features.*

- Structure setbacks
- Calculate the buildable area
- Calculate the maximum land coverage for the house
- Building heights and **daylight plane** (**Figure 24-4**)

The next step is to note the orientation of the building site by the (**See Figure 24-3**):

- **Compass orientation**
- Path of Southern sun
- Views
- Direction of noise
- Direction of usual breezes
- Occupant's wishes

The final step is a list of the needs (must haves) and the wants (flexible) of the occupants. It is often impossible to fulfill all the needs and wants of a client, so be ready for compromises. A typical example for one family may be:

- Bedroom facing the breeze
- Bedrooms on cool north side
- Two bedrooms
- One bathroom
- Adequate storage and closets
- Large U-shaped kitchen
- Dining area
- Service area blocking noise
- Garage on southwest corner to block the sun's heat

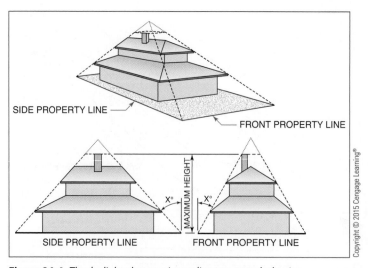

Figure 24-4 *The daylight plane zoning ordinance controls the size of upper floors.*

- Kitchen of east side to receive the morning sun
- Fireplace
- Two-car garage
- Separate entry (want)
- Half-bath in service area (want)
- Den (want)
- Separate laundry room (want)
- Exercise room (want)

Designing the Floor Plan

A floor plan has four major areas. They are the:

1. Entry
2. Living area—living room, dining room, family room, den, entertainment rooms
3. Sleeping area—bedrooms, closets, bathrooms
4. Service area—kitchen, laundry, utility room, work rooms, garage

A well-planned design will have the entry adjacent to the living, sleeping, and service areas (**Figure 24-5**). This will help to create an efficient and minimal traffic flow. Regardless of the basic shape of a floor plan, it should have an efficient traffic flow (**Figure 24-6**). It is also important to know if the plan is to be a **formal** (**Figure 24-7**) or **informal design** (**Figure 24-8**).

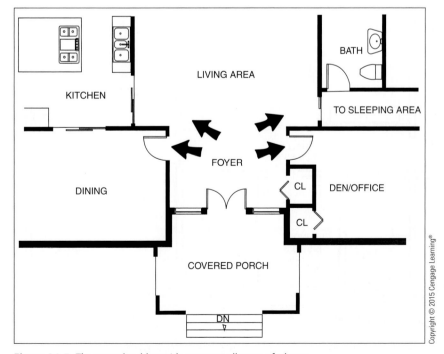

Figure 24-5 *The entry should provide access to all areas of a home.*

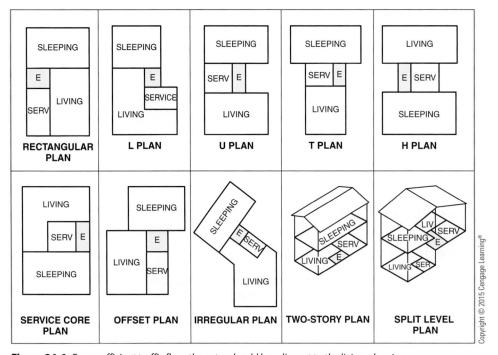

Figure 24-6 *For an efficient traffic flow, the entry should be adjacent to the living, sleeping, and service areas.*

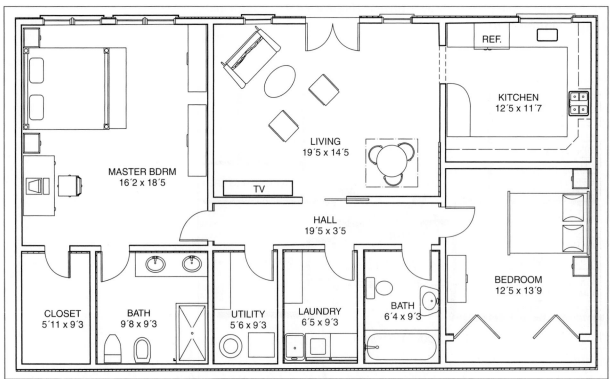

Figure 24-7 *A formal floor plan has all the rooms separated with walls.*

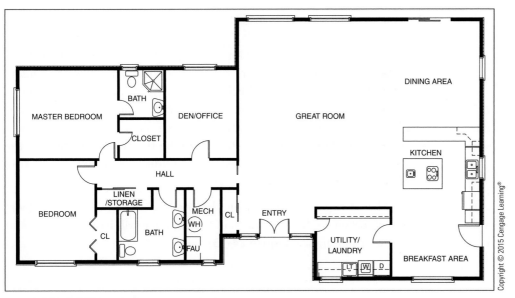

Figure 24-8 *An informal floor plan is an open plan.*

Room Planning

To plan the proper room sizes the architect/designer should know the amount and dimensions of the furniture and that will go into each room. A room too small will be crowded an uncomfortable. A room too large is a waste of the occupant's money, especially with what the price of construction is today. Typical room sizes are shown in **Figure 24-9**.

A GENERAL GUIDE TO ROOM SIZES			
BASIC ROOMS	**TYPICAL ROOM SIZES (FEET)**		
	SMALL	AVERAGE	LARGE
LIVING ROOM	12 x 18	16 x 20	22 x 28
DINING ROOM	10 x 12	12 x 15	15 x 18
KITCHEN	5 x 10	10 x 16	12 x 20
UTILITY ROOM	6 x 7	6 x 10	8 x 12
BEDROOM	10 x 10	12 x 12	14 x 16
BATHROOM	5 x 7	7 x 9	9 x 12
HALLS	3' WIDE	3'–6" WIDE	3'–9" WIDE
GARAGE	10 x 20	20 x 20	22 x 25
STORAGE WALL	6" DEEP	12" DEEP	18" DEEP
DEN	8 x 10	10 x 12	12 x 16
FAMILY ROOM	12 x 15	15 x 18	15 x 22
WARDROBE CLOSET	2 x 4	2 x 8	2 x 15
ONE ROD WALK-IN CLOSET	4 x 3	4 x 6	4 x 8
TWO ROD WALK-IN CLOSET	6 x 4	6 x 6	6 x 8
PORCH	6 x 8	8 x 12	12 x 20
ENTRY	6 x 6	8 x 10	8 x 15

Figure 24-9 *A general guide to room sizes for primary residences.*

Elevation Design

The architect/designer must know what architectural style the owners prefer. There are hundreds of different architecture styles and many that are mixed styles, called **eclectic**. Once determined, the elevations can be drawn. Elevation drawings are usually projected from the floor plan (**Figure 24-10**). The elevation drawings are coded by the compass orientation (**Figure 24-11**). Dimensioning the critical heights on the elevations will ensure that the construction will progress smoothly (**Figure 24-12**).

When designing the floor plan, the locations of the windows and doors are important so that the design of the elevation will have a pleasing and balanced appearance (**Figure 24-13**).

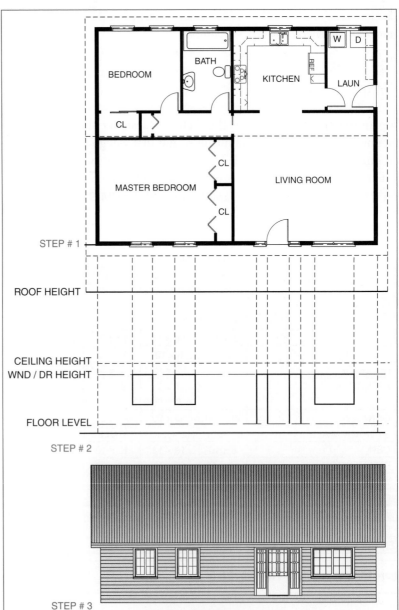

Figure 24-10 *The procedural steps to draw an exterior elevation.*

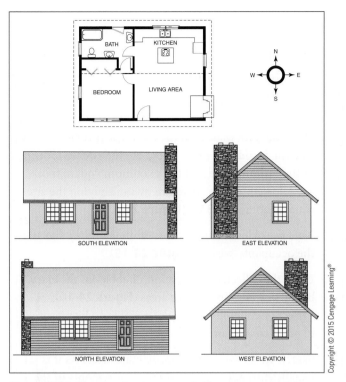

Figure 24-11 *Elevation drawings are coded by compass orientation.*

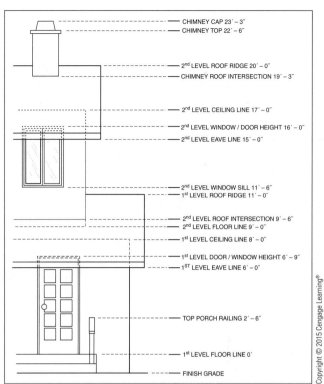

— CHIMNEY CAP 23´ – 3˝
— CHIMNEY TOP 22´ – 6˝

— 2nd LEVEL ROOF RIDGE 20´ – 0˝
— CHIMNEY ROOF INTERSECTION 19´ – 3˝

— 2nd LEVEL CEILING LINE 17´ – 0˝

— 2nd LEVEL WINDOW / DOOR HEIGHT 16´ – 0˝
— 2nd LEVEL EAVE LINE 15´ – 0˝

— 2nd LEVEL WINDOW SILL 11´ – 6˝
— 1st LEVEL ROOF RIDGE 11´ – 0˝

— 2nd LEVEL ROOF INTERSECTION 9´ – 6˝
— 2nd LEVEL FLOOR LINE 9´ – 0˝
— 1st LEVEL CEILING LINE 8´ – 0˝

— 1st LEVEL DOOR / WINDOW HEIGHT 6´ – 9˝
— 1ST LEVEL EAVE LINE 6´ – 0˝

— TOP PORCH RAILING 2´ – 6˝

— 1st LEVEL FLOOR LINE 0´

— FINISH GRADE

Figure 24-12 *Vertical dimensioning is critical for accurate construction.*

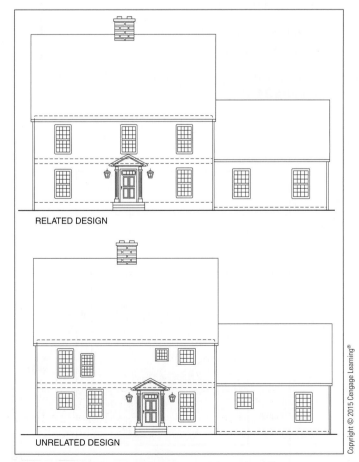

RELATED DESIGN

UNRELATED DESIGN

Figure 24-13 *The locations of windows and doors should have a pleasing appearance.*

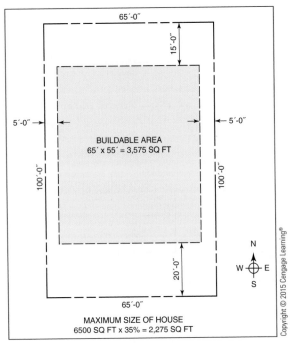

Figure 24-14 *Locate the setback lines from the property lines as prescribed by the local zoning codes. Calculate the square footage of the lot and the maximum square footage of the house: Square footage of lot x 35%. This is a typical zoning percentage for a city's residential building site.*

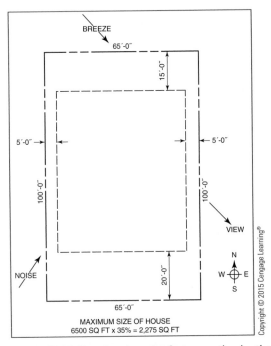

Figure 24-15 *Note the orientation features on the plan that will affect its design layout and list the homeowner's requirements.*
- *Two bedrooms*
- *Two baths*
- *Informal design*
- *Breeze into master bedroom*
- *The southwest corner of the house is the hottest area. Do not locate the bedrooms on this side.*
- *Large windows for the view from the living area.*
- *The service area and garage should block off the noise.*
- *Easy off-loading of groceries from the garage to the kitchen's refrigerator and pantry.*

The Design Process

After studying the gathered information you should now be ready to start the design process, which will consist of many rough sketches and changes after meeting with the occupants and builders.

Step 1. Draw the plot plan with the zoning setbacks to delineate the **buildable area**. Calculate the square footage of the building site and the maximum square footage allowed for the structure by multiplying the site's square footage by 35% (**Figure 24-14**).

Step 2. Note the **orientation** features for the building site and the occupants' wishes (**Figure 24-15**).

Step 3. Using the collected information, "bubble-in" the entry, living, sleeping, and service areas (**Figure 24-16**). Perfect orientation is often not possible, so be prepared to make concessions.

Step 4. Sketch-in the rooms to approximate scale. As it is usually difficult to meet all the requirements of the occupants, zoning and building department, and contractor, plan to meet with them to make the necessary changes (**Figure 24-17**).

Step 5. Complete the floor plan using a scale of $1/4'' = 1'-0$ (**Figure 24-18**).

Step 6. Complete the plot plan using a scale of $1/8'' = 1'-0$ (**Figure 24-19**).

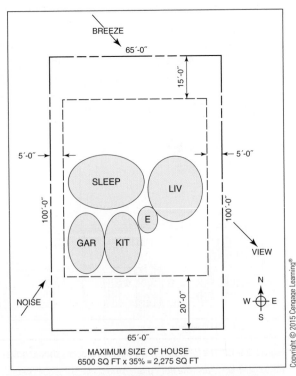

Figure 24-16 *Orient the general areas with a bubble sketch.*

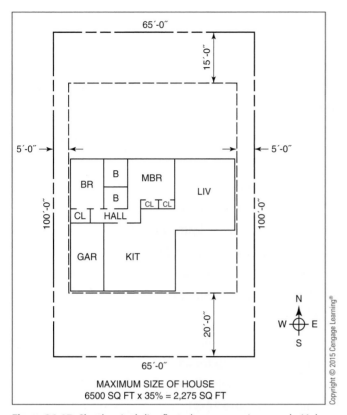

Figure 24-17 *Sketch a single line floor plan to approximate scale. Make as many changes as necessary for the final approval of the plan. Check the approximate square footage of the plan.*

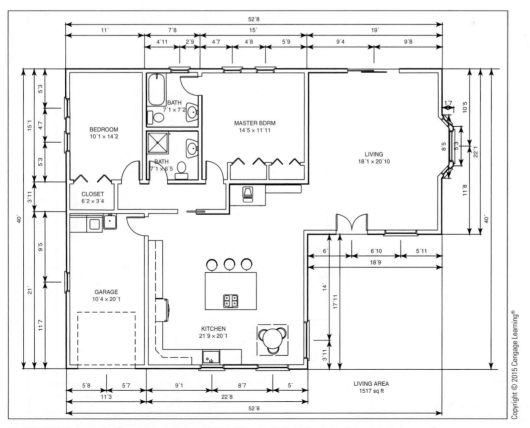

Figure 24-18 *Draw the floor plan using a scale of 1/4" = 1'–0". Double check the square footage of your design to ensure it is not over the maximum size.*

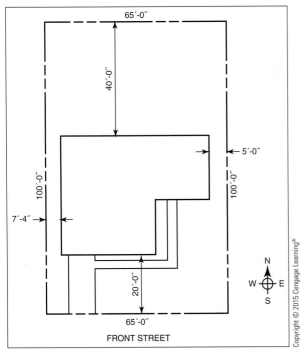

Figure 24-19 *Complete the plot plan using a scale of 1/8" = 1'–0".*

Sample Set of Residential Plans

An example of a minimal set of architectural plans is shown in **Figures 24-20** through **24-31**. You should use this plan set as a reference for your own architectural design work. Note that examples of two foundations are shown. Select one to use for your set of plans.

- Plot plan (**Figure 24-20**)
- Floor plan (**Figure 24-21**)
- Exterior elevations (**Figure 24-22**)
- Interior elevations (**Figure 24-23**)
- Roof plan (**Figure 24-24**)
- **T foundation** plan (**Figure 24-25**)
- **Slab foundation** plan (**Figure 24-26**)
- Electrical and heating plan (**Figure 24-27**)
- Framing plans (**Figure 24-28**)
- Construction details (**Figures 24-29** and **24-30**)
- Schedules (**Figure 24-31**)

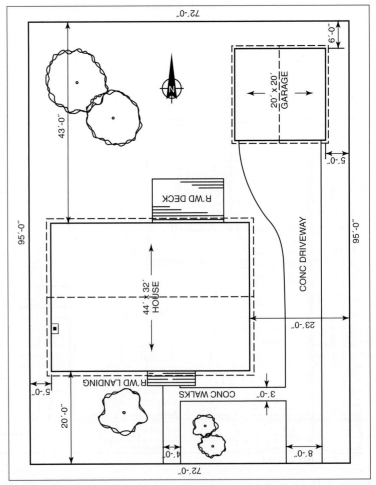

Figure 24-20 *Plot plan.*

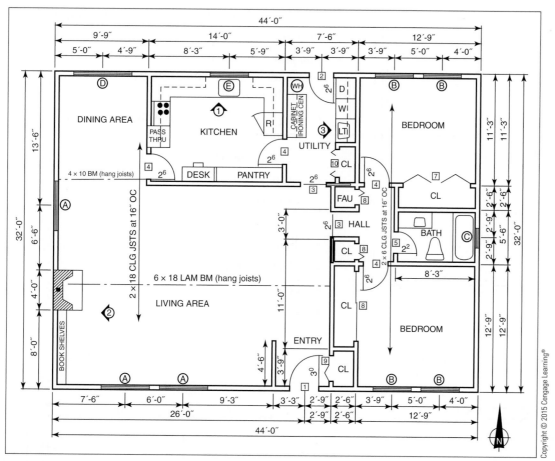

Figure 24-21 *Floor plan.*

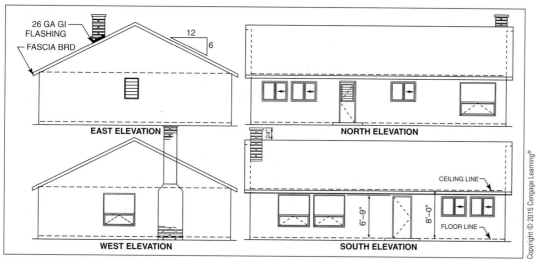

Figure 24-22 *Exterior elevations.*

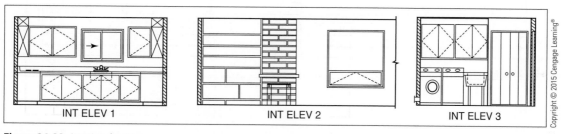

Figure 24-23 *Interior elevations.*

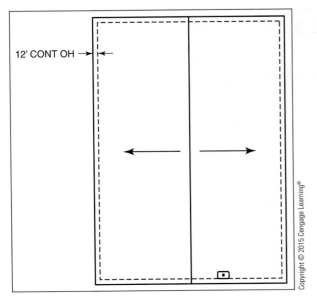

12' CONT OH

Figure 24-24 *Roof plan.*

A fully detailed set of architectural working drawings will include:

- *Plat plan*—A map of a large number of properties and surrounding streets (**Figure 24-32**). The following data should be included:
  - Property line dimensions with the compass orientation
  - Building setbacks
  - North arrow
  - Easements
  - Street widths
  - Street names
  - Block numbers
  - Lot numbers
  - Drawing scale

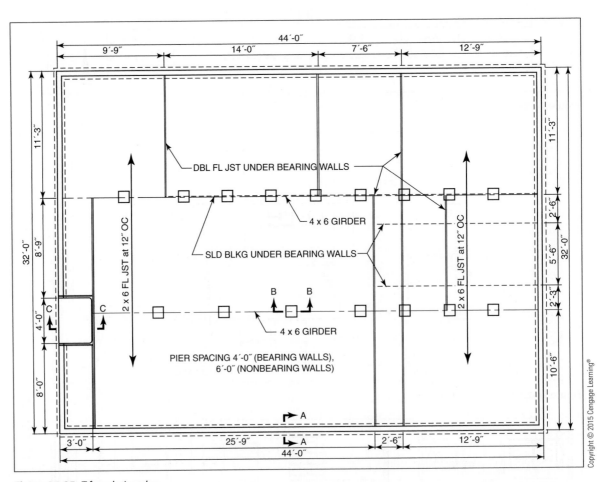

44'-0"
9'-9"   14'-0"   7'-6"   12'-9"

DBL FL JST UNDER BEARING WALLS

4 x 6 GIRDER

SLD BLKG UNDER BEARING WALLS

2 x 6 FL JST at 12" OC

2 x 6 FL JST at 12" OC

B   B

4 x 6 GIRDER

PIER SPACING 4'-0" (BEARING WALLS),
6'-0" (NONBEARING WALLS)

A

11'-3"   32'-0"   8'-9"   4'-0"   8'-0"

11'-3"   2'-6"   32'-0"   5'-6"   2'-3"   10'-6"

3'-0"   25'-9"   A   2'-6"   12'-9"
44'-0"

Figure 24-25 *T foundation plan.*

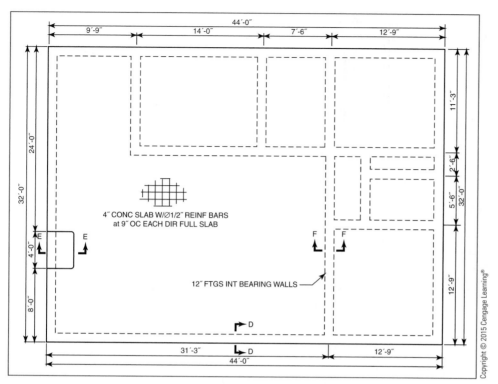

Figure 24-26 *Slab foundation plan.*

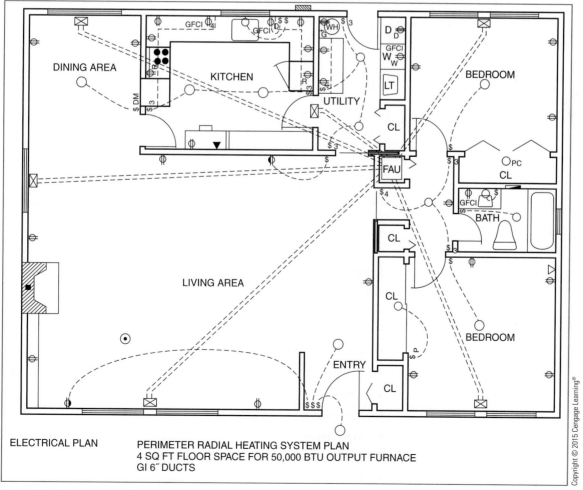

ELECTRICAL PLAN

PERIMETER RADIAL HEATING SYSTEM PLAN
4 SQ FT FLOOR SPACE FOR 50,000 BTU OUTPUT FURNACE
GI 6″ DUCTS

Figure 24-27 *Electrical and heating plan.*

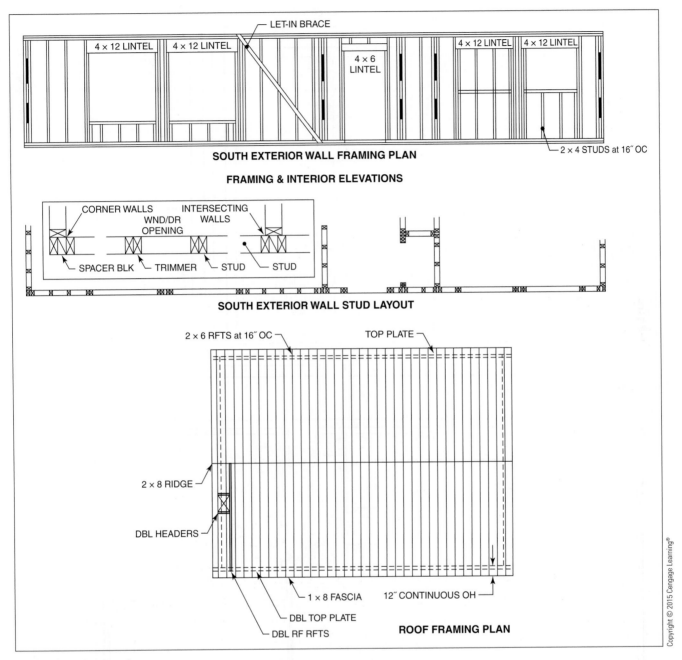

Figure 24-28 *Framing plan.*

- *Site plan*—A drawing describing the physical elements of a single piece of property (**Figure 24-33**). The following data should be included:
  - Elevation at all the property corners
  - **Contour lines** with the elevation heights
  - Property line dimensions with the compass orientation
  - Landscaping
  - Large outcrop of rocks
  - Waterways
  - Easements
  - Utility lines
  - Recontoured lines (dotted)

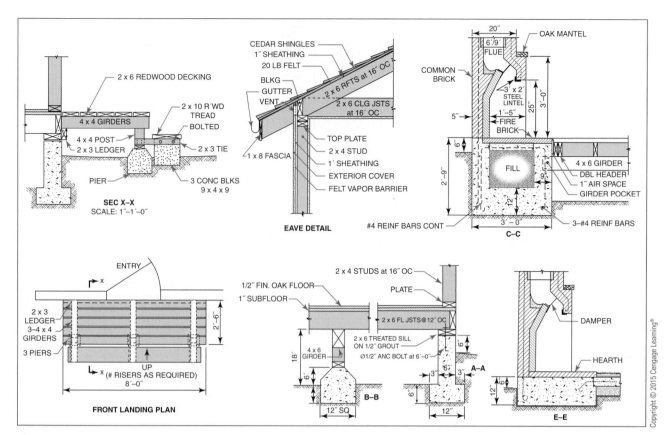

Figure 24-29 *Construction details.*

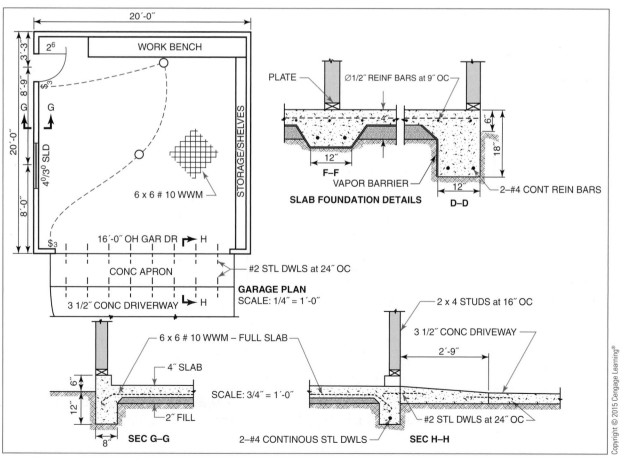

Figure 24-30 *Construction details.*

WINDOW SCHEDULE									
SYM	WIDTH	HEIGHT	MATERIAL	TYPE	SCREEN	QUAN	MANUF	CAT #	REMARKS
A	5´-0˝	5´-0˝	STEEL	AWNING	YES	3	CUSTOM WND MFG	ST-A5	
B	4´-0˝	3´-0˝	STEEL	SLIDING	YES	4	"	ST-S9	
C	2´-0˝	3´-0˝	STEEL	LOUVERED	NO	1	"	ST-L7	OBSCURED GLASS
D	6´-0˝	5´-0˝	STEEL	AWNING	YES	1	"	ST-A9	
E	4´-0˝	2´-9˝	STEEL	SLIDING	YES	1	"	ST-S7	

DOOR SCHEDULE									
SYM	WIDTH	HEIGHT	THICK.	MATERIAL	TYPE	SCRN	QUAN	MANUF	CAT #	REMARKS
1	3´-0˝	6´-8˝	1 7/8"	OAK	SOLID CORE	NO	1	CUSTOM DR MFG	EXT-75	4 COATS NAT OIL
2	2´-6˝	"	1 7/8"	FIR	SOLID CORE	YES	1	"	EXT-35	2´ × 3˝ LOUVERED WND
3	2´-6˝	"	1 1/2"	MAHOGANY	HOLLOW CORE	NO	2	"	SLGD-7	SLDG POCKET DOORS
4	2´-6˝	"	1 3/8"	MAHOGANY	HOLLOW CORE	NO	4	"	INT-9	2 COATS VARNISH
5	2´-2˝	"	1 3/8"	MAHOGANY	HOLLOW CORE	NO	1	"	INT-3	4 COATS VARNISH
6	3´-0˝	"	1 3/8"	MAHOGANY	HOLLOW CORE	NO	2	"	CL-80	2 COATS VARNISH
7	21˝	"	3/4"	PINE	LOUVERED	NO	4	"	LV-22	PAINT TRIM COLOR
8	7˝	"	3/4"	PINE	LOUVERED	NO	4	"	LV-10	PAINT TRIM COLOR
9	12˝	"	3/4"	PINE	LOUVERED	NO	4	"	LV-14	PAINT TRIM COLOR
10	6˝	"	3/4"	PINE	LOUVERED	NO	4	"	LV-09	PAINT TRIM COLOR

Figure 24-31 *Schedules.*

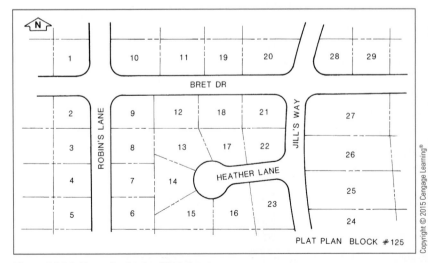

Figure 24-32 *Example of a simplified plat plan.*

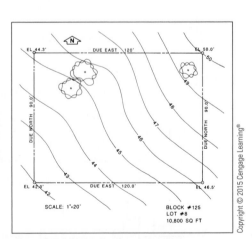

Figure 24-33 *Example of a simplified site plan.*

- *Plot plan*—A drawing of the building site with the structure/s oriented on the property. **Figure 24-34** is an example of a basic plot plan. **Figure 24-35** shows the zoning setbacks and the setbacks locating the buildings on the site.
  - The structure's **setback** from the property lines
  - Outline of the house and roof
  - Outline of all auxiliary buildings
  - Property line dimensions with the compass orientation
  - Elevation at the property corners
  - Decks, patios, walkways
  - Landscaping
  - Legal description
- *Floor plan*—A two-dimensional drawing of all the exterior and interior walls. Included details are:
  - Windows (**Figure 24-36**)
  - Doors (**Figure 24-37**)
  - Closets (**Figure 24-38**)
  - Kitchen and utility rooms (**Figure 24-39**)
  - Bathroom fixtures (**Figure 24-40**)
  - Built-in components (**Figure 24-41**)
  - Electrical symbols (**Figure 24-42 and 24-43**)
  - **Heating, ventilation, and air conditioning** (HVAC; **Figure 24-44**); may be a separate drawing
  - Plumbing lines (**Figure 24-45**); may be a separate drawing
  - Wastewater lines/miscellaneous (**Figure 24-46**)
  - Stairs
  - Sectional detail callouts
  - Interior elevation callouts
  - Roof outline
  - Attached decks and patios
- *Exterior elevations*—Two-dimensional drawings of all the exterior walls and roof. Included details are:
  - All exterior building materials
  - Roof covering
  - Callouts on all windows and doors
  - All required vertical dimensions
  - Roof outline
  - Roof slopes
  - Chimney
  - Plumbing vents

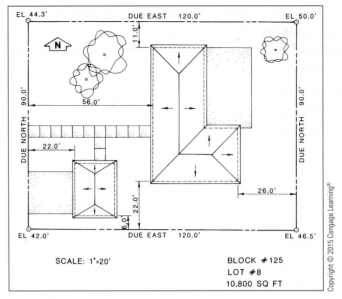

Figure 24-34 *Example of a simplified plot plan.*

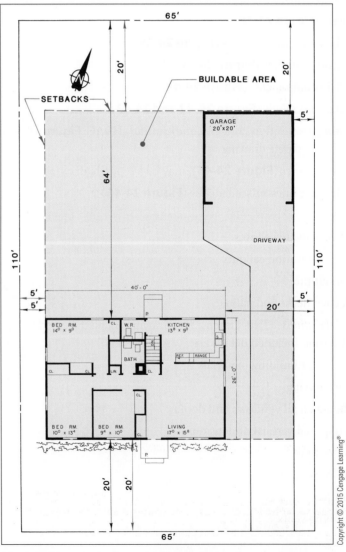

Figure 24-35 *Example of typical setback minimums for a structure on a building site.*

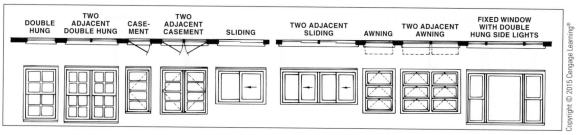

Figure 24-36 *Window styles and their symbols.*

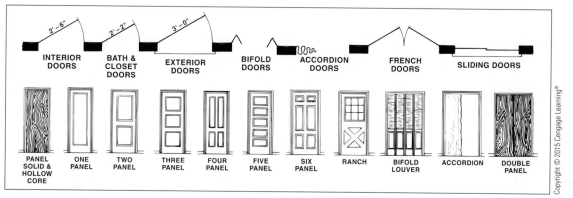

Figure 24-37 *Door styles and their symbols.*

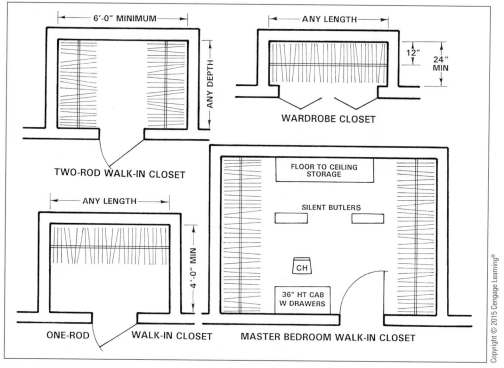

Figure 24-38 *Storage and closet areas must be adequate and easily accessible.*

SINK	REFRIGERATOR	COOK–TOP	FREE–STANDING RANGE	OVEN	DISHWASHER	BUILT–IN MIXER	LAUNDRY TRAY	WASHER	DRYER	KITCHEN CABINETS	IRONING BOARD	IRONER	SEWING MACHINE
S	R	●●●●		O	DW	⊠	L T	W	D	wall cab / floor cab			SM

Figure 24-39 *Kitchen and utility room symbols.* Copyright © 2015 Cengage Learning®

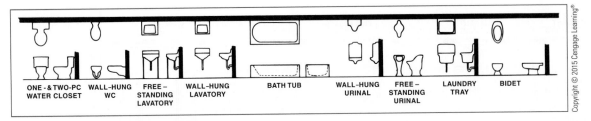

Figure 24-40 *Bathroom fixture floor plan and elevation symbols.*

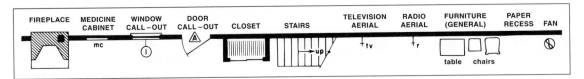

Figure 24-41 *Built-in component symbols.* Copyright © 2015 Cengage Learning®

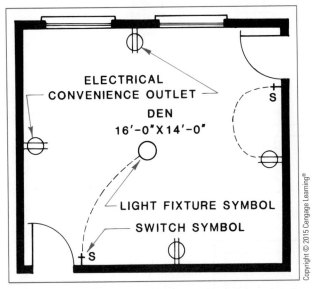

Figure 24-42 *Light switches must be accessible at the opening of each door and conform to local building codes for their placement.*

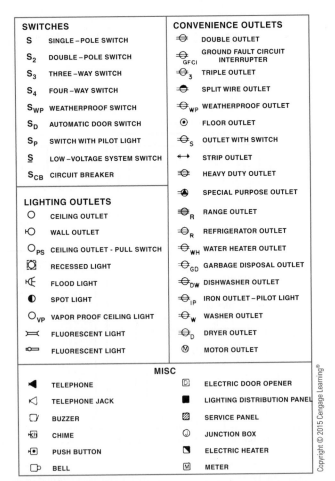

Figure 24-43 *Electrical symbols.*

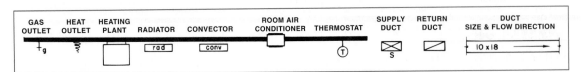

Figure 24-44 *HVAC (heating, ventilation, and air conditioning) symbols.* Copyright © 2015 Cengage Learning®

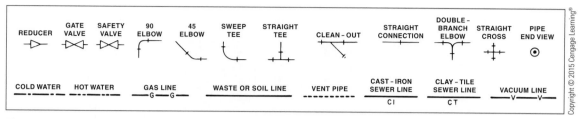

Figure 24-45 *Plumbing line symbols.*

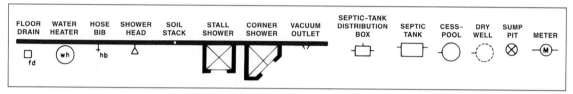

Figure 24-46 *Wastewater facility symbols.* Copyright © 2015 Cengage Learning®

- *Interior elevations*—Two-dimensional drawings of interior walls with built-ins and equipment that the builders must be aware of. Included details are:
  - Windows and doors and their callouts
  - All built-in components
  - Wall finishes
- *Roof plan*—An outline drawing of a birds-eye view of the roof. Included details are:
  - Dotted outline of the house
  - **Ridge board**
  - **Valleys**
  - Direction of roof slant
- *Foundation plan* (slab and T foundations)—A two-dimensional plan of all the concrete work and floor framing systems. Included are:
  - Outline drawing of all concrete walls and piers
  - Dotted lines specifying all footings
  - Notation of all steel **rebar** and anchor devices
  - Wood floor system for the T foundation
  - **Sectional details** through footings areas and fireplace
- *Framing plan*—Two-dimensional single-line or double-line drawings depicting all areas of the rough construction.
  - Floor framing
  - Wall framing
  - Roof framing
- *Construction details*—Two-dimensional drawings of any construction phase that is not defined in any of the other drawings. Usually drawn to a larger scale for easier reading.
  - **Longitudinal** and **transverse sections** (**Figure 24-47**)
  - Concrete and framing sectional details

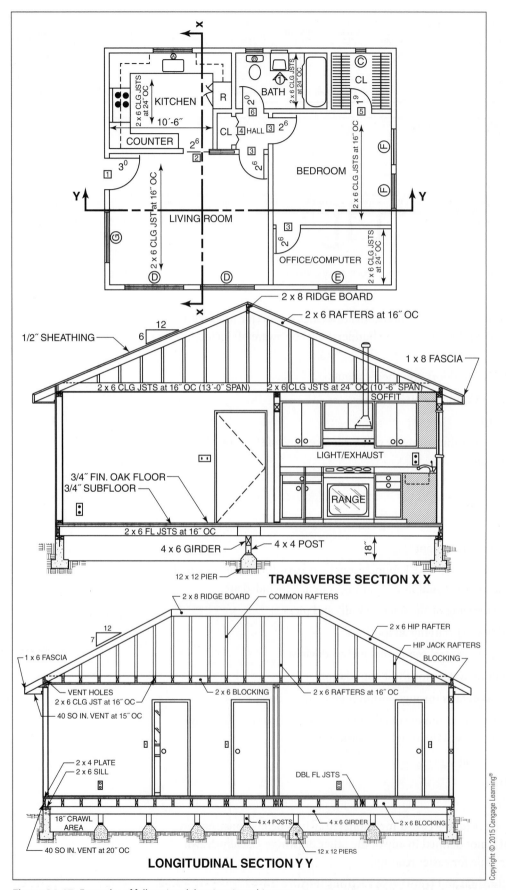

Figure 24-47 *Examples of full sectional drawings in architecture.*

- *Schedules and specifications*—These are charts and written instructions for all materials and installations procedures that are not listed in the set of working drawings:
  - Windows
  - Doors
  - Hardware and plumbing fittings
  - Electrical
  - Appliance
  - Floor and wall coverings and finishes
  - Written specs of any material work process not listed in schedules or drawings.

Procedural Steps to Draw a Floor Plan

Whether the floor plan is a small cabin or a very large structure, the procedure to lay out the drawing for the floor remains the same. Note the procedural drawing steps in **Figure 24-48 a-f**.

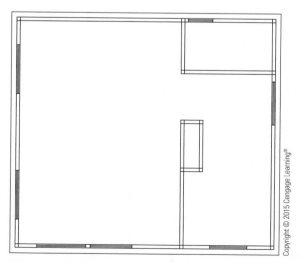

Figure 24-48a *Step 1. Draw exterior walls.*

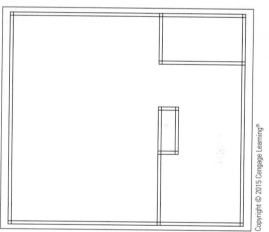

Figure 24-48b *Step 2. Draw interior walls.*

Figure 24-48c *Step 3. Draw windows.*

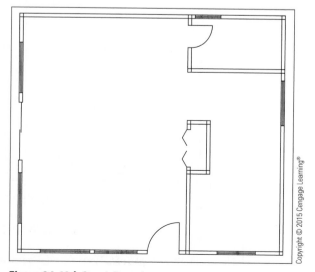

Figure 24-48d *Step 4. Draw doors.*

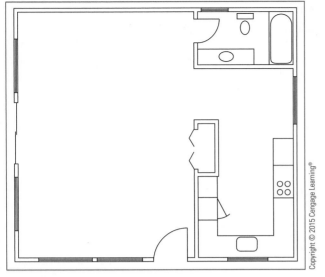

Figure 24-48e *Step 5. Draw cabinets, fixtures, and appliances.*

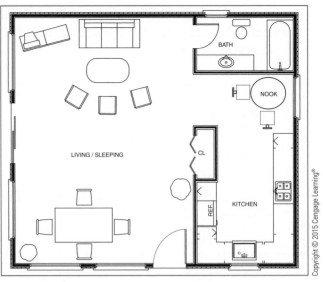

Figure 24-48f *Step 6. Draw furniture or place templates.*

DRAFTING EXERCISES

1. Discuss the design qualities of the good and poor designs in **Figure 24-49**.

2. Draw the interior elevation in the kitchen facing the range in **Figure 24-50**.

3. Draw the floor plan for the 30' × 24' home in **Figure 24-51**.

4. Study the math and purchasing conditions involved in **Figure 24-52** and design a home under 2,800 square feet.

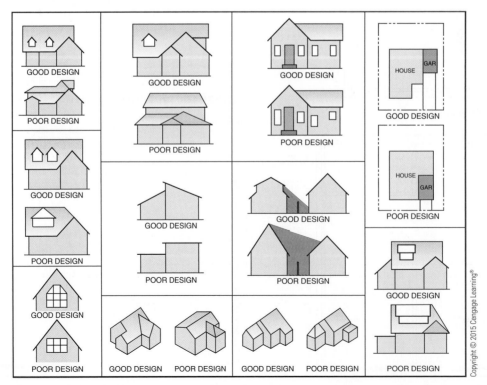

Figure 24-49 *Exercise #1.*

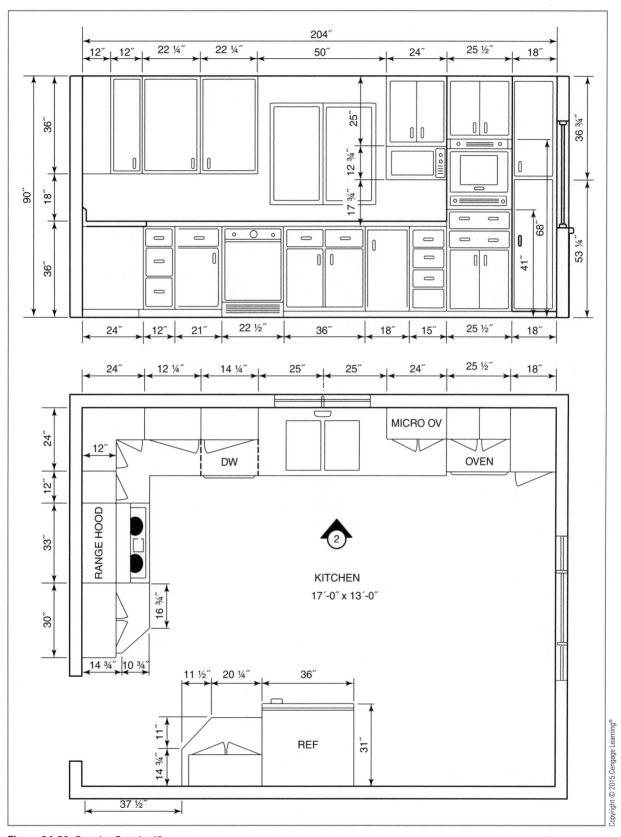

Figure 24-50 *Drawing Exercise #2.*

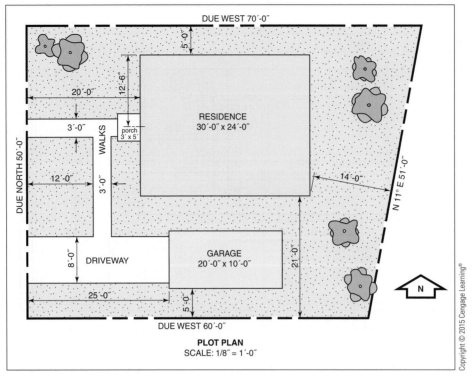

Figure 24-51 *Design Exercise #3.*

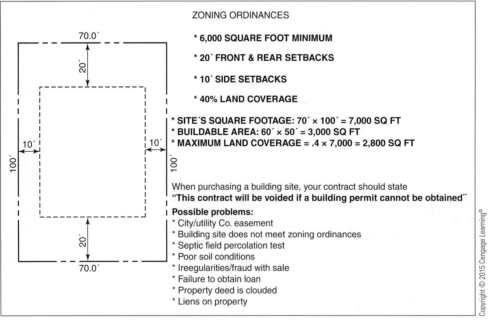

Figure 24-52 *Design Exercise #4.*

5. Design a floor plan, exterior elevations, and a roof plan for the farm styled house in **Figure 24-53**.

6. Draw a floor plan and the exterior elevations for A-frame house in **Figure 24-54**.

7. Draw a floor plan, exterior elevations, roof plan, and a slab foundation for the Southern Colonial house in **Figure 24-55**.

8. Design a new residence to replace your existing home.

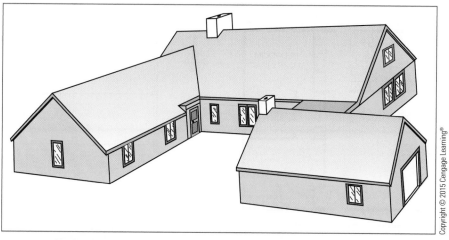

Figure 24-53 *Drawing Exercise #5.*

Figure 24-54 *Drawing Exercise #6.*

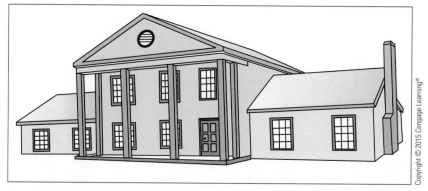

Figure 24-55 *Drawing Exercise #7.*

9. Design a mountain cabin to fit your needs.

10. Design a beach cabin to fit the needs of a friend.

11. Add to the floor plan in **Figure 24-56** a master bedroom and bath, a den, and an expanded living room. Draw the new elevations and roof plan.

12. Name the floor plan symbols in **Figure 24-57**.

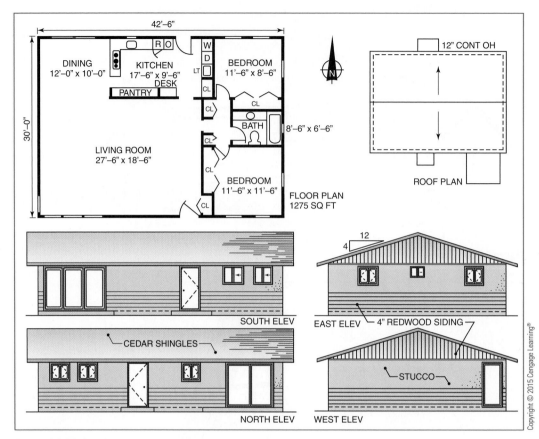

Figure 24-56 *Exercise #11.*

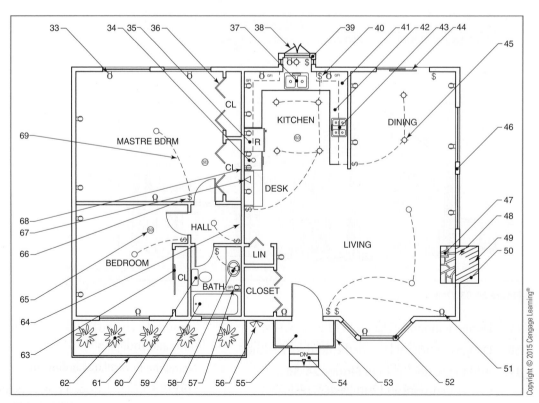

Figure 24-57 *Exercise #12.*

ADVANCED DESIGN PROJECT

The architectural students should not learn architecture by copying plans from a textbook. She/he should start their design with a real-life situation to develop critical thinking and reasoning to create an ideal home design for a client. Additional research outside of this chapter will be necessary to complete this design project. The floor plan of the following design exercise (**Figure 24-58**) should be drawn to a scale of ¼"= 1'-0". Select a smaller scale, depending on the size of your drawing format, for the plot plan drawing. The zoning setbacks are 15' front setback, 20' rear setback, and 5' side setbacks.

The client's needs are:

- The minimum size of the home is 1,400 square feet. The maximum size allowed by the 35% land overage is 3,170 square feet.
- Informal design
- Large master bedroom with bath in a quiet area
- Three additional bedrooms
- Two full bathrooms and a half-bath
- Attached double garage
- Breeze enters the master bedroom
- Locate the garage and service area in the "hot" southwest corner, also to block the noise.
- Large window in the living area for the view
- Easy off-loading of groceries from parking area to the kitchen's refrigerator and pantry

The client's wants are:

- A 10' × 10' storage room
- Hot tub in bedroom
- Additional work space in garage
- Wine storage closet
- Media room
- Music room

As your design progresses be certain to check all the design phases with the client (instructor). This eliminates a lot of redesigning.

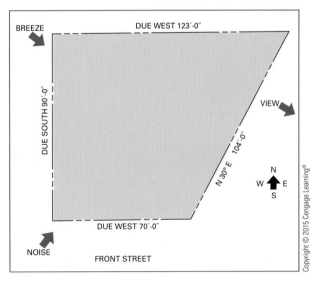

Figure 24-58 *Drawing Exercise #12.*

PROBLEM DESIGN AREAS

- Poor site orientation
- Keeping the plan inside the zoning setbacks
- Not following the clients wish list
- Rooms too small for furniture
- Rooms too large
- Long halls
- Poor traffic through the house
- Poor traffic flow through rooms
- Odd shaped rooms (should be rectangular at approximately a 3:5 ratio)

KEY TERMS

Buildable area

Compass orientation

Contour lines

Daylight plane

Eclectic

Formal design

Heating, ventilation, and air conditioning (HVAC)

Informal design

Longitudinal section

Orientation

Plat plan

Plot plan

Rebar

Ridge board

Sectional details

Setback

Site Plan

Slab foundation

T foundation

Transverse section

Valley

GLOSSARY

A

active solar Energy collected from the sun and mechanically transferred.

addendum The radial distance from the pitch circle to the top of a gear tooth.

addendum circle The outside diameter of a gear.

aeronautical The science and flight of all types of aircraft in the earth's atmosphere.

airline diagrams (baseline diagrams) An electronic diagram that is a shortcut way to quickly draw the components to the connectors by placing them close together.

aligned dimensioning The numerical dimension that is placed in line with the dimension line.

aligned section Sections that result from the offsetting or revolvement of a part to line up with the cutting plane.

allowance Acceptable clearance between mating parts.

alphanumeric keyboard A computer keyboard that allows an operator to input letters and numerals into a computer.

American Society of Mechanical Engineers (ASME) An organization that sets the standards and conventions for manufacturing working drawings.

American Wire Gage (AWG) A standard numerical system for measuring wire sizes.

angle of thread The included angle formed between the major diameter of a thread and the depth of thread.

angular gears Tow gears connected at an angle. They are not parallel or perpendicular.

arc welding Welding metals with an electric arc.

architect A professional designer of structures.

architect's scale A ruler with various drawing scales used to draw and/or to check architectural designs.

Architecture, Engineering, and Construction (AEC) A society that sets standards for architecture, engineering, and construction.

arrow line A leader with an arrowhead that points to a specific item.

assembly drawing A drawing that shows the relationship of various parts of an item when assembled.

assembly section A sectional working drawing of two or more mating parts.

astronautical The science and flight in outer space.

auxiliary plane A drawing plane whose auxiliary surface is rotated 90° from a multiview's angular surface.

auxiliary section A sectional working drawing of an auxiliary view.

auxiliary view A view of an object other than the normal six orthographic views. Primary auxiliary views are drawings projected directly from any of the six normal orthographic views. A secondary auxiliary view is a drawing projected from a primary auxiliary view.

axis A line around which parts rotate or are concentrically arranged.

axonometric drawing A pictorial drawing that shows the front, side, and top faces of an object in one view. The receding lines in axonometric drawings are always parallel. Isometric, dimetric, and trimetric drawings are types of axonometric drawings.

B

backing weld A weld used to add strength to a joint by making additional welds to the backside of the material being welded.

base circle The point where a gear's tooth starts its curve.

baseline dimensioning Dimensioning to a common datum point or plane.

basic (simple) dimension Also called a specified dimension. An exact value used to describe the size, shape, location of a feature, and from which tolerances can be specified.

basic hole size The standard size of any basic hole cutter.

basic shaft size The standard manufactured size of a shaft.

basic size The segment of the dimension from which limits are calculated.

bell and spigot connection The connection between two pipes. The straight spigot end of one pipe is inserted into the flared-out end of the other pipe. The joint is sealed with caulking or a compression ring.

belt drive Transmits the power with a belt between the shafts with a pulley.

bevel An inclined surface that is not constructed at a right angle to the adjoining surface.

bevel gear Connected at an angle less than 90°.

bilateral tolerance dimensions A tolerance that allows variation in both directions from a dimension.

bill of materials A list of materials specified for a project.

binary system This is any system of counting or measurements whose units are a power of two.

biomass The oldest source of renewable energy since humans learned how to make fire. It may consist of plants, animals, and/organisms.

block diagram Connected blocks labeled to represent sections of an electronic system.

blow molding A process where a hot liquid plastic is injected into a mold.

Blu-ray disc This is an optical disc storage system designed to replace the DVD format. The disc is 120 mm in diameter and 1.2 mm thick.

bolt circle A circular centerline that locates centers of holes positioned about a common center point.

boot A term that means "to load a software program into a computer."

bore To enlarge a hole with a hand or machine tool.

bow compass A compass used to draw circles and arcs up to 6" in diameter.

brazing The joining of two pieces of metal with a brass alloy filler that melts at about 800°F. This process is similar to soldering.

break line A line used to interrupt a drawing if an object will not fit on a drawing sheet.

broken-out section A small part of a sectional drawing that is removed and drawn at a larger scale. Also called a removed section.

buffer A device used for the temporary electronic storage of information.

buildable area The area within the building site that a structure may be placed.

building line An imaginary line on a plot beyond which the building cannot extend.

built-up body A jig built-up with separate segments.

bushing A hollow cylindrical sleeve bearing.

butt weld A welded joint that joins two pieces butted directly at their ends.

byte A single character of computer memory.

C

cabinet drawing An oblique drawing drawn with full size dimensions.

cabinet oblique drawing A pictorial drawing containing receding lines drawn at half-scale, at an angle of 30° or 45° from the horizontal.

cableform diagram A drawing of groups of wires bound together into a cable or harness.

calendaring The process of passing plastic sheets into rollers to produce flooring or gaskets.

callout A drawing note connected to a drawing with a leader and arrowhead.

cam A machine device used to transmit motion. It may change the speed, direction, or type of motion.

cam clamp A type of holder used to keep an item firmly in place in a jig.

cam follower A plunger whose reciprocating motion is produced by contact with the surface of a cam.

camshaft A rotating shaft that the cam is attached to.

Cartesian coordinate system Cartesian geometry is based on a two-dimensional set of points with an X- and Y-axes set at right angles to each other. They depict four imaginary quadrants where points may be located.

cartographer A person who designs and draws maps.

cast body A jig that is cast in one piece.

casting Pouring of molten metal into a sand or plaster mold.

cavalier drawing An oblique drawing whose front view is drawn with full-size dimensions, but whose receding sides are drawn at half-scale.

cavalier oblique drawing A pictorial drawing containing receding lines drawn to true scale.

cementing The process of permanently joining pipes.

center drill A drill used to drill the mounting holes in the end of a workpiece.

centerline A line that defines the common center of symmetrical objects.

central processing unit (CPU) The basic control center hardware for a computer system.

chain and sprocket drive Similar to a belt drive, but with a chain and sprocket wheels.

chain dimensions Continuous dimensions that connect aligned features of an object without reference to a common datum point or plane.

chain line A line used to indicate a specific area.

chamfer An angle across a corner that eliminates a sharp edge.

character A symbol used to represent data in an organized system.

chassis drawings The drawings used to construct the housing for electronic circuits and components.

checker A person who looks for errors and additional information required on working drawings.

chemical engineer A professional engineer that applies the science of chemicals to the industrial processes.

chemical machining Manufacturing process that uses a strong chemical to remove unwanted material.

chief/senior drafter The head drafter for working drawings.

chord A straight line joining two points on a curve.

chordal addendum Distance from the top of a spur gear tooth to the chordal thickness line.

chordal distance The distance of a line touching the diameter of a circle in two places.

chordal thickness The thickness of a spur gear tooth measured on a chord on the circumference of the pitch diameter.

circularity (roundness) In a geometric tolerance, it is the tolerance zone for a round object.

circular pitch The length of an arc along the pitch circle between the center of one gear tooth and the center of the next tooth.

circular thickness The length of an arc between the sides of a spur gear tooth measured on the pitch circle.

civil engineer A professional engineer who designs the environment around structures such as roads, bridges, tunnels.

civil engineer's scale A ruler with various drawing scales used to draw mechanical parts using the inch and decimal part of an inch.

clearance fit A clearance space between mating parts resulting from the limits of the dimensions.

collar A flange or ring attached to a shaft or pipe.

command Instructions transmitted to a computer by an operator.

commercial artist An artist who creates illustrations for advertisements and pictorial drawings depicting a manufactured items.

compact disc (CD) An optical disc that has deep and shallow pits on its surface, created by laser technology, allowing data read from and written to it.

compass orientation An arrow pointing north on the plot plan.

component One part of a multipart object or structure.

compression length The exact length of an object after its final compression.

computer-aided drafting (CAD) A computer system that uses a technical software program to generate working drawings.

computer-aided drafting operator A drafter who uses a computer system to draw working drawings and pictorials.

computer-aided engineering (CAE) A process that uses a direct link from the CADD system to the machine tool.

computer-aided manufacturing (CAM) Computer-augmented manufacturing; computer-automated manufacturing. A computer-aided software program that is read by manufacturing machines to produce an item. Uses a direct link from the CADD system to the machine tool.

computer-integrated manufacturing (CIM) Integrates all the CADD manufacturing systems into a single management system.

computer-numerical controlled (CNC) A computer's program that is connected to a manufacturing process. Uses a direct link from a high end, detailed CADD program to the machine tool.

concave Surfaces that curve inward.

concept An idea.

connection diagrams (wiring diagrams) The electronic drawings that contain all the details for electronic connections.

construction lines Light lines used for the layout of a drawing.

contact area The surface areas of two or more mating parts.

contour lines Curved lines that note the elevation, usually on a site plan.

convex Surfaces that curve outward.

coordinates Points of intersection of lines on a grid system.

counterbore To enlarge a hole to allow a screw or bolt head to be recessed below a surface.

countersink A recess made to allow the tapered head of a fastener to lie below a surface.

creativity The ability to develop new ideas for development.

crest The point of widest diameter of a screw thread that lies on a major diameter plane.

crosshatching Closely spaced parallel lines drawn obliquely to represent sectioned areas.

cursor A marker used to locate points on a video display screen (monitor) while working with a computer program.

cutaway view A partial sectional drawing of an object. Also called a broken-out section.

cutting plane line A line drawn on a view where a cut was made in order to define the location of the imaginary sectional plane.

cycloidal curve A curve formed when a point on a circumference is rotated 360°.

cylindricity Having the form or properties of a cylinder.

D

database A stored collection of computer information.

datum A line, point, or plane from which other locations are measured.

datum dimensioning A dimensioning system in which dimensions originate from a common datum.

datum identification box A symbol that locates a specific item and contains geometric tolerancing data. It is also called a datum target.

daylight plane A zoning code that restricts the area and height of multiple stories.

dedendum The radial distance from the pitch circle to the bottom of a spur gear tooth.

dedendum circle Defines the root diameter (bottom of the gear's tooth).

default Action initiated by a computer software program unless an operator specifies other actions.

degree (°) A unit of angular measurement.

descriptive geometry Techniques for drawing points, lines, and three-dimensional objects on a two-dimensional surface.

design assembly drawing A very refined assembly working drawing.

design size The size of a part after clearances and tolerances are applied.

design technician A drafter that completes working drawings, usually from an engineer's sketches.

detail drawing A dimensioned working drawing of a single part.

development drawing The drawing of a surface of an object unfolded on a flat plane. It is sometimes called a stretchout pattern. Development drawings are used in *pattern drafting*. Parallel line and radial line developments are used to develop patterns for objects that contain flat or single-curved surfaces. *Parallel line development* drawings are used to develop patterns for objects that contain parallel lines. *Radial line development* is used to develop patterns for objects that do not contain parallel lines. These methods cannot be used to develop patterns for warped surfaces. *Triangulation development* is a method of dividing a warped surface and other complex forms that cannot be developed with parallel or radial line development into a series of triangles, and transferring the true size of each triangle to a flat pattern.

developmental engineer A professional engineer who develops new products for production.

diagram assembly drawing Shows the form and location of all the subassemblies with multiview working drawings.

diameter The length of a straight line that passes through the center of a circle and terminates at each end on the circumference.

diametral pitch A ratio of the number of teeth on a gear per inch of pitch diameter.

diazo A printing process that produces black, blue, or brown lines on surfaces. The process exposes a drawing with a sensitized material exposed to an ultraviolet light, then processes it in an ammonia chamber.

digital circuit Electronic circuit that functions through the use of binary states encoded by binary numbers (1s and 0s).

digital readouts Devices that display numbers from the output of calculators and computers.

digitize To locate points and select computer commands with an input device.

digitizer An electronic tablet that serves as a drawing surface for the input of graphic data.

dimension line A thin line denoting the length of a dimension with arrowheads at the ends and numerical information placed at its center.

dimensioning Labeling radial or linear length (width, height, or depth) on a technical drawing.

dimetric drawing Pictorial drawing that contains two axes that form equal angles with the plane of projection.

direct current (DC) The basic current that electronic devices require.

direct dimension A method used when the distance between two points is critical to avoid tolerance build-up.

disc drive A computer device that allows the reading and writing of data to and from rotating storage media.

displacement diagram A drawing that shows the path of a cam follower.

dividers A device used to divide and transfer measurements or for scribing arcs on hard surfaces.

documentation Drawings or printed information that contains instructions for assembling, installing, operating, and servicing.

double-line piping drawing Drawing the outline of a pipe.

drafting/engineering supervisor A professional engineer or a drafter with many years of experience who coordinates the development of the working drawings and production of a product.

drafting pencil Mechanical or wood-encased graphite pencils used for manual drafting.

drill bushing A very hard bushing set into a jig to guide a drilling operation.

drum cam A cylinder with a groove in its surface that guides a follower through the groove to produce reciprocating motion.

drum plotter A cylindrical graphics pen plotter that accepts continuous-feed paper.

dual dimensions The use of the metric and the U.S. customary measuring systems on a drawing.

dump To transfer computer information from one system to another.

Dwell (rest) The period that the follower does not move during the rotation of the cam.

E

eclectic A mixture of architectural styles.

effective throat The depth of the penetration of a weld.

electrical/electronics engineer A professional engineer who designs and supervises the working drawings for and production of electrical and electronic components.

electric beam cutting Use of an electron beam to machine materials to shape.

electronics drawings Working drawings that depict electronic circuits and electronic items.

elementary diagrams (see Schematic diagrams) Most frequently used diagrams to simply and quickly show the connection and function of each circuit in a device.

element A distinct group of lines, shapes, and text, as defined by a computer operator; also referred to as entities. In development drawings, imaginary lines used to draw the pattern.

elevation drawing A drawing of the exterior or interior walls of a building. It is a perpendicular, or upright, projection from the floor plan to show vertical architectural or design details.

ellipses A closed curve in the form of a symmetrical oval.

ellipse template A thin plastic template used to draw ellipses.

engineering change documents Documents used to implement a specific change in a production drawing. Engineering Change Request (ECR), Engineering Change Notice (ECN), and Engineering Change Order (ECO).

engineering drawing A technical drawing in any field of engineering.

enlarged section A segment of a sectional working drawing, drawn at a larger scale.

enter To input information into a computer.

environmental engineer A professional engineer who improves the environment by reducing the pollution of the soil, water, and atmosphere.

environmental study Study and research aimed at developing a healthy and safe environment.

equilateral triangle A three-sided figure with three equal interior angles.

erection assembly drawing Provides the drawing information for the construction and/or erection of a building product.

ergonomics The study of human space and movement needs.

exploded view A pictorial drawing that shows all parts disassembled, but in relation to each other.

extension line Thin lines marking the ends of a part that the dimension line touches.

external thread The outside thread of a product such as a bolt.

extrusion The process of forcing soft plastic in a mold through a die.

F

face angle The angle between the top of a gear tooth and the gear axis. Also called a bevel angle.

face cam Has the follower facing the face of the cam rather the circumference of the cam.

fall The period that the follower drops during the rotation of the cam.

FAO Finish all over.

fastener A device used to hold two or more objects together.

feature An element of a part or object.

feature control frame Contains data for each controlled specification and includes the characteristic symbol, tolerance value, and datum reference letter.

feet button Used to raise and level the part to be machined in the jig.

fillet The curved interior intersection between two or more surfaces.

fillet weld The weld for two perpendicular objects. The weld is similar to a triangle in shape.

film An indestructible polyester film used for as a paper base for drawings.

finish Applications of materials to the surface of objects.

finish symbol A small symbol in the form of a check mark stating how smooth a surface must be.

finite-element analysis Also called FEM is a practical application to analyze with numerical techniques to solve solutions for math and physics. It is used in product design for the testing of materials.

first-angle projection A multiview drawing of the front view, top view, and left-hand side view.

fit The relationship between mating parts.

fittings (pipe) The parts of a piping system used for joining pipe lengths together such as bends, elbows, tees, etc.

fixed drill bushing A component forced into a hole in the jig that holds the item firmly during machine operations.

fixtures In architecture, electrical devices that can be secured to the ceiling or walls. In manufacturing: a tool used to position and clamp an item for a machine operation.

flange A circular rim around a pipe or shaft.

flange connections Pipe fittings with a flange that can be bolted.

flange weld The welding of the edges of two flat pieces bent at a right angle.

flank The lower side of a gear's tooth.

flash drive This is a small, removable storage device used to back up computer data. It is about 35 mm long and 3 mm thick.

flat-faced follower A cam follower that has a flat face adjacent to the cam.

flat pattern A development or layout drawing that shows the true shape of an object before bending.

flat springs An arched or flat metal spring such as a leaf spring for automobiles.

flat surface pattern A two-dimensional development drawing.

flatbed plotter A pen plotter in which the drawing paper is placed on a flat surface and the pen carriage moves over the surface in a manner similar to handheld pens.

flatness In geometric tolerancing, it is the condition of a surface where a tolerance may be applied.

floor plan A horizontal, sectioned, working drawing showing all walls, windows, doors, and architectural details with the use of drawing symbols.

flow lines Lines representing the direction of a sequence of operations.

flute A groove on drills, reamers, and taps.

fold (bend) line Line of intersection between two planes in a multiview projection, also called a reference line. Represents the position of a fold to be made on flat materials such as paper, sheet metal, and plastic.

footing An extension at the bottom of a building foundation wall that distributes the load into the ground.

force fit The joining of two mating parts in a manner that requires pressure to force the pieces permanently together.

foreshortened To show lines or objects shorter than their true lengths. Foreshortened lines are not perpendicular to the line of sight.

forging A process of shaping metal by forcing it into a die or hammering malleable metals.

form tolerancing An allowable variation of a feature from the precise form as shown on a drawing.

formal design A symmetrical floor plan with all rooms separated by walls.

foundation The basic structure on which a building rests.

free fit Liberal tolerances between mating parts that allow free movement without binding.

free length spring The length of a spring when here is no pressure or stress applied.

friction wheel Transmits power when two smooth gear wheels are pressed together.

frontal plane In descriptive geometry, it is the front plane (view) of an object.

full-divided scale A scale that contains full subdivision lines throughout the entire length of the scale.

full isometric section An isometric drawing of a sectional drawing of the whole object.

full section A sectional drawing based on a cutting plane line that extends completely through an object.

fusion welding A welding process that results in the mixing of molten metals.

G

gauge Numerical standard for the thickness of sheet metal and wire.

gas welding A welding process produced from one or more burning gases. A filler metal and pressure may or may not be used.

gear A toothed wheel used to transmit power or motion between shafts.

gear axis The center of gear drive shaft.

gear ratio The mathematical ratio of two mating gears.

general assembly drawing Includes all the multiview working drawings of all parts, material lists, and data needed for the manufacturing of the product.

general tolerance note When all tolerances are identical, the tolerance may be listed once in a note.

geologic features Land features such as water, contours, outcropping, and landscaping.

geometric characteristic symbol The modifying (shorthand) symbols for geometric tolerances.

geometric construction The drawing of geometric forms.

geometric dimensioning and tolerancing The assignment of tolerances that prescribes allowable variances in shape, angle, straightness, and position of features and parts.

geothermal energy This is the heat that is in the earth. It can be collected and processed and used for our energy needs.

graphical user interface (GUI) A user interface that allows a user to interact with electronic devices using images (icons) rather than text commands.

graphics tablet An electronic flat surface on which a stylus or mouse is moved to transmit information, digitize points, or pick commands.

grid paper A paper drawing medium with light nonreproducible lines used to help with freehand sketching.

grid size (CAD) The size of the grid or series of light dots that appears on the screen to aid the operator. The grid will not print on the final drawing.

groove weld A weld that deposits metal in a groove between two members.

guide lines Light rules used for freehand lettering.

H

half isometric section An isometric drawing showing only half a section of an object.

half-section A sectional drawing based on a cutting plane line that cuts through one-quarter of an object. A half-section reveals half of the interior and half of the exterior.

handwriting-activation device The software and hardware that allows a computer to read the handwriting or lettering on a tablet connected to a computer.

hard copy The output of a computer printed or plotted on paper or film.

hard disk drive A storage device containing magnetic media; it may be a peripheral unit or built directly into a computer's CPU. A hard disk holds more data than a microdisk. Data can be stored to it and retrieved from it quickly.

hardware The physical components of a computer system.

harmonic motion Describes the smooth transition of the follower during the rotation of a cam.

harness drawing A drawing showing the layout or pattern for making an electrical harness or cable.

helical gear A gear that has two parallel drive shafts. The gear's teeth are at an oblique angle to the shafts.

helix An evenly spaced spiral curve around a cylinder.

hem The additional material added to an edge of a pattern to add strength for a connection.

herringbone gear A gear that has two sets of teeth that form a V shape.

hexagon A polygon with six equal sides.

hidden line Dashed lines that represent object lines that fall behind an object's surface.

highway diagram A point-to-point diagram used to simplify electrical interconnecting complex lines, sometimes called a trunkline diagram. An electronic drawing in which all the wires are bundled into a thick line called a path. The wires are separated near their connection with a thin line called a feeder.

hinged planes The imaginary folded planes used in orthographic and auxiliary projections.

horizon In perspective drawings, the horizon is at the viewers eye level and have the vanishing points located on it.

horizontal plane In descriptive geometry, it is the projected top plane of the frontal plane.

hydroelectricity This is the energy produced by moving water that spins a turbine, which in turn spins a generator that produces electrical power.

hydroforming A process of spraying a very high-pressure fluid onto a piece of metal over a form.

I

inclined Any line or plane located at an angle other than 90° from an orthographic plane.

inclined surface The angular surface of an object.

increment The spacing of a weld.

industrial designer A creative person who designs external packaging with new ideas and materials.

industrial engineer An engineer who designs and optimizes products in many different areas in engineering and the sciences.

Industrial Fasteners Institute (IFI) The organization that sets the standards for fasteners, such as nuts and bolts, for the aerospace, industrial, and automotive industries.

informal design An asymmetrical (unbalanced) floor plan, with open planning for the living areas and kitchen.

injection molding A process in which molten plastic is injected into a mold.

ink-jet printer A printer that sprays tiny drops of ink from a cartridge onto paper to form text or drawings.

input The insertion of information into a computer.

input device A physical component used to insert data into a computer system.

installation assembly drawing Provides the information on how to assemble a product.

Institute of Electrical and Electronics Engineers (IEEE) A worldwide organization dedicated to advancing technological innovation in the fields of electronics and electrical engineering.

instructor of engineering/drafting A professional engineer or educator who teaches classes of industrial working drawings.

integrated circuit (IC) A semiconductor device that is a miniaturized electronic circuit. It is also called a microchip.

interchangeable Parts manufactured with consistent dimensional standards that can be used with a variety of parent objects.

interference fit When parts have negative clearances and must be forced together. Used when mating parts are designed to be permanently attached.

internal bus A component inside a computer that relays information within the computer or to peripheral hardware.

internal thread The threads that are inside an object such as a nut.

International Electrotechnical Commission (IEC) An organization that publishes, worldwide, technical guidelines about the production and storage of electricity.

involute curve A spiral curve that follows a point on a string as it unwinds from a cylinder or other shape.

irregular curve A curved, plastic drawing instrument used to help draw smooth curves. It is also called a French curve.

isometric A pictorial drawing based on the 30° angle.

isometric axis The basic layout of an isometric drawing containing three axes 120° apart.

isometric circle An isometric circle will appear as a 30° or 35° ellipse in an isometric drawing.

isometric drawing A pictorial drawing that contains receding planes at 30° from the horizon.

isometric lines In an isometric drawing, all vertical and 30° lines are considered isometric lines.

J

jig A device that holds a workpiece and guides the cutting tool.

junior detailer A beginning level drafter with good drafting techniques that produces simple working drawings and makes changes or corrections of finished working drawings.

junior drafter A beginning level drafter with good drafting techniques and a background of CAC.

K

key A device embedded partially into two mating parts to prevent movement.

keyboard An alphanumeric keyboard similar to a typewriter's keyboard. It is used to input data in a computer system.

keyseat A slot that holds a key.

kilo (k) Metric prefix for one thousand (10^3).

knurling The machine roughing of a surface to make it nonslip.

L

land coverage The square footage of land that a structure will cover. Also called a footprint.

laser machining A process in which a high-frequency laser beam is used to cut materials.

laser printer A printer that uses a focused laser beam to copy text and drawings.

lay The direction of the dominant surface pattern made by machining operations.

layers Separate overlay drawings using the same base drawing.

layout assembly drawing Usually a rough sketch of a new design concepts used to provide an idea of the appearance and use of a new product.

lead The lateral distance a screw thread moves in one revolution.

lead per revolution The distance a threaded fastener will advance with one full revolution.

leader Used to connect a specific area on a drawing to a dimension or notation.

Leadership in Energy and Environmental Design (LEED) The organization that rates and sets standards for green construction.

least material condition (LMC) The least amount of material possible in the size of a part. The LMC of an external feature is its lower limit. The LMC of an internal feature is its upper limit.

left-hand thread These threaded fasteners will advance when turned counterclockwise.

light-emitting diode (LED) A semiconductor diode that emits a light when conducting an electrical current.

limits Acceptable variances of a dimension or, on a CAD monitor, the size of the drawing area.

limit tolerance The maximum distance of tolerance notation.

line conventions Standardization of lines used on technical drawings by line weight and style.

line precedence Used to describe which line has priority. In a working drawing, when lines overlap, the heavier line takes precedence.

lines of vision Imaginary lines connecting the item being drawn with the picture plane.

location dimensions Dimensions that show the exact location of parts of an object.

locator A pinlike device that firmly positions and holds a workpiece in a jig.

logic diagram Illustration used to design electronic microcircuitry.

longitudinal section A full section through the length of a structure.

M

machining tool Any power-driven tool.

magnetic tape A magnetized strip of plastic film on which information is stored.

mainframe system A central processing unit networked to individual satellite terminals (workstations).

major diameter The outside diameter of the threaded portion of a fastener measured from crest to crest.

market potential The study and prediction of how a product may sell.

master An original drawing or intermediate print from which reproductions are produced.

mating parts Two or more parts are assembled together.

maximum material condition (MMC) The greatest amount of material possible given the size of a part. The MMC of an external feature is its upper limit. The MMC of an internal feature is its lower limit.

mechanical engineer An engineer who applies the principles of physics and materials for manufacturing and mechanical systems.

mechanical engineer's scale A ruler similar to the civil engineer's scale.

melt-through weld This is a weld used when complete penetration is required from one side. It is used with all types of groove welds.

menu Listings of specialized tasks in a software program.

metric dimensioning A form of dimensioning in which all working drawing dimensions are shown only with millimeters (mm).

metric system The international measurement system based on units in multiples of 10.

micro (μ) Metric prefix for one millionth (10^6).

microbial fuel cells A bioelectrochemical system that creates an electric current by copying bacterial interactions in nature.

microcomputer A computer that uses a microprocessor as a basic CPU element. A microcomputer is a stand-alone system. It can process information without a link to a mainframe computer.

microfilm Film used to store photographically reduced drawings.

microprocessor A miniaturized electronic circuit device used in personal computers.

milli Metric prefix for one thousandth (10^3).

millimeter (mm) One-thousandth of a meter.

MIL STD Military Standards.

minor diameter The diameter of the thread measured from root to root, perpendicular to the axis.

minus tolerancing dimensions The negative variation of a dimension.

miter gears Gears that intersect at a right angle.

modem A device used to transmit data over telephone lines.

modified uniform motion Creates a smooth transition between the rise, fall, and rest for a rotating cam's follower.

molecular/nanotechnology engineer A professional engineer who designs working products from extremely small particles such molecules.

monitor A display screen. Monitors may be monochrome (one color, usually green or amber) or may have full-color capabilities.

motherboard Also known as a logic board. It is a printed circuit board (PCB) used in all computer systems. It holds most of the electronic components required for a computer. It provides connectors for other hardware peripherals.

mouse A graphics input device that is moved across a flat surface to control the movement of a cursor on the screen.

multiple-thread fastener Threaded fastener that will advance (lead) further with one full revolution. May also be a double- or triple-thread lead.

multiview drawing Views of an object projected onto two or more orthographic planes.

Mylar Polyester plastic drafting film.

N

National Coarse (NC) American Standard coarse screw thread series.

National Fine (NF) American Standard fine screw thread series.

networked When stand-alone microcomputer systems are connected to a minicomputer or mainframe system in order to use the larger system's database.

nominal size The designation of the stock size of a standard material before machining the surfaces.

nonisometric lines Lines that not vertical or at a 90° angle are not true size in an isometric drawing.

normal surface True-size surfaces that are parallel or perpendicular to the plane of projection.

notations Notes that specify materials and any other necessary manufacturing information on a drawing.

numerical control (NC) Control of a machine process with a computer program. Sometimes called computerized numerical control (CNC).

O

object line A heavy solid line used on a drawing to represent the outline of an object.

oblique drawing A pictorial drawing with a front view and angular receding lines.

octagon An eight-sided geometric figure with each corner forming a 135° angle. Each center angle is 45°.

offset section A sectional drawing created by a cutting plane bent at right angles to features as though they were in the same plane.

ogee curve Two reversed connected curves.

one-point perspective A perspective drawing using only one vanishing point.

opaque Material that light cannot penetrate.

opaque drawing paper A semi-transparent paper used for working drawings. It is available as a tracing paper (inexpensive) or vellum, which is a high-quality drawing medium.

open-divided scale A scale on which only one major unit is graduated with a full-divided unit. It is adjacent to zero.

operation assembly drawing A working drawing that depicts the moving parts of an assembled product.

optical disc A thin disc that is coated with plastic that stores data in shallow pits etched in the surface. A laser beam is used to read the digital data.

optical image scanner A hardware input device that is usually connected to a computer. It can copy text and drawings into a digital file that may be manipulated with a computer program.

orientation The placement of a structure on building site to take advantage of the sun, view, winds, and noise.

orthographic projection In a two-dimensional multiview working drawing, the other views are projected from the front view at a 90° projection.

outline assembly drawing A minimal drawing that shows the major outlines to describe the product.

output Processed computer data transmitted to a monitor, hard-copy device, storage media, a control device, or another computer.

output device Any device that a computer uses as a receiving device such as a printer, monitor, or storage media.

overall dimensions Dimensions that describe the total depth, height, or width of an object.

oxyacetylene welding The welding process that uses a mixture of oxygen and acetylene for the heat source.

P

parallel Lines that are equal distance.

parallel port It is an older method for connecting hardware peripherals. Computers now use the USB port for most peripherals.

parchment paper A waterproof and grease-resistant paper that may be used to produce working drawings.

Parthenon A Greek temple built 438 B.C. that is an example of the Doric style of architecture.

passive solar A heating and cooling system that uses the energy of the sun without the use of mechanical aids.

pattern A drawing or a template prepared to the exact size and shape of a workpiece. Also, a model from which sand molds are made.

pattern development drawing A drawing that shows the outline of all connected surfaces of an object when laid flat.

pentagon A five-sided geometric figure with each corner forming a 118° angle. Each center angle is 72°.

perfluorinated compounds (PFC) Toxic chemicals that are found in many household products.

peripherals The supplemental equipment used in conjunction with, but not as part of, the computer system.

personal computer (PC) A microcomputer, also called a desktop computer.

perspective drawing A pictorial drawing that contains receding lines that converge at vanishing points on the horizon. A perspective drawing can use one, two, or three vanishing points.

phantom line Line used to show the alternate position of an object or matching part without interfering with the main drawing.

photocopier Also called a copy machine. It makes copies of documents by using a technology called xerography, which is a dry process that sues heat, static electricity, and a black powder called toner.

photodrafting The use of photographs for a working drawing.

photovoltaic panel A unit that converts sunlight into an electric current.

pictorial assembly drawing A photolike illustration of the finished product.

pictorial drawing A drawing that shows the width, height, and depth of an object in one view.

picture plane Represents the vertical plane upon which the object is viewed.

pin graphics The drafting system that separates segments of working drawings into separate drawing formats and keeps all the segments aligned.

pinion gear The smaller gear in a set of mated gears.

piping drawing The working drawings of a pipe system drawn in isometric or with orthographic projection. A pipe may be represented by a single line or a double line describing the thickness of both the pipe and the fitting.

pitch A uniform distance from a point on one part to a corresponding point on an adjacent part.

pitch circle An imaginary circle concentric with a gear axis that passes through the thickest point on a gear tooth.

pitch diameter (spur gear) The diameter of a pitch circle.

pitch point The point of intersection between a gear's tooth face and pitch circle.

pixels The series of dots that make up the picture on the monitor.

plane of projection In a two-dimensional multiview working drawing, all views are shown on a 90° projection plane, but are rotated to a flat drawing surface.

plat plan An architectural drawing showing the relationship of the property to the adjacent community.

plate cam A basic cam plate with a follower with the follower adjacent to the circumference of the cam.

plug weld The weld of two flat pieces with the weld passing through one piece.

plot plan An architectural drawing showing the location of all structures on the property.

plus tolerancing dimensions The maximum variation of a dimension.

pointed follower A cam follower that has a point adjacent to the cam.

point-to-point diagram An electronic drawing that shows components, terminals, and connecting wires.

polar coordinate dimension The use of both linear and angular dimensions for locating features.

polygon A multisided closed form.

positional tolerancing The allowable variation of a feature from the precise position shown on a drawing.

powder metallurgy Process in which a powder is pressed into a form and heated.

power supply unit (PSU) Supplies power to the computer by converting alternating current (AC) to a low-voltage direct current (DC) for the internal components.

primary auxiliary view The first auxiliary drawing projected from a multiview drawing.

primary revolution The first revolved drawing from a multiview drawing.

primary storage The main storage for data inside of a computer.

printed circuit Copper electronic circuits bonded to the surface of an insulated board.

printed circuit board (PCB) The electronic printed circuit assembly, including the insulated board. Replaces wiring with conductors printed on a board.

printer A device that converts computer data into printed alphanumeric or graphic images. Printers are usually ink-jet or laser operated.

printing operator The person who makes copies from original documents and drawings.

prism A solid with base intersections that are parallel polygons and with sides that are parallelograms.

product-data management (PDM) Supervision of a product through all of its operations from concept to finished product.

profile plane In descriptive geometry, it is the projected side plane of the frontal plane.

project engineer A professional engineer who coordinates all the phases of research, drafting , and manufacturing.

projection weld Similar to a spot weld. A small projection is made at the location of the weld. Under pressure and heat the two pieces are welded together.

prompt Instructions on the monitor.

prototype An original functional model of a product constructed prior to mass production. The first finished sample of a new product.

protractor A drawing instrument used to measure angles in degrees.

puck A graphics input device that is moved across a graphics tablet to control the movement of the cursor on the monitor.

Q

quality assurance Ensuring that the quality of a product is satisfactory.

R

rack A flat bar containing gear teeth that engage with the teeth in a pinion gear.

rack and pinion gear Has a flat-toothed gear rack that is adjacent to a pinion gear.

radial line The line elements used for radial line pattern drawings.

radial line development Objects without parallel lines use a stretchout arc and radial lines to develop a pattern drawing.

radius The distance from the center of a circle to its circumference. It is one-half of the diameter.

random-access memory (RAM) A form of temporary computer memory storage that holds the data input from the operator and the software program being used.

rapid prototyping (RP) Also called 3D printing. See *stereolithography.*

read-only memory (ROM) A computer memory system that stores instructions permanently, allowing the computer to read the software programs.

rebar Steel reinforcement bar placed in concrete.

rectangular coordinate dimension A method of linear dimensioning that locates features, such as surfaces, edges, centerlines, and the like, from a baseline.

reference designation Part numbering is used to identify components.

reference line A line used to locate features on a working drawing.

regardless of feature size (RFS) A condition that requires a tolerance of position regardless of where the feature lies within an object.

relief A groove between surfaces to provide clearance for machining.

removed section A sectional view removed from the area of the cutting plane and positioned in another location.

rendering Adding embellishments to a drawing to provide a more realistic appearance.

resistance welding Welding by using the resistance of metals to the passage of an electric current to produce fusion heat.

revolution drawing A drawing of the repositioned view after the primary orthographic drawing has been rotated. The first view projected from a rotated primary orthographic view is the *primary* revolution. Rotating a primary revolution and projecting a new view is a successive revolution.

revolved section A sectional view that is revolved 90° and perpendicular with the plane of projection.

ridge The upper construction member in slanted roof.

right-hand thread A standard threaded fastener that advances when turned clockwise.

ring gears Circular gears. The large ring gear has its teeth inside. The smaller gear that is inside the larger ring has its teeth on the outside.

rise (1) The vertical height of a roof, of a step, or a flight of stairs. (2) The distance the follower will rise during the rotation of the cam.

rivet A fastener with a head and a shaft. The rivet is inserted in a hole and the end flattened to join two or more pieces of material.

robotics This refers to the action of a programmable manipulator (machine) that will perform a variety of tasks.

robotics engineer A professional engineer that designs mechanical robots for engineering and general uses.

roller follower A cam follower that is a rolling wheel adjacent to the cam.

root The deepest point or valley of a thread.

root opening (root gap) The distance between members at the root of a weld joint.

rotational molding A process in which a plastic pellet is spun in a die to achieve its desired form.

roughness-width cutoff The distance from the highest to lowest points on a textured surface.

round A curved exterior corner of two surfaces.

round follower A cam follower that is rounded at the contact point of the cam.

run The width of a step or the horizontal distance covered by a flight of stairs. This term is also used to describe the horizontal length of a rafter.

runout The area of intersection of two differently shaped surfaces.

S

scale drawing A drawing prepared proportionally smaller or larger than the object it represents.

schematic diagram (elementary diagram) A drawing of an electronic circuit showing symbols for electronic components and lines representing connecting conductors.

schematic symbol Drawings are made easier by using simple schematic icons to represent components.

screw clamp A clamp that is screwed into the jig to hold the workpiece.

screwed fitting Threaded pipe fittings.

screw thread series The two types of threads are Unified National Course (UNC) and Unified National Fine (UNF).

seam weld A continuous spot weld with a break.

secondary auxiliary view The next auxiliary drawing from a primary auxiliary drawing.

secondary storage Also called auxiliary storage and may be separate from the computer. It stores software and data, on a semipermanent basis, that is not currently in the computer's primary storage or memory. Its data may be saved on a hard disk, CD, DVD, or on any other type of removable media.

section drawing A working drawing depicting the interior of an object.

section line Angular line in a section drawing depicting solid surfaces.

sectional details Small section drawings for construction instruction.

sectional view A drawing that shows the interior of an object as it would appear if cut in half or quartered.

senior detailer An experience drafter who works directly with the design engineers.

serial port A serial communication connector that transfers data in or out, one bit at a time, in a computer system. Serial ports are being replaced with UBS ports as was the parallel port.

setback A zoning ordinance specifying the minimum distance a structure may be placed in relation the property lines.

shaft A cylinder to which rotating machine parts may be attached to transmit power and motion.

simple outline symbol (pictorial) The pictorial electronic symbols closely represent the component.

single-line piping drawing Pipes that are represented with a single line.

single thread The 360° revolution will advance (lead) only the distance of one thread (pitch).

SI International System of Units. The metric system of measurement as adopted by the General Conference of Weights and Measures.

site plan An architectural drawing depicting an area or plot of land with defined property lines.

size dimension Describes the size of each geometric form on a working drawing.

slab foundation A concrete floor and foundation system poured directly on the ground.

slope diagram A drawing that indicates a comparison between the horizontal run and the rise of a roof.

slot weld An elongated plug weld.

smart structure An environmentally safe home using electronic devices, insulation, nontoxic sustainable building materials, solar energy, natural daylight, and efficient orientation.

snap interval The distance the cursor will jump (snap) on the display screen.

software Computer programs that contain specific instructions for the functioning of a computer system.

soldering Done with a lead filler at a much lower temperature than welding.

solid model The creation of a three-dimensional drawing of the exterior of an object on a CAD system with no hidden features revealed.

solid-phase forming Similar to metal forging. Soft plastic is forced into a die.

solid-state welding The process of heating two metal pieces to their melting point, then holding them together under pressure until the hot atoms for both pieces mix together and cool.

span The distance between structural building supports.

specification A detailed description of components, parts, or materials that includes size, color, manufacturer's number, and sometimes cost.

spiral bevel gear Have curved teeth that provides more power than conventional gear teeth.

spline One of a number of keyways cut around a shaft. Flexible strips used for drawing curves.

spot weld Used for very thin materials. Electrical resistance process is used to fuse the material.

spring A helical coil of wire that will yield to the force of contraction or expansion.

spur gear Conventional gear with the teeth perpendicular to the gear.

stamping A process where thin metal is formed between two dies with pressure.

stand-alone A computer system that's operates by itself. It is not connected to a mainframe.

standards Rules (conventions) for working drawings.

station point Represents the location of the observer for a perspective drawing.

stereolithography Also called 3D printing. A process that produces a three-dimensional prototype with instructions from a CADD program. The prototype is gradually built up with thin layers with a plastics type material.

stitch line Line used in a working drawing that indicates sewing or stitching.

stretchout A pattern development drawing showing the exact shape of a flat material before forming into a three-dimensional shape.

stretchout arc The arc used to develop pattern drawings for objects without a parallel surface.

stretchout lines The stretched out, straight line of the circumference of a cylinder or circle.

straightness In geometric tolerancing, it is the condition of how straight a surface or axis may be in relation to its surface's length.

stylus pen An electronic touch pen that can activate commands by touching them on the monitor or tablet.

subassembly An assembled part that is a part of a larger assembly.

subassembly drawing Working drawings of only the smaller segments of a product.

successive revolutions The continual revolution drawings projected from the last revolution drawing.

surface model A CAD drawing showing only exterior surfaces.

surface texture The waviness, roughness, lay, and flaws of a surface.

surface weld The surface build-up of a metal surface using a metal filler and heat.

Surfacing (building up) The surface of a material may be made thicker with heat and a filler.

survey drawing A site drawing of a property containing dimensions, contours, compass orientation, and all physical features on the land.

survey plan An architectural drawing showing the geological aspects of the property.

sustainability Using only replaceable materials for construction, preserving the environment and natural resources.

symbol A graphic form used to represent a standard object, part, material, or component on a drawing.

symmetrical An object that has the same configuration on both sides.

T

tab The extra material added to an edge to strengthen a connection of two surfaces.

tabulated outline dimensioning Dimensions from perpendicular datum planes listed in a table.

T foundation A concrete foundation that looks like an inverted T.

tangent A line that intersects a circle or an arc at a point 90° from the radius.

tape drive This is a data storage device that reads and writes data on a narrow, flexible magnetic tape.

taper A wedge or conical shape that increases or decreases in size at a uniform rate.

taper reamer A reamer that produces tapered holes.

T square A drawing instrument in the shape of a T used to draw horizontal lines.

technical illustration A pictorial drawing used to interpret technical working drawings.

technical illustrator A drafter with artistic skills that produces two-dimensional working drawings into three-dimensional drawings.

template A flat form used as a guide to draw symbols and holes.

tertiary storage This includes any removable storage device that is connected to a computer. It is used in industry because of its high capacity for data and its low cost.

thin wall section Used when materials are too thin to add a section lining. Thin wall sections remain solid.

third-angle projection A multiview drawing of the front, top, and right-side views.

thread depth The distance from the crest (major diameter) to the root (minor diameter) measured perpendicular to the axis.

thread forms The standard thread form is a V shape. There are a dozen other forms that threads can take.

three-dimensional drawing A pictorial drawing using three vanishing points. Usually used for tall objects.

three-dimensional geometric forms Forms possessing height, width, and depth.

three-point perspective A perspective drawing using three vanishing points.

title block General information about the working drawing listed along the border of the drawing.

toggle-head clamps A clamp with a locking lever, used to hold a workpiece securely in position for machining.

tolerance The amount of variation allowed in the dimension of an object. Plus and minus tolerancing dimensions indicate the tolerance range above and below the basic dimension.

tolerance zone The total permissible variation of size or location of an item.

tool designer A professional engineer or experienced designer that designs the tools required for the manufacturing of products.

tool engineering The designing of tools and machinery required for manufacturing.

tooth face The surface of the gear's tooth that applies the pressure on the adjacent gear's tooth.

torsion springs Springs designed to transmit energy by a turning or twisting action.

torque Rotating or twisting force.

tracer A beginning level drafter, with good drafting skills and a knowledge of CAD, who reproduces and/or does simple working drawings.

tracing paper An inexpensive, transparent paper use for sketching and drafting.

train Meshed gears in a series.

transition fit Mating parts that may have either a clearance or interference fit.

transverse section A full section through the width of a structure.

triangle A three-sided drawing instrument used to draw various angles.

triangulation development A development drawing of a curved (warped) surface that is drawn into a series of triangles.

trimetric drawing A pictorial drawing in which the three principal axes are unequally foreshortened.

true length (true size) The condition that results when the line of sight is perpendicular to surfaces or lines.

true position The exact position of a feature.

true size edge The edge view of a surface that is actual size.

true size surface Any surface of line that shows the true shape in a working drawing.

truncated An object with the apex, vertex, or end removed by an angular plane.

two-dimensional drawing A flat drawing that shows only length and height of an item.

two-dimensional geometric forms Forms possessing height and width.

two-dimensional software 2D CAD programs that can only function on the X- (horizontal) and Y- (vertical) axes, and not on the Z-axis.

two-point perspective A pictorial drawing using two vanishing points on the horizon.

typical A term used to describe details or conditions that are identical in many parts of a drawing.

U

ultrasonic machining While the work piece is suspended in an abrasive fluid, a high-frequency vibration shapes the workpiece.

undercut To allow an overhanging edge or a cut with inwardly sloping sides.

unidirectional dimensioning All isometric letters, numerals, and arrowheads are positioned vertically. A variation of this is *vertical plane dimensioning*, where all dimensions are drawn in the isometric plane.

Unified National Fine (UNF) A thread standard for threaded fasteners.

uniform accelerated motion A parabolic curve is used to smoothly speed up the cam drive.

uniform decelerated motion A parabolic curve is used to smoothly decrease the speed of the cam drive.

uniform motion A cam motion that is uniform but not smooth at the change of directions.

unilateral tolerance dimension The tolerance size of a dimension that goes only in one direction (plus or minus).

upgrade To improve the quality of a home or a product.

USB port A standardized connector for most computer peripherals, replacing serial and parallel ports.

U.S. customary system Units based upon the inch and pound commonly used in the United States.

V

valleys The meeting of two roof surfaces that form a valley.

value designation The value of an electronic component expressed as volts, amperes, watts, or ohms.

valves Fittings that control the flow of materials through a pipe.

vampire energy Also called standby power. It draws a very small amount of energy to keep your electronic devices warm and easy to activate. Because there are so many electronic devices in use today, they consume about 10% of our total energy output.

vanishing point The point at which the receding lines of a perspective drawing meet.

vector A line that has value and direction and defined by its two extremes.

vellum Translucent drafting medium.

vernier scale A movable scale attached to a fixed scale that contains subdivisions of the fixed scale.

vertex A point in the intersection of two or more sides.

virtual reality system A lifelike reality presented with a computer with special hardware and software programs.

voice activation device To be able to activate a command on a computer by speaking the command required.

volatile organic compounds (VOCs) Toxic gases that are emitted from thousands of products such as building materials, paints, cleaning fluids, furnishings, office equipment, and so forth.

W

warped surfaces Surfaces that are curved in two directions.

water-jet cutting A high-velocity jet of water with an added abrasive will cut and dissolve the work piece to form.

weld symbol A shorthand icon for a type of weld.

welded connections Pipe fittings that are permanently welded together.

welding process The metallurgic combination of heat, energy, and materials to create welded bond between two metals.

whole depth Full height of a gear tooth, which is equal to the sum of the addendum and the dedendum.

windmills/wind turbines These turn a turbine, which turns a generator that creates electrical energy.

wireframe Three-dimensional computer models in which all visible and hidden edges appear as lines.

Woodruff key Crescent-shaped flat key.

working assembly drawing A working drawing of the finished product with enough data to complete its manufacture.

working depth Distance a gear tooth extends into a mating space; two times the addendum.

working drawing Two-dimensional drawing that provides all information needed to manufacture or construct a product.

workpiece A piece of material that has machining operations performed on it.

workstation A location where workers perform specific job tasks.

worm The driver of a worm gear that contains at least one complete tooth (endless screw) around the pitch surface.

worm gears Gears with teeth that curve to mesh with and be driven by a worm.

X-axis The horizontal axis in a rectangular coordinate system.

Y-axis The vertical axis in a rectangular coordinate system.

Z

Z-axis The axis in a three-dimensional Cartesian coordinate system that is perpendicular to the X- and Y-axes.

zoning The legal restrictions on size, location, and type of structures to be built in a designated area.

zoning ordinance The building code laws that controls the size and location of a structure on a building site.

INDEX